£49.00

Climate Change and
Human Impact on the Landscape

Climate Change and Human Impact on the Landscape

Studies in palaeoecology and environmental archaeology

Edited by

F.M. Chambers

Environmental Research Unit
Keele University
UK

CHAPMAN & HALL
London · Glasgow · New York · Tokyo · Melbourne · Madras

Published by Chapman & Hall, 2–6 Boundary Row, London SE1 8HN

Chapman & Hall, 2–6 Boundary Row, London SE1 8HN, UK

Blackie Academic & Professional, Wester Cleddens Road, Bishopbriggs, Glasgow G64 2NZ, UK

Chapman & Hall, 29 West 35th Street, New York NY10001, USA

Chapman & Hall Japan, Thomson Publishing Japan, Hirakawacho Nemoto Building, 6F, 1–7–11 Hirakawa-cho, Chiyoda-ku, Tokyo 102, Japan

Chapman & Hall Australia, Thomas Nelson Australia, 102 Dodds Street, South Melbourne, Victoria 3205, Australia

Chapman & Hall India, R. Seshadri, 32 Second Main Road, CIT East, Madras 600 035, India

First edition 1993
© 1993 Chapman & Hall

Typeset in 10/12pt Palatino by Pure Tech Corporation, India
Printed in Great Britain by Clays Ltd, St Ives plc

ISBN 0 412 46200 1

Apart from any fair dealing for the purposes of research or private study, or criticism or review, as permitted under the UK Copyright Designs and Patents Act, 1988, this publication may not be reproduced, stored, or transmitted, in any form or by any means, without the prior permission in writing of the publishers, or in the case of reprographic reproduction only in accordance with the terms of the licences issued by the Copyright Licensing Agency in the UK, or in accordance with the terms of licences issued by the appropriate Reproduction Rights Organization outside the UK. Enquiries concerning reproduction outside the terms stated here should be sent to the publishers at the London address printed on this page.

The publisher makes no representation, express or implied, with regard to the accuracy of the information contained in this book and cannot accept any legal responsibility or liability for any errors or omissions that may be made.

A catalogue record for this book is available from the British Library

Library of Congress Cataloging-in-Publication data available

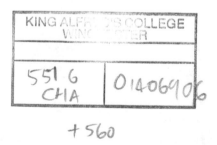

∞ Printed on permanent acid-free text paper, manufactured in accordance with the proposed ANSI/NISO Z 39.48–199X and ANSI Z 39.48–1984

Contents

Contributors	xiii
Acknowledgements	xv
Preface	xvii
Editor's Notes	xix

Part One: Precision and Accuracy in Studies of Climatic Change and Human Impact

Introduction

1	**Precision, concepts, controversies: Alan Smith's contributions to vegetational history and palaeoecology** H.J.B. Birks	**5**
	1.1 Introduction	5
	1.2 Precision	5
	1.3 Concepts	8
	1.4 Controversies	9
	1.5 Personal reflections	10
2	**Forward to the past: changing approaches to Quaternary palaeoecology** Frank Oldfield	**13**
	2.1 Introduction	13
	2.2 Inductive investigation in Quaternary palaeoecology	14
	2.3 Deductive studies: the acid rain research	16
	2.4 The future	19
3	**Radiocarbon dating and the palynologist: a realistic approach to precision and accuracy** Jon R. Pilcher	**23**
	3.1 Introduction	23
	3.2 The pioneer phase	23

3.3	The routine phase	24
3.4	The questioning phase	25
3.5	Is there a future?	29
3.6	A hypothetical project	29

4 Great oaks from little acorns. . . : precision and accuracy in Irish dendrochronology — 33
M.G.L. Baillie

4.1	Introduction	33
4.2	First case study	34
	4.2.1 An Ulster crannog	34
4.3	Discussion	35
	4.3.1 Precise dating	35
	4.3.2 Climatic and environmental effects	35
	4.3.3 The 'suck-in' effect	36
	4.3.4 Proxy data: quantitative climatic records	36
4.4	Second case study	36
	4.4.1 An attempt to apply dendrochronology to the dating of the inner ditch at Haughey's Fort	36
	4.4.2 Attempts to 'date' Q-7971	38
	4.4.3 Inferences	38
4.5	Third case study	38
	4.5.1 Early 16th century defoliations in Ulster	38
	4.5.2 Localized effects and lessons therefrom	40
4.6	Conclusions	41

Part Two: Climatic Change on the Landscape

Introduction

5 Peat bogs as sources of proxy climatic data: past approaches and future research — 47
Jeff Blackford

5.1	Introduction	47
5.2	Past approaches	47
	5.2.1 The Blytt–Sernander scheme	47
	5.2.2 Recurrence surfaces	50
	5.2.3 The search for new methods	50
5.3	The present position	52
5.4	Future research	52
	5.4.1 Dating techniques	52
	5.4.2 Sites	53
	5.4.3 Methods	55
5.5	Conclusions	55

6 Forest response to Holocene climatic change: equilibrium or non-equilibrium — 57
Richard Bradshaw

	6.1	Introduction	57
	6.2	Case studies	58
		6.2.1 Lake Michigan and the migration of *Fagus* in the USA	58
		6.2.2 The western limits of *Pinus sylvestris* in Europe	60
		6.2.3 The boreal–nemoral ecotone in central Sweden	63
	6.3	Conclusions	65

7 Isolating the climatic factors in early- and mid-Holocene palaeobotanical records from Scotland 67
J. John Lowe

7.1	Introduction	67
7.2	Pollen records and the Holocene vegetational history of Scotland	68
	7.2.1 Pioneer phase (*c.* 10 300 to 8500 BP)	68
	7.2.2 Afforestation phase (8500 to 5000 BP)	70
	7.2.3 Deforestation phase (5000 BP to present)	72
7.3	Pollen stratigraphy and palaeoclimatic inferences	72
	7.3.1 Blytt–Sernander scheme	73
	7.3.2 Time-lagged responses	73
	7.3.3 Indicator-species approach	74
	7.3.4 Conclusions	74
7.4	Isolating the climatic factors in Holocene palaeobotanical records	75
	7.4.1 Early Holocene climatic revertance	75
	7.4.2 Holocene range limits of pine	77
	7.4.3 Variations in pine macrofossil abundance	80
	7.4.4 Conclusions	80

8 Radiocarbon dating of arctic-alpine palaeosols and the reconstruction of Holocene palaeoenvironmental change 83
John A. Matthews

8.1	Introduction	83
8.2	Field sites	84
8.3	Problems of soil dating and the importance of laboratory pretreatment	84
8.4	Time elapsed since burial	88
8.5	Time since the onset of soil formation	90
8.6	Holocene glacier and climatic variations	92
	8.6.1 Climatic implications	92
8.7	Soil history at Haugabreen and Vestre Memurubreen	93
	8.7.1 Podsol development	93
	8.7.2 Brown Soil	94
8.8	Vegetation history	94
	8.8.1 Haugabreen (podsol) sites	94
	8.8.2 Vestre Memurubreen (Brown Soil) sites	96
8.9	Conclusions	96

Part Three: Evidence for Human Impact

Introduction

9 Earliest palynological records of human impact on the world's vegetation — 101
D. Walker and G. Singh

   9.1 Data — 101
   9.2 Commentary — 104
      9.2.1 The nature of human impact — 104
      9.2.2 Gaps on the map — 106
      9.2.3 Time span — 106
      9.2.4 Oldest disturbance — 108
   9.3 Celebration — 108

10 Vegetation change during the Mesolithic in the British Isles: some amplifications — 109
I.G. Simmons

   10.1 Introduction — 109
   10.2 The background — 109
      10.2.1 The earlier Mesolithic — 109
      10.2.2 The later Mesolithic — 110
      10.2.3 The Mesolithic–Neolithic transition — 111
   10.3 Two more-detailed examples — 112
      10.3.1 Bonfield Gill and North Gill, North York Moors — 112
   10.4 A wider context — 117

11 The development of high moorland on Dartmoor: fire and the influence of Mesolithic activity on vegetation change — 119
Chris Caseldine and Jackie Hatton

   11.1 Introduction — 119
   11.2 The Black Ridge Brook model — 122
   11.3 Pinswell — 124
      11.3.1 Site location and pollen and charcoal results — 124
      11.3.2 Vegetation change and peat initiation — 127
   11.4 Comparison with Black Ridge Brook and other Dartmoor sites — 130

12 Models of mid-Holocene forest farming for north-west Europe — 133
Kevin J. Edwards

   12.1 Introduction — 133
   12.2 Models of early farming activity — 136
      12.2.1 The landnam model — 136
      12.2.2 The leaf-foddering model — 138
      12.2.3 The expansion–regression model — 140
      12.2.4 The forest-utilization model — 141
   12.3 Conclusions — 144

13 The influence of human communities on the English chalklands from the Mesolithic to the Iron Age: the molluscan evidence — 147
J.G. Evans

  13.1 Introduction — 147
  13.2 Molluscan analysis — 149
  13.3 The Mesolithic period — 150
  13.4 The earlier Neolithic — 151
  13.5 The later Neolithic — 152
  13.6 The Bronze Age and Iron Age — 153
  13.7 Research strategy — 154

14 Mesolithic, early Neolithic, and later prehistoric impacts on vegetation at a riverine site in Derbyshire, England — 157
Patricia E.J. Wiltshire and Kevin J. Edwards

  14.1 Introduction — 157
  14.2 The site — 158
  14.3 Methods and presentation of results — 159
    14.3.1 Field sampling — 159
    14.3.2 Pollen analysis — 159
    14.3.3 Radiocarbon dating — 163
    14.3.4 Statistical analysis — 163
  14.4 Vegetation history — 163
    14.4.1 General — 163
    14.4.2 Reconstruction — 164
  14.5 Discussion — 167

15 Holocene (Flandrian) vegetation change and human activity in the Carneddau area of upland mid-Wales — 169
M.J.C. Walker

  15.1 Introduction — 169
  15.2 Study area — 170
  15.3 Research strategy — 171
  15.4 Methods — 172
  15.5 Regional pollen assemblage zones — 172
  15.6 Vegetation and landscape change in the Carneddau region — 176
    15.6.1 Early- and mid-Flandrian woodland development — 176
    15.6.2 Mid- and late-Flandrian woodland decline: evidence of human impact — 180
  15.7 Conclusions — 182

16 Early land use and vegetation history at Derryinver Hill, Renvyle Peninsula, Co. Galway, Ireland — 185
Karen Molloy and Michael O'Connell

  16.1 Introduction — 185
  16.2 Sites investigated — 188
  16.3 Methods — 189

16.4	Results and interpretation: short profiles on Derryinver Hill	189
	16.4.1 Description of soil profiles	189
	16.4.2 Pollen and macrofossil analysis	189
	16.4.3 Chronology, and significance in terms of land use, of events recorded in profiles DYR I, II, III and VI	196
16.5	Discussion	198

Part Four: Climatic Change and Human Impact: Relationship and Interaction

Introduction

17 Rapid early-Holocene migration and high abundance of hazel (*Corylus avellana* L.): alternative hypotheses — 205
Brian Huntley

17.1	Introduction	205
17.2	The hypotheses	206
	17.2.1 Succession and soil development	206
	17.2.2 Migrational lag	207
	17.2.3 The position of 'glacial refugia'	209
	17.2.4 Late-glacial expansion	209
	17.2.5 Human assistance	210
	17.2.6 Climate	211
	17.2.7 Plateaux in the radiocarbon timescale	213
17.3	Discussion	213
	17.3.1 Rapid expansion of geographical range	213
	17.3.2 Anomalous early-Holocene abundance	214
17.4	Conclusions	215

18 The origin of blanket mire, revisited — 217
Peter D. Moore

18.1	Introduction	217
18.2	Mechanisms of blanket mire inception	219
18.3	Extent and intensity of prehistoric forest modification	221
18.4	Replacement of forest by blanket mire	223
18.5	Conclusions	225

19 Climatic change and human impact during the late Holocene in northern Britain — 225
Keith E. Barber, Lisa Dumayne and Rob Stoneman

19.1	Introduction	225
19.2	The late-Holocene environment of northern Cumbria	226
19.3	The Roman impact on the landscape of the frontier zone	229
	19.3.1 Historical accounts	229
	19.3.2 Palynological evidence	230
19.4	Testing and extending the record of climatic change	233
19.5	Conclusions	235

20 Palaeoecology of floating bogs and landscape change in the Great Lakes drainage basin of North America   237
Barry G. Warner

    20.1   Introduction   237
    20.2   Early investigations   239
    20.3   Ecological variation   240
    20.4   Distribution   241
    20.5   Hydrological variation   242
    20.6   Stratigraphic variation   243
    20.7   Human activities   243

21 Late Quaternary climatic change and human impact: commentary and conclusions   247
F.M. Chambers

    21.1   Introduction   247
    21.2   Precision and accuracy   248
           21.2.1   Precision and accuracy in taxonomy and sampling   248
           21.2.2   Precision and accuracy in dating   248
    21.3   Climatic change and vegetational response   249
           21.3.1   Climate changes of the last 500 000 years   249
           21.3.2   The last cold stage   249
           21.3.3   Holocene temperate forest development   251
           21.3.4   Volcanic activity and effects on climate   253
           21.3.5   Proxy records of climate from peat stratigraphy   254
           21.3.6   Conflicting evidence of past climatic change   254
    21.4   Assessment of evidence for human impact   255
           21.4.1   Pleistocene human impact   255
           21.4.2   Holocene human impact: a recurrent theme   255
           21.4.3   Mesolithic human impact in north-west Europe   256
    21.5   Concluding remarks   256
           21.5.1   Timescales of change   256
           21.5.2   Nature conservation and environmental management   258

Appendix   261

References   263

Index   295

Contributors

Professor M.G.L. Baillie,	Palaeoecology Centre, School of Geosciences, Queen's University, Belfast BT7 1NN, Northern Ireland.
Dr Keith E. Barber,	Department of Geography, University of Southampton, Southampton S09 5NH, UK.
Professor H.J.B. Birks,	Botanical Institute, University of Bergen, Allegaten 41, N-5007 Bergen, Norway.
Dr Jeff Blackford,	School of Geography, University of Birmingham, Edgbaston, Birmingham B15 2TT, UK.
Dr Richard Bradshaw,	Forestry Faculty, Swedish University of Agricultural Sciences, Box 49, S-230 53 Alnarp, Sweden.
Dr Chris Caseldine,	Department of Geography, University of Exeter, Amory Building, Rennes Drive, Exeter EX4 4RJ, UK.
Dr F.M. Chambers,	Environmental Research Unit, Keele University, Keele, Staffs ST5 5BG, UK.
Lisa Dumayne,	Department of Geography, University College of Swansea, Singleton Park, Swansea SA2 8PP, UK.
Dr Kevin J. Edwards,	School of Geography, University of Birmingham, Edgbaston, Birmingham B15 2TT, UK.
Dr J.G. Evans,	School of History and Archaeology, University of Wales College of Cardiff, PO Box 909, Cardiff CF1 3XU, UK.
Jackie Hatton,	Department of Geography, University of Exeter, Amory Building, Rennes Drive, Exeter EX4 4RJ, UK.
Dr Brian Huntley,	Environmental Research Centre, University of Durham, Department of Biological Sciences, South Road, Durham DH1 3LE, UK.

Professor J. John Lowe,	Centre for Quaternary Research, Department of Geography, Royal Holloway, University of London, Egham, Surrey TW20 0EX, UK.
Dr John A. Matthews,	ESPEC, Department of Geology, University of Wales College of Cardiff, PO Box 914, Cardiff CF1 3YE, UK.
Dr Karen Molloy,	Department of Botany, University College, Galway, Ireland.
Dr Peter D. Moore,	Division of Biosphere Sciences, King's College London, Campden Hill Road, London W8 7AH, UK.
Dr Michael O'Connell,	Department of Botany, University College, Galway, Ireland.
Professor Frank Oldfield,	Department of Geography, University of Liverpool, Roxby Building, PO Box 147, Liverpool L69 3BX, UK.
Professor Jon R. Pilcher,	Palaeoecology Centre, School of Geosciences, Queen's University, Belfast BT7 1NN, Northern Ireland.
Professor I.G. Simmons,	Department of Geography, University of Durham, Science Laboratories, South Road, Durham DH1 3LE, UK.
Dr G. Singh (deceased),	formerly of Department of Biogeography and Geomorphology, Research School of Pacific Studies, The Australian National University, GPO Box 4, Canberra ACT 2601 Australia.
Rob Stoneman,	Department of Geography, University of Southampton, Southampton SO9 5NH, UK.
Professor D. Walker,	Research School of Pacific Studies, The Australian National University, GPO Box 4, Canberra ACT 2601, Australia.
Dr M.J.C. Walker,	Department of Geography, St David's University College, University of Wales, Lampeter SA40 7ED, UK.
Dr Barry G. Warner,	Department of Geography, University of Waterloo, Waterloo, Ontario N2L 3G1, Canada.
Patricia E.J. Wiltshire,	Department of Human Environment, Institute of Archaeology, University College London, 31–34 Gordon Square, London WC1H 0PY, UK.

Acknowledgements

I thank contributors for responding to heavy-handed editorial comment, Helena Chambers and Pauline Jones for assistance with text and references, John Birks, Richard Bradshaw, Keith Briffa, Stephen Briggs, Ed Cloutman and Kevin Edwards for advice, Valerie Hall, Jonathan Lageard, Peter Moore and Mike Walker for additional help; the referees (who shall remain anonymous) for comments to the editor on individual chapters, Bob Carling of Chapman & Hall for assistance, and Andrew Lawrence for re-drafting some of the figures.

On behalf of the contributors, may I express deep regret that Dr Gurdip Singh died suddenly in November 1990, not long after his joint manuscript was submitted.

Preface

I am pleased to present this volume of invited reviews and research case studies, produced to mark the retirement of Professor A.G. Smith – one of the leading researchers in Holocene palaeoecology. A.G. Smith took his first degree at the University of Sheffield, graduating in 1951 with a first-class honours degree in Botany. His doctorate was awarded in 1956 for a study in late-Quaternary vegetational history, based in the Sub-Department of Quaternary Research at the University of Cambridge, under the supervision of the late Sir Harry Godwin, FRS. He then researched and taught at Queen's University, Belfast, from 1954, leading the Nuffield Quaternary Research Unit there, becoming Co-Director of the Palaeoecology Laboratory from 1964. He was appointed Professor and Head of the Department of Botany (later, Plant Science) at University College, Cardiff, in 1973, and retired from the School of Pure and Applied Biology at the renamed University of Wales College, Cardiff, in August 1991.

Although his principal interests have been concerned with the post-glacial environmental history of the British Isles, Professor Smith has significantly influenced many researchers elsewhere in their interpretation of biological and other evidence for human modification of the natural environment. He helped to promote the widespread use of radiocarbon dating of peat and other sediments to provide a timescale for vegetational and environmental change, and was instrumental in setting up the world-renowned Palaeoecology and Dendrochronology Laboratories at Belfast, and the Cardiff Radiocarbon Dating Laboratory. His major interests in Holocene palaeoecology have been in the fields of human impact on vegetation, particularly in the Mesolithic and Neolithic, and of the thresholds or inertia involved in climatic forcing of vegetational change. His approach has been interdisciplinary, and he has regularly collaborated with archaeologists, biologists, geographers and geologists.

This volume of invited contributions has been compiled in celebration of his work. The contributors include former colleagues of Professor Smith, and researchers who have been influenced by his supervision or by his writing; many are themselves internationally renowned figures in research. I am pleased that they all agreed to contribute to this compendium of current views and recent research practice in areas of late-Quaternary palaeoecology and environmental archaeology.

The text is aimed at researchers, staff and final-year undergraduates in geography, environmental science, environmental biology, environmental archaeology, and Quaternary geology; at Masters students in environmental archaeology, prehistory and archaeology, and Quaternary science; and at anyone else with an enquiring mind who has an interest in Quaternary science, Holocene palaeoecology, environmental archaeology, climatic history, vegetational history and landscape history. I hope it will prove a stimulus to research and teaching, and a valuable source of ideas and references for staff and students alike. Finally, I hope this volume will prove of more than mere academic interest for the staff and supporters of national and regional conservation agencies and archaeological trusts, who are faced with the unenviable tasks of managing or excavating and restoring the landscapes that have emerged through the combination of climatic change and human impact over the last ten thousand and more years.

Editor's notes

NEWCOMERS AND NON-SPECIALISTS

Newcomers to palaeoecology and environmental archaeology – and readers interested in the background to contemporary climatic change or in environmental history and management – may find it helpful to read Chapter 21 first, to guide them to relevant chapters.

STANDARDIZATION

Nomenclature

Despite attempts at standardization, scientific terminology has proved remarkably resistant to fossilization by international convention: terms change their meaning, or are substituted over time; at any one time, there may be a wide range of contemporary usage. Recognizing this, I have permitted a range of terminology: post-glacial and Holocene, cultural periods (Mesolithic, Neolithic, etc.) and chronozones (Flandrian I); phases, zonules and pollen assemblage zones. In Chapter 17, 'migration' has been used where other authors might prefer 'spread'. Had 'spread' been substituted, legitimate nuances of meaning may have been lost. As Science cannot progress without either diversity or debate, I have permitted authors a degree of latitude in their use of terminology.

Readers may note that the gender-specific term 'man' has been avoided, and, for reasons that will be clear from the beginning of Part Three, I have also sought to avoid the adjective 'anthropogenic' (except in direct quotations).

Radiocarbon dating conventions

Different ways of quoting radiocarbon ages have emerged over the past two decades, and the norm in one science (e.g. Quaternary geology) has not been mirrored in another (e.g. archaeology). Journals have adopted (and amended) their own house-styles. Despite the wishes of the international radiocarbon community (see *Radiocarbon*), multicultural methods of citing radiocarbon ages continue to be used. Some major journals, such as *Journal of Ecology, Journal of Quaternary Science, Nature, Quaternary Research, Science* and *The Holocene*) require or prefer radiocarbon ages to be quoted uncalibrated (as BP – before present, where present is AD 1950). Others, notably *Antiquity, Archaeological Journal, Journal of Archaeological Science* and *Proceedings of the Prehistoric Society*, have not adhered to this convention. The situation has become muddled and, despite recent changes in policy by some journals, it remains so. For a volume as multidisciplinary as this, addressing as it does a range of questions and issues, including calibration itself, it proved impractical to insist on one (and only one) system.

Scientists from various disciplines have contributed to this book. Some are used to quoting radiocarbon ages on a bc/ad timescale, i.e. by subtracting the uncalibrated radiocarbon age from AD 1950. This practice is not recommended by the international radiocarbon community. Nevertheless, in certain fields, much scientific literature *has* used this system. I was advised to avoid lower-case bp, formerly used by archaeological journals for uncalibrated ages (even though bp has the advantage of removing all ambiguity), and to use BP instead – as used by the majority of the scientific journals listed above (cf. Bowman, 1990). However, to use *only* BP when also discussing documentary evidence (in calendar years AD) from the Roman or medieval periods in Britain, or the post-Columbian period in the New World, would be perverse.

In this volume, following international convention, the (rounded) Libby estimate of 5570 ± 30 years for the half-life of radiocarbon has been adopted, i.e. all uncalibrated ages are quoted in conventional radiocarbon years BP. However, to facilitate comparison with published literature by British chalkland archaeologists, uncalibrated radiocarbon ages BC have been added in parentheses in Chapter 13. Similarly, in Chapters 15 and 19, for the benefit of British archaeologists, I have allowed the use in parentheses of bc and ad for some uncalibrated radiocarbon ages. The resulting figures may be more comprehensible to those readers than the unfamiliar BP. For reference, Fig. 3.1 contains a timescale showing equivalent ages BP and bc/ad.

The existence of a so-called high-precision radiocarbon calibration curve has meant that quotation of calibrated ages has recently become more common. Unfortunately, 'plateaux' in the radiocarbon calibration curve result in rather imprecise calibrated age ranges for certain periods, though this does emphasize their uncertainty. Chapters 3, 4 and 8 tackle some of the problems that arise from calibration, and by necessity, authors of these chapters have quoted both calibrated and uncalibrated ages. Calibration is also attempted in Chapters 15 (see Table 15.1) and 16. A calibration table is printed in Appendix A for easy conversion from BP to Cal. BC and Cal. AD.

Hence, readers should note that in this volume:

1. AD or BC indicates a calendar (historical) date.
2. All radiocarbon dates (or estimated ages) cited BP, bc or ad are uncalibrated, and expressed in conventional radiocarbon years. Dated samples have their statistical precision given to one standard deviation, unless explicitly stated otherwise.

3. Cal. BC or Cal. AD signifies a radiocarbon date (or interpolated radiocarbon age), calibrated using the Belfast–Seattle chronology of Stuiver & Pearson (1986) and Pearson & Stuiver (1986). Tables 5.2, 8.2 and 15.1 also include other calibration methods.
4. There is reference in Chapter 13 to a thermoluminescence age. These are routinely cited as years BP (with present taken as AD 1980).

Part One

Precision and Accuracy in Studies of Climatic
Change and Human Impact

Introduction

Instrumental meteorological records began a mere 300 years ago in parts of Europe, and much more recently in most parts of the world. For any earlier period, students of climatic change depend upon 'proxy' records of climate. Many of these proxy records derive from palaeoecological research. For studies of human impact upon vegetation and landscape, even during the historical period (of written records), the researcher will be heavily dependent upon palaeoecological or environmental archaeological techniques. These have been developed and refined over the years.

For the Holocene (the last 10 000 years), the principal terrestrial techniques have been pollen analysis and radiocarbon dating. However, radiocarbon 'dates' are not equivalent to calendar dates. Attempts have thus been made to calibrate radiocarbon ages using tree-ring dating (dendrochronology).

Underlying this section is the search for precision (exactness, or repeatability) and accuracy (correctness) of approaches, data collection, measurement, and analysis in palaeoecological research. Implicit are the complementary and often conflicting needs for rigour in scientific method and approach, but for this to be coupled with new ideas, new hypotheses – the requirement that for Science to advance, some scientists need to break with convention.

In Chapter 1 Professor H.J.B. Birks considers the contributions of A.G. Smith to research in vegetational history and palaeoecology. In Chapter 2 Professor Frank Oldfield explores changing approaches in research in Quaternary palaeoecology of the past few decades. In Chapter 3 Professor Jon R. Pilcher considers the problems that have emerged in the calibration of radiocarbon dates and suggests a realistic approach for palynologists and others reliant on radiocarbon dates for chronology, and finally, in Chapter 4, Professor M.G.L. Baillie explores some of the achievements and problems in Irish dendrochronology and dendroclimatology.

1

Precision, concepts, controversies: Alan Smith's contributions to vegetational history and palaeoecology

H.J.B. Birks

SUMMARY

1. A review of the published contributions of A.G. Smith to our understanding of vegetational history and palaeoecology in the British Isles is presented under three broad headings – precision, concepts, and controversies.
2. Examples of his different contributions within these headings are discussed.
3. Some personal reflections are included.

Key words: A.G. Smith, Holocene, pollen analysis, precision, inertia, threshold.

1.1 INTRODUCTION

Although I was never formally taught or supervised by Alan Smith, my early interest in pollen analysis and vegetational history in the early 1960s was strongly stimulated by Alan's encouragement and keen interest in my schoolboy attempts to count pollen at Chat Moss. This encouragement, along with the considerable help and guidance I also received from the late Sir Harry Godwin, Frank Oldfield and John Tallis, played a major role in developing my life-long fascination for bogs and lakes and my attempts at trying to unravel the secrets of their past. I am therefore delighted to have this opportunity to review and evaluate Alan Smith's work, to discuss its influence on our research approaches and on our understanding of the Holocene vegetational history of the British Isles, and to comment on its long-lasting contribution to Quaternary palaeoecology. I will discuss Alan Smith's work under three main headings – Precision, Concepts, and Controversies.

1.2 PRECISION

The single most distinctive characteristic of Alan Smith's work is his concern for precision, whether it be precision in pollen morphology, precision in sampling stratigraphically or geographically, or precision in chronology. In this age of increasing syntheses and rapid summaries of pollen-analytical data, with isopoll and isochrone maps, analogue matches, rates of change of smoothed and interpolated data, detrended correspondence analysis axes, and the like, it is surprisingly easy to forget that the basis of any synthesis is good data. With the increasing number of broad-scale continental or sub-continental syntheses of pollen data involving more and more sites and fewer and fewer taxa, it is increasingly important to remember that almost all developments in our understanding of vegetational history have come from improved precision in pollen identification, sampling resolution, chronology, or spatial scale. The work of our Danish colleagues such as the late Johs. Iversen, the late J. Troels-Smith, Svend Th. Andersen, Bent Aaby, Bent Fredskild and Bent Odgaard clearly shows the close positive correlation between precision of investigation and our understanding of vegetational history. The acquisition of reliable, repeatable pollen-analytical data remains a labour-intensive activity, despite early promises from developments in image analysis, automated counting, and other pattern-recognition techniques. Detailed pollen analyses continue to demand taxonomic, stratigraphical, spatial, and chronological precision.

The building-up in the mid-1960s of the 'Alan Smith' collection of pollen and spore reference slides of virtually the entire native British flora, all prepared in a uniform way and mounted in silicone oil, was a clear expression of Alan Smith's concern for **taxonomic** precision. It was, for a long time, the only pollen reference material in the Cambridge Botany School mounted in silicone oil and thus the only source of reliable size measurements. For people like Hilary Birks, Rona Pitman (née Peck), Dick Sims, and myself, who began to use silicone oil in Cambridge in the late 1960s (rather than glycerine jelly with its undesirable features) for reliable and reproducible pollen counts, the 'Alan Smith slides', as Miss R. Andrew called them, were in constant use. The building-up of such a large and comprehensive reference collection must have been a mammoth task, and it is something that those of us who used it so much will always be grateful for.

A second clear demonstration of Alan Smith's overriding concern for precision in palaeoecology is his extremely detailed close-interval **stratigraphical** sampling, usually involving large (by British standards!) pollen counts. Such stratigraphically precise studies were made by Alan Smith and his associates in an attempt to solve critical problems in pollen stratigraphy and vegetational history. The classic investigations on the early 'landnam' phases at Fallahogy, Co. Down, with samples at 2 cm intervals (Smith, 1958a; Smith & Willis, 1961–2; Smith, 1964), on the Mesolithic layers at Newferry, Co. Antrim, at 1–2 cm intervals (Smith & Collins, 1971; Smith, 1981a, 1984), on early Neolithic horizons at Ballynagilly, Co. Tyrone, at 1 cm intervals (Pilcher et al., 1971; Pilcher & Smith, 1979), and on near-contiguous samples from the horizons under the Pubble barrow in Co. Londonderry (Smith et al., 1981) provide clear demonstrations of the importance of such stratigraphical precision. These studies show how extremely rewarding such detailed analyses can be in elucidating particular problems. It is easy, however, to forget the huge numbers of hours of pollen counting represented in one such study. The diversity and complexity of the palynological and, by inference, the vegetational changes revealed by such detailed sampling at Newferry or Ballynagilly provide exciting challenges to many ecological ideas about resilience, stability, and integrity of vegetation, challenges not yet being addressed by ecologists or vegetation historians.

Today, such fine-detailed studies are termed 'fine-resolution pollen analysis' (Simmons, this volume) and this approach is regarded by some as one of the new exciting developments in Quaternary pollen analysis. Exciting – yes; new – no!

A third important component of Alan Smith's concern about precision in vegetational history is his attempts at achieving some **spatial** precision in discussions of vegetational history at the local, site scale. The study of Waun-Fignen-Felen, a small upland bog on Black Mountain, South Wales, involves seven detailed stratigraphical transects, over 90 corings, 13 pollen diagrams, and maps of the local distribution of selected pollen-stratigraphical changes (Smith & Cloutman, 1988). It is a unique contribution to our appreciation of the almost unbelievable complexity of local vegetational changes. Current palynological or plant ecological theory does not readily accommodate this degree of spatial and temporal complexity, and possibly future models of pollen dispersal and vegetation dynamics may need to incorporate a substantial stochastic component if the diverse patterns in time and space at Waun-Fignen-Felen are to be interpreted in any detail.

The palaeovegetational reconstructions for an area 600×600 m for six separate times in the Holocene and the demonstration of the extreme variability in the expression of a presumed regional palynological change, the elm decline, within a single small basin are salutary reminders in this time of broad-scale syntheses that there is an enormous and largely unappreciated inherent variance within a site, variance that should not be dismissed as local site noise and that cannot be removed simply by including 200 sites from southern Spain to the Lofoten islands. Smith & Cloutman (1988, p. 203) note that 'the lessons of this exercise are that exact definition of elm declines demands close sampling, that there is variability, and that local events may possibly be recorded'. Their study, still poorly known, must rank as one of the most detailed studies of the vegetational history of a site ever attempted.

A similar but less detailed approach was successfully used by Cloutman & Smith (1988) in their elegant reconstructions of the local and regional vegetational history at Star Carr and Seamer Carr in the Vale of Pickering, North Yorkshire.

The fourth aspect of precision that characterizes Alan Smith's work concerns *chronology*. Sound correlation is the basis for comparison and correlation in vegetational history and palaeoecology at all scales. Alan Smith and his associates have made several extremely important contributions to chronological precision. Alan's first contribution was his classic study with Eric Willis (Smith & Willis, 1961–2) at Fallahogy when they attempted, for the first time, to establish the duration of an early 'landnam' phase. This approach was extended by Jon Pilcher (Pilcher & Smith, 1979) and associates (Pilcher *et al.*, 1971). The famous radiocarbon laboratory at Belfast began whilst Alan Smith was at Queen's University. This laboratory is closely associated with high-precision radiocarbon-dating (e.g. Pearson *et al.*, 1977) and with high-precision calibration of the radiocarbon timescale (e.g. Pilcher & Baillie, 1978; Pearson *et al.*, 1986; Pearson & Stuiver, 1986) using the long northern Irish Holocene tree-ring chronology developed in Belfast (Smith *et al.*, 1972; Pilcher *et al.*, 1977).

It was Alan Smith's close concern with precision in chronology and with the critical evaluation of radiocarbon dates that resulted in his masterly survey with Jon Pilcher of radiocarbon dates and pollen-zone boundaries in the British Isles (Smith & Pilcher, 1973). Their review was one of the first unambiguous demonstrations of the non-synchronous nature of the major pollen-zone boundaries in the British Isles. This analysis paved the way for many new studies and approaches in British vegetational history, including: (1) a radically different approach to pollen zonation, with emphasis on local (or site) and regional pollen assemblage zones, and a greater concern for explicit definitions and zone demarcation, sometimes by constrained numerical classification techniques; (2) a consideration of the nature and purpose of stratigraphical divisions within the Holocene; (3) alternative approaches to pollen zonation, such as splitting of individual pollen-taxon sequences; (4) pollen and vegetation maps of the British Isles for 5000 BP; and (5) patterns of tree spreading and expansion in the British Isles. This work of Smith & Pilcher (1973) was such an important contribution that, like so many other influential papers, it is so absorbed into the general thinking of palynologists that it is rarely cited nowadays. Such classic papers make a mockery of the use of citation indices in evaluation of scientists and their work!

A close correlate with Alan Smith's overriding concern for precision is the extreme thoroughness of his review papers. His synthesis of the Atlantic/Sub-Boreal transition (Smith, 1961a) remains one of the most thorough and thoughtful reviews of the possible causative processes for the mid-Holocene elm decline. Similarly Alan Smith's reviews of Irish vegetational history (Smith, 1970a), of Neolithic and Bronze Age impacts in northern Ireland (Smith, 1975), and of Neolithic impacts in the British Isles (Smith, 1981b) are invaluable, thorough and critical syntheses.

Although I do not agree with all of it, Alan Smith's review (Smith, 1970b) on the possible influence of Mesolithic and Neolithic people on British vegetation remains one of the most stimulating (and, to me, most controversial!) articles written about Holocene vegetation history in the British Isles in the last two or three decades. I shall return to the importance of this review in section 1.4 (Controversies).

1.3 CONCEPTS

Of the 50 or so papers by Alan Smith, the one that I have used most often, cited most frequently, re-read most, and encouraged others to read and think about is his 1965 paper 'Problems of inertia and threshold related to post-Glacial habitat changes' (Smith, 1965). It was delivered at the most stimulating meeting I have ever attended, a Royal Society Discussion Meeting organized by the late W.H. Pearsall in March 1964, with lectures by the late John Mackereth, Winifred Anne Pennington, Ron Pearson, Judy Turner, and others. After 27 years, Alan Smith's contribution still stands out as an intellectual 'high' for me. Why?

In his lecture and 'inertia and threshold' paper (Smith, 1965), Alan Smith laid the foundation for several important conceptual developments in vegetational history. He proposed that 'differences of migration rate and the rates of pedogenesis should be given more weight'. He asked 'whether we can attach any climatic significance to the pollen zone transitions up to the beginning of the Atlantic period' and warned that 'it is clearly to argue in a circle to use a chronology based on vegetational evidence'. He postulated that 'the effect of a climatic change will always depend on whether or not the changing local climate or microclimate could cross the threshold for a vegetational change' and discussed how 'a time lag in the response of vegetation to a climatic change serves to emphasize that the vegetational development in different areas may have proceeded at different rates according to local edaphic or microclimatic conditions'. He suggested that if we could 'attempt to define regions in which similar vegetational events had taken place at the same time', then 'the study of the post-Glacial phytogeography of our native trees could begin on a new basis'. He discussed how we might 'discriminate between the importance of migration rates and rates of establishment of various species' and introduced into palaeoecology the concept of inertia. He emphasized the idea of Harry Godwin that 'representatives of the forest trees were always present in locally favourable habitats and that when conditions changed in their favour, extension from these centres was rapid and effective'. Smith notes that 'doubtless it will be those areas which now appear to be exceptions to a general rule that will furnish evidence for these processes'. He proposed that 'the present pollen zonation scheme gives no prominence to transitions... it does not assist in the examination of ecological processes', that 'we shall have to attempt to construct pollen diagrams on an absolute basis', and that 'we must explore all possible methods of estimating past climatic conditions that do not depend on argument from present distributions of plants and animals'. He concluded by suggesting that the concepts of threshold and inertia 'may help us in coming closer to an understanding of the complex changes of the post-Glacial period by indicating the need for a change of emphasis in our approach'.

I have deliberately quoted extensively from Smith (1965) to show that it contains a treasure-trove of ideas and concepts, a treasure that in over 25 years has been barely explored. Re-reading Smith (1965) in 1991 in light of the many papers that have appeared in the last five to ten years on pollen (vegetation)–climate equilibrium, rates and processes of tree migration, rates of population expansion, rates of palynological change, alternatives to pollen zonation, palaeoclimatic reconstructions from pollen data, how to interpret late-Quaternary palynological data, history of *Alnus glutinosa*, and the like, emphasizes how prophetic it was. It presented a major research agenda for improving our understanding of Holocene forest dynamics. In view of its extraordinary richness of important ideas and concepts, it is very surprising that Smith (1965) is so rarely cited by European or North American vegetation historians, despite the (re)current interests in tree migrations, response times, lags, rates of change and palaeoclimatic reconstructions.

In terms of conceptual contributions to Quaternary palaeoecology, Smith (1965) and Smith & Pilcher (1973) represent long-lasting and major contributions that have strongly influenced my own research interests and approaches and those of some of my former students and research associates.

1.4 CONTROVERSIES

During his scientific career, Alan Smith has not been frightened to present new, even revolutionary (*sensu* Kuhn, 1970) ideas about the underlying causes of Holocene vegetational history. In Smith (1970b), he proposed that *Corylus avellana* may be fire resistant and thus that the early Holocene expansion of *Corylus* pollen may have resulted from Mesolithic disturbances and its remarkable early-Holocene abundance may have represented a 'fire climax'. In the same paper and also in Smith (1984), Alan Smith suggested that the establishment of *Alnus glutinosa* in mid-Holocene times may have been facilitated by widespread destruction of forest cover by Mesolithic people. He proposed that the environmental impact of Mesolithic people on British vegetation may have been considerably greater than previously thought.

This is not the place to repeat my reasons for not agreeing with these ideas, as I have presented them elsewhere. Alan Smith knows my reservations. Indeed on the latest reprint I received from him (Smith *et al.*, 1989), Alan has written 'John: you won't believe this one'. He was right – I do not believe it!

Leaving aside these disagreements, I believe that Alan Smith's ideas have been particularly fruitful ones because they have initiated a large amount of interesting, detailed work by others and have stimulated much critical thought about Holocene forest history. I suspect that much of the fascinating palynological work done, for example, on early and mid-Holocene deposits on the North York Moors and in parts of the Pennines and Scotland, that the detailed charcoal stratigraphies being elucidated for sites in, for example, Scotland and East Anglia, and that the different syntheses published on the Holocene history of *Alnus glutinosa* might not have been done but for Alan Smith's stimulating ideas. Appropriately, the most cited paper written by Alan Smith is Smith (1970b), not surprisingly because it abounds in ideas that invite testing.

Although I feel that the generality of many of the ideas presented in Smith (1970b, 1984) has now been falsified, this is immaterial. In the Popperian view of science, a view that I adhere to (see also Oldfield, this volume), science proceeds not by being 'proved right' but by posing falsifiable hypotheses and attempting to reject them rather than by seeking confirmatory evidence for favoured explanations or stratigraphical schemes that, for various reasons, have become constructs like a religion. Alan Smith's ideas on the early and mid-Holocene have served an invaluable catalytic role by stimulating research efforts and defining new research problems.

The critical resolution of some of the problems associated with interpreting the early-Holocene patterns of change in *Corylus* and *Alnus* highlights our inabilities to interpret unambiguously rather major pollen-stratigraphical changes. These difficulties should encourage caution in assuming cause-and-effect relationships between, for example, charcoal and a pollen change at one site when at another site the same palynological change may occur but charcoal is absent, or at a site where charcoal is present but there are no palynological changes (see Caseldine & Hatton, this volume).

Hypothesis testing in palaeoecology and vegetational history is notoriously difficult because many interesting palynological changes may be so localized that they are isolated singular phenomena. A statement about their causes cannot often be tested with data from another site. Indeed some of the changes may not be reproducible in time or space (e.g. Smith & Cloutman, 1988). Coherent but untestable narrative explanations may thus be the best we can hope for in such situations, although Popper has argued that singular historical events are not precluded from explanation by falsifiable hypotheses simply because they are singular, historical events. I am increasingly impressed by the accumulating and increasingly detailed evidence to implicate Mesolithic people as disturbance agents in early-Holocene vegetation, particularly in upland Britain. I doubt, however, that we will ever be able to frame falsifiable hypotheses about Mesolithic human impact (Simmons, this volume). If palynology and palaeoecology are to retain their scientific credibility, we should be modest about the generality of our statements concerning the possible causes of particular palynological changes in discussion with, for example, archaeologists or climatologists. We must admit that we are presenting local, untestable narratives and not general tested or testable theories.

1.5 PERSONAL REFLECTIONS

In this chapter I have not discussed Alan Smith's work with Gurdip Singh and others on sea-level changes in Co. Down (Singh & Smith, 1966, 1973; Dresser *et al.*, 1973), his work on Holocene vegetation history in the Gwent Levels (Smith & Morgan, 1989), or other detailed studies, for example at Cannons Lough, Co. Down (Smith, 1961b).

I conclude by saying that I fondly remember my first contact with Alan Smith. It was 8 February 1962. On that day, three reprints from Alan Smith arrived. They were Smith (1958b, 1958c and 1959). As someone trying to learn about pollen analysis and how to interpret simple pollen-analytical results, I read and re-read those three papers many times. They are still models of clarity and detail and of clear, readable pollen diagrams and stratigraphical sections. Besides the three reprints, Alan Smith enclosed a friendly letter in which he tactfully enquired if I might be interested in a job at Queen's University, Belfast, as he was trying to build up a pollen group there. I was flattered but had to decline because I had to take my final school examinations that summer and the Scholarship Examination for Cambridge later that year!

It is with very considerable pleasure that I dedicate this chapter to Alan Smith and say thanks to him for his many contributions to vegetational history and pollen analysis and for his encouragement over the years. I wish him many years of happy retirement.

Acknowledgements

I am grateful to Hazel Juggins for her help in preparing this essay and to Hilary Birks and Frank Chambers for their critical reading of it.

Editor's note: Most of A.G. Smith's publications are listed in the references at the end of the volume. The most recent are Smith & Goddard (1991) and Smith & Green (in press). A chronological list of A.G. Smith's publications can be obtained from the editor.

2

Forward to the past: changing approaches to Quaternary palaeoecology

Frank Oldfield

SUMMARY

1. This chapter considers some of the ways in which, over the last decades, research in Quaternary palaeoecology has reflected changes in the inferential structures within which problems are defined and research pursued.
2. A sequence is outlined from inductive, through deductive, to what are here termed projective modes of problem definition and research development.
3. The implications of each are explored and illustrated, and some of their possible strengths and weaknesses considered.

Key words: Quaternary, palaeoecology, philosophy

2.1 INTRODUCTION

The aim of this chapter is to explore structures and modes of reasoning that have been prevalent in Quaternary palaeoecology over the last few decades. Most reviews deal with methodology and results. This seeks to delve beneath them and to consider instead the ways in which, explicitly or more often implicitly, changing research priorities have both moulded and grown out of changing inferential structures.

The main contention presented here is that Quaternary palaeoecology, having developed from a long period of predominantly inductive activity into a recent phase during which deductive reasoning has played an increasingly convincing and important role, is now becoming strongly influenced, if not dominated, by what I here term a **projective** mode. The logical and scientific consequences of this bear further, albeit speculative, examination. As an unrepentant Popperian, I have welcomed the emergence of more rigorous deductive frameworks within the subject and I fear for their partial demise. So my starting point for comparative evaluation is a conceptual model giving my own version of a Popperian, essentially deductive, framework for scientific investigation (Fig. 2.1). It is incomplete, not least because it stresses *activity*

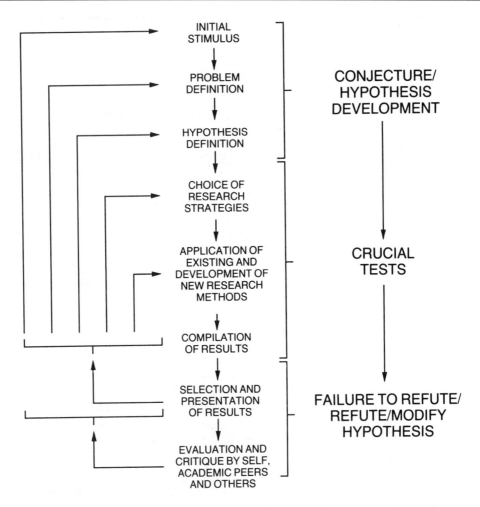

Fig. 2.1 A partial and simplified conceptual model of the deductive mode of research. The generalized terms to the right of the model are those of Popper (1963); the return arrows to the left are intended to represent the way in which results from one phase of deductive research feed into the next.

without adequately defining important contextual aspects such as the *frame of reference* within which stimuli arise and, related to this, the nature of the *assumptions* that feed in to the business of problem definition, the specification of hypotheses and the choice of research strategies. Nevertheless it is, for me, a kind of idealized template with which to articulate and compare alternative investigative frameworks.

2.2 INDUCTIVE INVESTIGATION IN QUATERNARY PALAEOECOLOGY

A large proportion of the research of Quaternary palaeoecologists has begun, not with the route to hypothesis development and testing outlined in Fig. 2.1, but from the adoption by individuals or groups, of some area of ignorance as their area of concern. Areas of ignorance have included, at different times and stages, themes such as the Holocene and Pleistocene antecedents of the present British flora, the nature and timing of vegetation changes in the Late Devensian or early Holocene, the causes of the elm decline, the environmental impact of Mesolithic peoples in upland Britain, the course of land- and sea-level change on a regional scale, or the pattern of terrestrial environmental change during interglacials. The availability of sites, techniques and of predominantly academic incentives to explore such issues, has led to the development, through comparative evaluation and synthesis, of what are in effect coherent myths, long before it is possible within these to identify potentially refutable hypotheses. I use the term myth in an entirely non-pejorative sense, for as Popper (1963, p. 38) makes clear, 'a myth may contain important anticipations of scientific theories'. By way of illustration, I have only to reflect on the translation from myth to established canon, during the span of my own career, of the notions of 'continental drift' and of 'Milankovitch cycles'.

Adopting the framework provided in Fig. 2.1 as much as seems reasonable, we may consider more fully what the concepts of stimulus, strategy, methodological development, and evaluation and critique imply within the more inductive inferential framework. Stimulus may come, for example, from the discovery of a new site or section; from a passionate interest in a particular time interval, sequence of past events or region or landscape type of exceptional geological or ecological interest or aesthetic beauty; from a perceived need to 'fill in' spatial or temporal lacunae in the record; or from a concern with the nature of long-term processes. Work undertaken in this way often seems to exist in a state of tension between several partially antithetic poles of emphasis, such as, for example, between site-specific value as against generalizable significance; between illumination of the past for its own sake rather than understanding of contemporary environmental problems; between predominantly descriptive or, in its stead, more process-oriented or functional appraisal; or between the application of established techniques in preference to the development of new ones.

Seminal empirical work within this type of framework can stem from the insight with which the investigator reconciles these tensions through a skilful union of sites, problems and techniques in the light of the existing record and current wisdom, and in response to issues of more fundamental theoretical interest. Examples of such work spanning the last three decades would include primary research shedding light on substantive questions as diverse as rates of plant migration (Watts, 1973), fire incidence and periodicity (Swain, 1973), and the role of plant pathogens in the mid-Holocene (Davis, 1981), as well as on methodological issues like the problems of radiocarbon dating in fossil soils (Dresser, 1970) the nature of pollen dispersal (Tauber, 1965), the limitations of relative pollen diagrams (Davis, 1964) and the value of statistical rigour in palaeoecological interpretation (Birks, 1985).

Equally important in the development of Quaternary research are the attempts at empirically based synthesis without which the inductive approach would tend towards formlessness and lack of direction. These provide the synoptic coherence from which explanatory 'myths' arise, as well as the conceptual framework within which further research may advance the subject. There are inevitable parallels with Wright Mills' (1959) discourse on 'overarching theory' and 'abstracted empiricism'. Quaternary palaeoecologists may generally eschew the former, but they are not beyond generating a fair amount of the latter, much of which gains shape and significance at a later date, if at all, through processes of critical selection, comparison and aggregation.

Most of the above falls some way outside Popperian notions of testing and refutation. Rejection may occur, but less often through processes of deductive logic than through imputation or acknowledgement of technical inadequacies, limited data or misplaced reliance on ambivalent methodologies. Favourable peer-group recognition for research arises not so much from failure to refute any hypothesis it may propose, as from its relationship to current research concerns, fashions or paradigms, the quality of the basic data, the apparent explanatory power over and above the purely descriptive value of the work in areas of more theoretical or synoptic concern and the contribution it makes to enhancing techniques and developing attractive methodologies. In Popper's (1963, p. 98) terms, 'degree of corroboration' is a more relevant concept here than is the notion of formal refutation. He also states that 'by trying hard and making many mistakes, we may sometimes, if we are lucky, succeed in hitting upon a story, an explanation, which "saves the phenomena"'. This has a familiar ring for the palaeoecologist.

Two problems can be signalled at this stage in anticipation of their renewed consideration later. Much of the eventual value of high-quality empirical work in the type of predominantly inductive mode discussed here is *latent*. Its full acceptance can lag behind its completion and publication, for it must often await a sufficient accumulation of related data, an improvement in interpretative insights and skills, or a shift in research priorities to a frame of reference congenial to its favourable reappraisal. There is thus an important, and indeed familiar and inevitable, sense in which the conceptual framework of the time will determine the perceived value of a piece of research. This presages the second problem, for interpreting palaeoecological data is rarely a matter of unambiguous, objective certainty. The search for the creation of necessary coherent myths often requires degrees of reinforcement (Watkins, 1971) and selectivity that carry a heavy toll in terms of vulnerability and scepticism.

2.3 DEDUCTIVE STUDIES: THE ACID RAIN RESEARCH

Over the last two decades especially, an increasing proportion of Quaternary palaeoecology has had a distinctly more deductive flavour. This had required the identification of a conjunction of environmental contexts, techniques and conceptual models that allow reasonably rigorous retrospective testing of hypotheses whether about 'applied' problems such as the timing and origin of eutrophic conditions in freshwater lakes (Battarbee, 1978, 1986) or 'purer' concerns, for example with the nature of peat-bog regeneration (Barber, 1981).

Much of the work recently published on the history of surface-water acidification on both sides of the Atlantic (Battarbee *et al.*, 1990; Charles *et al.*, 1990) can be used to illustrate the type of Popperian, deductive framework of scientific reasoning outlined in Fig. 2.1. The initial stimulus for the work arose from confirmation of the current acidification of lake waters and of the economic and political consequences, and also from the realization that the causes of acidification were in doubt, that questions of timing and relation to other ecological, environmental or geochemical changes in the past were crucial to resolving the doubt, and that the palaeolimnological approach had reached the stage of addressing such issues successfully and promised to help to resolve outstanding doubts about timing. Note that, whereas the scientific stimuli central to the inductive mode as described above are usually defined from within the discipline and are thus, in Passmore's (1974) terms, 'problems in ecology', the greatest stimulus here is external to the whole scientific framework. It is the perception of an 'ecological problem' that, as Passmore points out, exists outside the discipline and is by definition also actually or potentially an economic, political, cultural and technological problem. However, as we shall see, this external agenda setting, of itself, does not weaken the inferential framework available, need not distort or diminish the quality of the research completed in response to its stimulus, and certainly does not force the neglect of, or in any way devalue, the corpus of inductively derived material available at the beginning of the research. Nevertheless, it asks fresh questions of it and builds from it in new ways.

The central definition of the task for palaeolimnologists was the reconstructing and dating of pH variations in selected acidified lakes from the evidence contained in their sediments. At the outset, demonstrating the potential of the palaeolimnological approach depended on an immense body of pre-existing knowledge, especially about the pH preferences of diatom taxa in contemporary waters. Without this prior work and the painstaking taxonomy that necessarily underpinned it, the palaeolimnological contribution to acidification studies would have been hard to envisage, but despite this, the questions posed and the statistical confidence required, forced the researchers involved into a new wave of more rigorously quantitative analyses of present-day diatom populations and their pH ranges.

The research task lent itself to a classic, deductive, hypothesis-testing approach, for a range of hypotheses had been proposed, the main ones being that lake-water acidification was the result of 1. natural long-term or more short-term recurrent processes operating within the timescale of the whole Holocene, 2. of recent catchment afforestation by conifer species, 3. of catchment soil and water acidification consequent on the abandonment, changed or reduced management of upland areas, or 4. of industrialization and especially the combustion of fossil fuels in power generation. Overtones of political and economic expedience, of bias and of entrenched preconceptions have, at times, made the results of the work bitterly controversial, especially in relation to the potential role of the land-use hypothesis (Rosenqvist, 1977).

I would claim that failure to agree on the nature and definition of the hypotheses being tested and failure on the part of the land-use protagonists to appreciate fully Popper's logical distinction between the structures of reasoning associated with corroboration and falsification have helped to fuel the controversy. For example, if the hypothesis being tested is 'soil and water changes arising from farm abandonment or the spread of moorland in lake catchments can lead to water acidification', the hypothesis can be falsified only by a total absence of evidence to support it at *all* sites. Moreover the procedure to 'test' the hypothesis can be inductive, the search for sites selective and the lines of inference essentially corroborative. In Popper's terms, such a formulation hardly constitutes a hypothesis, partly because the probability of falsification is so low.

The inferential framework changes if we substitute the hypothesis that 'soil and water changes arising from farm abandonment or the spread of moorland in lake catchments are alone sufficient to account for most instances of lake water acidification'. This opens up the opportunity to falsify by regarding the history of lake and catchment changes as potential partially controlled retrospective experiments. Not only can paired catchments with and without land-use change be directly compared, but the hypothesis can be tested in a wide variety of contexts wherever land-use change can be excluded or controlled for, within the limits of the methods available.

This last qualification sounds a necessary note of caution, for *post hoc* experimental control in the 'real' environment is never perfect. We are 'coaxing history to conduct experiments' (Deevey, 1969, p. 40). Not only has history lacked the foresight to oblige us completely, but our methods will always fall short of the task, even where she has done her level best. This means that retrospective falsification tests have to be carried out in as many as possible of the contexts that favour them. The results of testing the hypothesis *as defined* are unambiguous, partly because of the definition and partly because of the consistency of the results across a wide area using different approaches to eliminate or to characterize possible land-use effects.

In summary, my contention is that the controversy surrounding this aspect of the acidification debate relates not merely to vested interests, prejudice or a genuine conflict of evidence, but also to a conflict of inferential frameworks and modes of investigation as between inductive, corroborative reasoning and the search for deductive falsification. The power of the latter framework surely prevails, but, as foreshadowed in the earlier section and illustrated here, the two modes are strongly interactive. For example, the inductive corroborative approach can develop the coherent myths within which the 'palaeoexperimentalists' can delve for opportunities to define, test, refute and modify hypotheses.

In exploring the crucial importance of explicit and unambiguous hypothesis specification, we have anticipated somewhat the issue of **strategy choice**, for the crux of the palaeolimnological approach to the problem has been to select sites for study that allow the testing and selective refutation of the competing hypotheses (e.g. Battarbee *et al.*, 1990). Achieving tests of sufficient rigour has depended on and promoted the development and application of a great variety of techniques, both well established and newly conceived. Three of the most important benefits of the research have been 1. an acknowledgement of the immense value of preceding inductive studies, 2. the stimulation of improved insights into many aspects of the basic science involved through research into subjects as diverse as the ecophysiology of diatoms, the chronology of recent sedimentation and the statistics of transfer functions, and 3. the welding together of a diverse range of techniques into a coherent methodology, despite the constraints imposed by problems of sediment sampling and the low volume of core material available for multidisciplinary analyses.

The upshot of all this coordinated work has been the refutation of all but one of the hypotheses about the *general* causes of acidification (as distinct from the contributory causes at particular sites). Built into the programme of research are many strategies designed to limit and if possible eliminate the dangers of spurious reinforcement. These take the form of multiple testing of all hypotheses in a range of environmental contexts, reconstructions of pH change using a variety of models and alternative groups of organisms, and the avoidance of single lines of evidence either for chronologies or for contaminant deposition histories. The empirical evidence is internally consistent and overwhelming, the rigour of the deductive reasoning unquestionable and the broad conclusions inescapable. The conclusions are sufficiently robust and quantitative to provide an adequate empirical basis for evaluating simulation models of acidification designed to improve predictive capability, and this foreshadows one of the main goals of the 'projective' mode considered in the next section. The approach is now being refined to address a series of redefined questions through testing modified hypotheses relating to issues such as catchment and lake recovery following declining sulphur dioxide emissions, the additional impact of catchment afforestation on lake-water acidification and the long-term effects of liming. Returning to Fig. 2.1, we may regard these second-generation studies as following the return arrows within the conceptual model.

A word of qualification and of warning concludes this section. Palaeoexperiments inevitably lack the rigour of true experimental science. Whilst a combination of replication, statistical validation, ingenuity and intellectual honesty can limit and constrain spurious reinforcement, the circumstantial nature of so much *post hoc* evidence, and the judgemental nature of critical aspects of sampling and interpretation will still influence the conclusions drawn. Under these circumstances, deductive frameworks alone are no guarantee against misinterpretation, as this author knows to his cost.

2.4 THE FUTURE

There is increasing evidence for the potential importance of human-induced forcing in the course of global change. A growing scientific consensus has shifted the balance of research priorities and resourcing, and bids fair to establish a major new research agenda, not simply for the duration of a passing fad, but for several decades.

The research agenda, from which evolved the deductive successes outlined in the previous section, required the Quaternary palaeoecologist to establish the immediate antecedents of environmental problems once they had been detected and, all too often belatedly acknowledged as worthy of concern, not merely for their scientific interest but above all for their economic, social and even political implications. Now, the research agenda is being set not by what has come to light as a result of changes in the past, but by projections of what *may* happen, given certain assumptions, in the future. This 'projective' rather than *'post hoc'* question-setting structure casts every research problem we may consider in a new and different light, irrespective of whether our concern is an environmental (hence inevitably economic, technological and policy) problem, or a 'problem in environmental science' (i.e. one set within the internal concerns of the research discipline). It is perhaps too soon to identify fully and unambiguously how things look in this new and different light, but not too soon to sketch out a few impressions.

The primary stimulus for the new, 'projective' research agenda arises from

1. Apprehension about the course and consequences of climatic change in the near future resulting from 'greenhouse' warming;
2. Realization that the causes of short-term climatic change are poorly understood, as are the likely impacts on other environmental systems;
3. Acknowledgment that many recent environmental trends are inadequately understood;
4. Indications that partial analogues for a warmer world or for periods of rapid warming may have existed in the geologically recent past;
5. Some hope, in the light of 4., that proxy data relating to past periods may be of value in helping to determine the causes and consequences of future change.

Of course, such considerations are strongly mediated by mechanisms of research funding, reflecting the extent to which policy priorities have understandably shifted in response to anxieties about the future.

At the outset, two somewhat opposed notions already foreshadowed in an earlier section strike me forcibly: latency and reinforcement. Piecing together the basis for reconstructing past climates from proxy records requires the establishment of rigorously evaluated databases. If the pattern of current research investment is any indication of future value, they may well be among the most significant contemporary research developments in the field. They will often be built, at least initially, on data acquired from past research not directly related either to the present research agenda or to the task in hand for the database builder. Nevertheless, the quality of the data is of critical concern and its current significance gives 'posthumous' validation to the best products of the inductive approach, and especially to its emphasis on the retrieval of high-quality detailed information rather than on the more intellectually economical selection and presentation of no more than is necessary to test hypotheses. It illustrates well the notion of latency in research data and reinforces the case for the comprehensive archiving of samples as well as data.

It should also be a warning both against being over-dismissive of the inductive approach at its best and against valuing palaeoecological works solely by the perceived priorities and enforced expediencies of the present. Set alongside this realization is a fear of moving away from inferential frameworks that permit rigorous tests of falsification, into (it is tempting to write **back** into) the mutual inference systems so familiar in Quaternary science and so open to beguiling and persuasive massage by the reinforcement syndrome (Watkins, 1971). In place of scientific rigour we are in danger of substituting a search for mutual compatibility between overlapping uncertainties. Where the sources of error are solely statistical, this is conceptually valid and progressively refinable. Most of the uncertainties both in modelling and in reconstructing past climate from proxy records, however, stem from errors and biases that are not statistically controlled. Such inferential frameworks tax the objectivity of the investigator to the limit and, sometimes, to destruction, since the only way to communicate and promote the value of the work externally, or even to find personal intellectual reward, is to highlight or even impose a degree of coherence or convergence at best hopeful, at worst spurious.

In the case of research on climatic change and its possible future course and consequences, many additional factors compound the difficulties, irrespective of whether our concern is the primary one of predicting the course of climatic change at global and regional level, or the secondary one of articulating the possible consequences of these changes on other environmental systems. For example, in relation to the primary concern, the palaeoenvironmental record contains very few situations where global scale and fine temporal resolution can be combined; in this regard, the polar ice fields may be unique. The task that lies ahead, therefore, must include a (global in so far as possible) synthesis of fine-temporal-resolution, small-spatial-scale proxy data sets, chronologically harmonized with precision and accuracy, and reconciled within mutually compatible databases. No wonder we look for some short cuts and for the rewards of dawning coherence, spurious or otherwise, on the way.

The problems inherent in the current projective research agenda are compounded by many further considerations. For example, the criteria of relevance that can be derived from concerns about future change are inevitably more ambiguous than those that arise out of defined contemporary problems already experienced. We therefore have too little objective indication of what the important questions are. Tests of mutual consistency between model simulations and reconstructions derived from proxy records often lack the recourse to verifiable reality that might independently validate the one, before it, in turn, has to do service in validating the other. The imprecise and unvalidated nature of current global climate models actually reduces the demands made on palaeoenvironmental research in terms of accuracy and precision. Coupled with the emphasis on *global* scale concerns, it tempts the researcher away from 'old-fashioned' detailed site studies towards more superficial worldwide syntheses.

The many imperfections in our understanding of the links between climate and the past and present distribution of organisms, communities and ecosystems, which inhibited an earlier generation of Quaternary palaeoecologists from making confident palaeoclimatic inferences from their data, have not been eliminated by any means. The incentive to exploit possible links has outstripped their comprehension. Only time will tell whether it has also outstripped their utility. In addition there is the complex nature of climate systems on all spatial and temporal scales, and the inadequacy in functional terms of the descriptions that we have inherited for it from earlier science and now use to express the climatic relationships of particular organisms or groups. In summary, the question 'what is going to happen next?' is very different from 'how do we come to have this mess?'.

The task ahead is truly overwhelming. The best I can offer in conclusion are questions, homilies and exhortations born more of bewilderment than of insight. There are some obvious dangers: that the rather monolithic task of building essential databases will concentrate power and influence too narrowly; that the fluency of future scenarios rooted in nothing more fertile than mutual reinforcement and hoped-for positive effects on funding will distort perceptions and responses; that external pressures responding to political timescales will generate destructive impatience in response to necessary caution and will further reduce funding for fundamental research; and that palaeoecologists will suffer a little too much from the self-assertive view that they can use current fears and opportunities to support what they can already do and want to get on with, rather than to develop new approaches that may address more closely the emerging research agenda. Nobody in the field can fail to have seen recent signs of the sort of intellectual imperialism that this last attitude can engender.

It is harder to take a detached view about the future than it is about the past, for there is a basic sense in which the past is entirely academic, and the future, by definition, is not. I believe we should try to see many of the transitory consensuses, visions of dawning coherence or convergence and of tenuous, reinforcement-based, syntheses in a more detached and existential light. Yet, the need for some detachment must not blind us to the probability that out of these imperfect and conflicting syntheses and scenarios may, perhaps inevitably will, emerge new integrative canons of greater scientific and intellectual power.

Finally, within the vast scope of the new research agenda, there are many opportunities for cultivating and extending the deductive, Popperian approach. Techniques of palaeoenvironmental reconstruction can be tested against the documented record of change for the last few decades and centuries and so refined and validated before being used to shed light on the more remote past. Techniques can be used independently to reconstruct palaeoenvironmental conditions for particular periods and places; these can then be compared, with due regard to the dangers of mutual reinforcement. Models of the relationships between changing climate and the response of other systems – geomorphic, biotic and human – can be tested by adopting the same 'palaeoexperimental' approach outlined in the previous section. All these types of activity can include reasonably rigorous tests of falsification. Within the broad research agenda we now have, they should ensure the continued health of the deductive mode and they should make crucial contributions to our response to a bracing task. This task must include responses not only to currently fashionable and directly future-oriented issues, but to fundamental questions about the nature, pattern and pace of past environmental and ecological change *per se*.

Acknowledgements

I am especially grateful to Keith Barber, Rick Battarbee, Frank Chambers, Andrew Morse and Andrew Plater for their comments on this paper. It is offered in thanks for Alan Smith's open and generous sharing of results and experience at critical stages in my career.

3

Radiocarbon dating and the palynologist: a realistic approach to precision and accuracy

Jon R. Pilcher

SUMMARY

1. The relationship between pollen analysis and radiocarbon dating has passed through three phases: pioneer, routine, and critical. These phases are examined.
2. Conventional radiocarbon dating is no longer adequate to answer the questions of rates of change, lags and leads, and absolute time intervals that are now required for climate modelling.
3. Errors of past radiocarbon dating are assessed and new approaches are suggested to allow radiocarbon dating to continue to make a valuable contribution.

Key words: radiocarbon dating, accuracy, precision, errors, wiggle matching

3.1 INTRODUCTION

Radiocarbon dating and palynology have been closely linked since the early days of radiocarbon dating. The relationship has passed through various phases: the pioneer stage when pollen analysts could obtain the first 'absolute dates'; the routine phase in which everything was dated because that was the fashion; and then the critical phase in which the meaning of dates was examined. This paper examines the three phases and then shows how a modern approach to radiocarbon dating will allow palynology to make its contribution in a world where precise calendrical timescales are increasingly important.

3.2 THE PIONEER PHASE

In the early days of palynology, the interpretation of vegetational change was easier as there was no absolute timescale. Assumptions of synchroneity of events and conclusions concerning the climatic causes of vegetational change appeared plausible on the evidence available. Take for example the Boreal–Atlantic transition. In 1949 Jessen wrote:

The climatic conditions that favoured the alder and made possible its rapid increase in Ireland in the beginning of Atlantic time must also have been felt in other parts of north-west Europe as the alder-curve in England, Holland, north-west Germany and Denmark commonly shows a marked rise in the lower part of Atlantic time. This rise gives further support to the view that the level indicated for the top of the Boreal zone is synchronous in the chronological schemes drawn up for the countries mentioned. (Jessen, 1949; p. 230)

In the pioneer phase of radiocarbon dating, attempts were made to test ideas such as those of Jessen by dating specific pollen-zone boundaries in various places. Godwin published the first Cambridge date list in 1959 and this list included the dating of zone boundaries at Scaleby Moss (Godwin & Willis, 1959).

Within a few years the radiocarbon timescale for all the major vegetational changes of the late- and post-glacial in Britain was established. In the Croonian Lecture to the Royal Society, Godwin presented a radiocarbon chronology for climatic and vegetational change in the Quaternary (Godwin, 1960). This included dating of such key sites as Upton Warren, Fladbury, nine late-glacial sites and a range of post-glacial sites from all over Europe, including some from Ireland.

By about 1965 a chronological framework was established both for British Isles archaeology and for the major post-glacial vegetational changes of north-west Europe. The concept of synchronous vegetational change was compatible with the new dates and it was generally assumed that these changes were caused by climatic change. However, Smith stated: 'While the radiocarbon age estimations so far obtained do not on the whole confute the assumptions [of synchroneity], there are a number of exceptions' (Smith, 1965, p. 331). This demonstrates that some cracks were starting to show, but in general the views of Jessen were still accepted.

3.3 THE ROUTINE PHASE

The routine phase started in about 1965 with the establishment of a radiocarbon laboratory at Belfast by Alan Smith and Martin Jope and by a change in philosophy:

At this stage in our study of ecological changes it is tempting to attempt to discover by means of radiocarbon assay whether the already-defined zone boundaries are synchronous; I suggest however, that we should approach the problem from a different angle; we may find it more advantageous to make serial radiocarbon determinations through deposits subjected to pollen analysis, and in suitable cases to chemical analysis, but independently of zone boundaries. We could then attempt to define regions in which similar vegetational events had taken place at the same time and the study of the post-Glacial phytogeography of our native trees could begin on a new basis. (Smith, 1965: pp. 335–6)

This was the ideal and it was with this philosophy that several long profiles were dated in the 'routine phase' by the Belfast laboratory (e.g. Pilcher, 1969; Goddard, 1971; Pilcher, 1973; Holland, 1975; Pilcher & Smith, 1979; Pilcher & Larmour, 1982; Francis, 1987). The results of this approach were very encouraging. Many of the bogs and lakes showed uniform deposition rates (Fig. 3.1) and much use was made of dates interpolated from deposition-rate curves rather than individual date determinations.

By 1973 there were sufficient radiocarbon-dated pollen diagrams from Ireland for a first synthesis to be attempted along the lines of Godwin's 1960 paper (Smith & Pilcher, 1973). The radiocarbon dates appeared to show a marked lack of synchroneity in most of the traditional pollen-zone boundaries with the elm decline standing out as the exception (Fig. 3.2).

3.4 THE QUESTIONING PHASE

The collation of dates for Smith & Pilcher (1973) brought to light discrepancies in how the presence, absence, increase or decrease of a species was defined. This paper thus ushered in the start of the questioning phase.

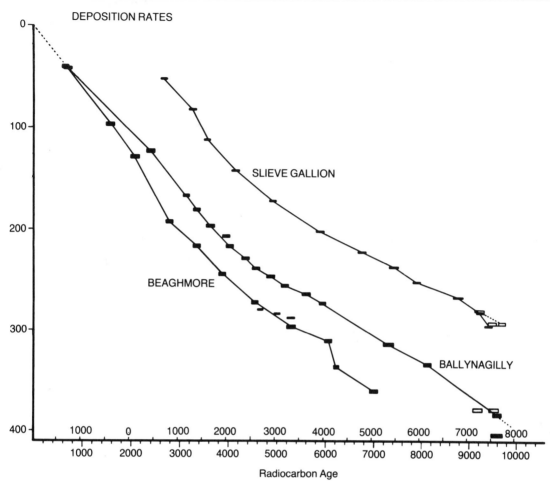

Fig. 3.1 Three deposition-rate curves from bogs in the north of Ireland, measured in the Belfast radiocarbon dating laboratory, showing uniform deposition rates and no inversions. The conventional, uncalibrated, radiocarbon age is shown as ad/bc (above axis) and BP (below axis). Depths are in cm.

Even where the pollen recruitment processes are well understood, there will always be some uncertainty in trying to guess the point at which a plant is locally present rather than contributing to the pollen rain from some distance.

Alan Smith's research student Quentin Dresser tackled the problem of age differences among the different components of peat (Dresser, 1970; Smith *et al.*, 1971). Dresser used physical and chemical separations to divide peats and muds into different fractions, and then radiocarbon dated these using the existing Belfast gas counting technology. The research showed up a number of inconsistencies, particularly in blanket peats, and left a number of unanswered questions. Dresser's study suffered from the lack of a base-line chronology against which to test the various peat fraction dates. There was no 'right' answer for the date of the samples analysed. In some cases Dresser picked the oldest fraction on the basis that most contamination was likely to be by downward percolation of younger material. One outcome from this work was the development of a technique for

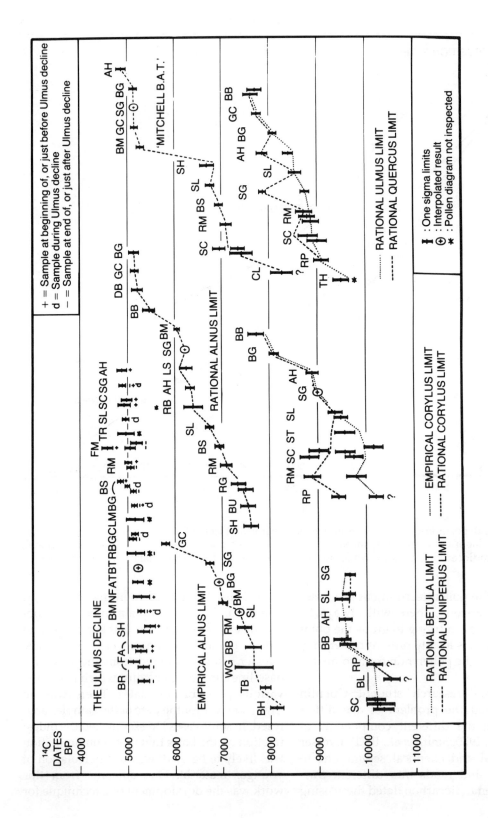

Fig. 3.2 Distribution of radiocarbon dates for conventional pollen zone boundaries in the British Isles (from Smith & Pilcher, 1973). If the dates for the elm decline are taken as multiple estimates of a synchronous event, the standard deviation of the dates is ± 290 years. For the Irish samples the standard deviation is ± 234 years.

extracting a 'fine particulate' fraction of peat (< 250 μm) that was a compromise between the ideal of dating the pollen grains and the practicalities of sample size.

It would now be possible to tackle this again using the latest technology and obtain a much greater insight into peat component ages, but this will only be possible using both the highest precision conventional dating and also accelerator mass spectrometer (AMS) dating on discrete fractions of known botanical composition such as seeds or pollen (Andree et al., 1986). In addition, there is now the possibility of testing a range of chemical and physical fractions of peat that enclose a tephra layer of known age (for example, the historically recorded AD 1104 eruption or the tree-ring dated 1159 BC eruption of Hekla in Iceland (Hekla 3), tephra of which can be found in Ireland). The tephra will provide a baseline date against which all the measurements can be compared. In many projects, however, it is the date of events in a pollen diagram that one is seeking, so the date of the pollen itself may provide the only truly valid chronology.

The third part of the questioning process came about with the advent of dendrochronology. With considerable foresight, Alan Smith allowed dendrochronology to get started at Belfast. It was a reluctant decision because it was clearly going to divert resources away from pollen analysis and it was not at all clear that it was going to be of any value – either as a dating method or as a component of the broad-based environmental research of the Belfast Palaeoecology Laboratory. However, Suess (1970) had just published his long radiocarbon calibration curve and the many questions that this posed made the construction of an alternative tree-ring chronology of international importance.

The tree-ring research programme benefited from the presence of an in-house radiocarbon laboratory. Since the age of each bog oak collected for this project was unknown, radiocarbon dating was used to place small sections of tree-ring chronology in the right relative order (Smith et al., 1972). At the same time as the tree-ring chronologies were growing, Gordon Pearson was refining the radiocarbon measurement process towards a goal of a precision better than ± 20 years and, more importantly an accuracy (lack of bias) that was equally good (Pearson, 1979).

By 1977 the Belfast laboratory had completed long sections of tree-ring chronology (Pilcher et al., 1977) and was able to publish its first section of radiocarbon calibration (Pearson et al., 1977). Similar work was in progress in Seattle, and in 1986 an internationally agreed calibration for the last 4000 years was published (Pearson & Stuiver 1986; Stuiver & Pearson, 1986) with the Belfast section extending back to 7000 years (Pearson et al., 1986). A calibration back to about 9000 years should appear soon.

Several authors are now starting to use calibrated dates in a realistic way (e.g. O'Connell, 1990b), but it is clear that a calibration curve with wiggles severely limits the ability of the method to resolve small time differences and to answer the phytogeographic questions of plant migrations that the pollen analyst might be asking. In spite of the fact that the calibration curves have been published for five years, many authors are still not taking them into account, even where calibration may profoundly affect the interpretation (see, for example, Gear & Huntley, 1991; Bennett et al., 1990a).

As tree-ring chronology building proceeded, a large number of wood samples that had been radiocarbon dated by various laboratories were then absolutely dated by dendrochronology. For the first time the radiocarbon method was subjected to a rigorous independent test using samples whose age was known to the year of formation. It did not fare well (Baillie, 1990a). Of 62 dates, two were out by 1000 years. With these outliers removed, the others could be taken as reasonable estimates of the true age if the standard deviations were each rather more than 200 years. Further evidence for the inaccuracies of radiocarbon dates come from two international intercalibration studies, the first in 1982 (International Study Group, 1982) and, more recently, in a more comprehensive study (Scott *et al.*, 1990). There was only a slight improvement between the two tests and both showed that radiocarbon laboratories were in many cases showing both poor reproducibility and considerable bias. From all these sources, one must conclude that all existing radiocarbon measurements are reasonable estimates of the nearest half millennium in which the sample falls.

It is worth making a reassessment of the errors associated with the dating of long pollen profiles in the Belfast radiocarbon laboratory and elsewhere. The samples for these profiles were usually measured in a single batch and often sequentially. Thus many of the bias errors, particularly those associated with the measurement of background and standard, will have remained constant during the batch. Thus the individual samples will be much more accurate *relative to each other* than on an absolute scale. This probably explains why there are relatively few inversions and anomalies in the deposition rate curves – except in those relating to the late-glacial, which is a different problem (Pilcher, 1991).

The second factor that helped to give the illusion of accuracy is that the sample width was typically about 4 cm, giving a time span of about 100 years for the sample. This would tend to iron out some of the sharper fluctuations seen in the calibration curve and again reduce the chance of inversions. The problem occurs when one attempts to correlate one such radiocarbon-dated profile with another, because here the absolute dates matter and the bias affecting one batch of dates will influence any comparison.

One other way of investigating the errors of these dates of the 1960s and 1970s is to look at the elm decline dates. For the sake of this argument we will use the elm decline simply as an event that has been extensively radiocarbon dated and make no comment on its real origins or time spread. If the elm decline dates are taken as replicate measurements of a single instantaneous event this allows us to assess the total errors associated with the dating, including the error associated with the interpretation of the pollen diagrams. If this is done (using only the Irish dates given in Edwards, 1985), the mean is 5055 BP and standard deviation 225 years. The full set of British Isles dates in Smith & Pilcher (1973) have a standard deviation of 290 years. Gandreau & Webb (1985) give 300 years for the standard deviation of the hemlock decline in the USA, and for the pine decline dates in Smith & Pilcher (1973), the standard deviation is 234 years. The significant thing about these calculations is that the standard deviations are very close to the estimates based on the wood samples described above. To take a positive view of these results, the ability of radiocarbon laboratories to replicate dates cannot have been routinely much worse than about ± 225 years.

From these results we know that for a large range of sites and several laboratories, the real errors of radiocarbon measurement cannot have been better than a mean figure of ± 200 years, but are also unlikely to be much worse than about ± 225.

Thus at the end of the questioning phase we have the following conclusions.

1. The routine radiocarbon dates available from most laboratories have been shown to have inaccuracies that only allow the calendrical date to be specified to the nearest half millennium. Attempts to interpret time differences or real ages closer than 500 years by conventional dates are simply not valid. Conventional radiocarbon dating is thus of little value in studies of migration rates, vegetation lags, etc.

2. Calibration may increase the age span in calendar years where a single radiocarbon content may relate to a number of different

real ages (as for example in the periods 400 to 800 BC and AD 1650–1900).
3. There is no consistent basis for picking events to date in pollen diagrams. In particular, there is no consistent basis for deciding when a species has arrived in a particular location.
4. The various components of peat and lake muds are likely to have different real ages (and radiocarbon ages) and these ages may differ from that of the pollen.

If this is accepted then there are now very few sites or situations in the British Isles where the palynologist could justify the use of conventional radiocarbon dating for the Holocene. We already know the answers to within the best precision that is obtainable.

3.5 IS THERE A FUTURE?

The chronological questions that the palynologist should now be asking are different from the questions being asked a decade ago. With the current emphasis on climate change, the interest is now turning to rates of vegetational change. How fast did the climate change at the end of the late-glacial? How quickly do species migrate? How much lag is there between vegetation and climate? We now need to be able to look at rates of change and we need to be able to assess time differences of a few centuries rather than the half millennium that could be resolved by the radiocarbon dates generally available up to now. Let us look at how this might be achieved.

3.6 A HYPOTHETICAL PROJECT

Suppose a pollen analyst wishes to determine the time difference between the decline of pine in the lowlands and highlands of a particular region of the British Isles. First let us consider the choice of sites. Uncertainties in the origin of organic matter in lake sediments argue against their use in this type of study, and the very slow deposition rate at the base of blanket peats might make them unsuitable. The choice has to be a valley bog or raised bog.

Decisions must next be made about the method of radiocarbon dating. There are two possibilities. Either one must use AMS dating and a fraction of the material that is biologically distinct (such as the pollen) or one must go for high-precision dating. The AMS dating has the great attraction of being able to date very small samples such as the pollen itself or specific seeds (Andree et al., 1986; Cwynar & Watts, 1989). Ideally an AMS laboratory would like about 5 mg of carbon, which is equivalent to about 15–20 mg of dry pollen. If we assume that, at best, 0.1% of a dry peat sample is pollen by weight, 20 mg will be found in 20 g dry peat or 200 g wet peat. Thus a peat sample of 10 cm × 10 cm by 2 cm thickness would be adequate. The 2 cm thickness in a raised bog should represent no more than 50 years. The precision of the AMS date will be ± 60–80 years. The intercalibration study mentioned above suggests that errors quoted by AMS laboratories are generally realistic; thus there will be a 95% chance of the date lying within ± 160 years and a reasonable chance of getting the real age (after calibration) correct to within a 400-year band.

If this is still not adequate to answer the questions being asked, we must turn to high-precision dating. In theory this could give a precision of better than ± 20 years. However, the requirements are for an ideal of 16 g of carbon. This represents about 200 g of dry peat or 2 kg of wet peat. If we wish to carry out some extraction of rootlets and to remove humic acid we should start with 4 kg (the ideal of extracting pollen is clearly impossible in this case), and this quantity can be obtained from a slice 1 cm thick and 40 cm × 40 cm. However, there will be serious

Table 3.1 Effect of different sizes of peat sample on radiocarbon dating error

Sample	Cutting error (years)	Thickness age span (years)	Radiocarbon date error (2σ) (years)
10 × 5 × 2 cm (1 g C)	10	± 25	± 200
20 × 10 × 2 cm (4 g C)	25	± 25	± 100
20 × 20 × 2 cm (8 g C)	25	± 25	± 80
40 × 20 × 2 cm (16 g C)	50	± 25	± 40
40 × 40 × 1 cm (16 g C)	100	± 12.5	± 40

problems of non-horizontal microstratigraphy in attempting to cut such a slice. On the other hand, if each 1 cm represents 25 years we will lose precision by increasing the sample thickness. If we take every 1 cm deviation in cutting the slice to cost 25 years of error and weigh these errors against the precision that various sample sizes are capable of yielding (Table 3.1), we can select the best compromise for a particular site.

Thus, to keep the sampling error reasonable we might compromise on a 2 cm slice and compromise on total carbon of half the optimum amount of 16 g. The combined errors of this option give ± 100 years for 95% confidence limits. In other words the date will fall within a 200-year band rather than the 400-year band of the AMS result. This is better than the AMS can offer, but is it good enough? In many cases the answer is probably no. We might be looking for age differences of only 200 years in total. It is possible to measure small samples to a better precision than indicated above by using longer counting times, but to halve the precision requires four times the counting time or the cost of three additional samples. If ± 100 years is still not good enough, and for our

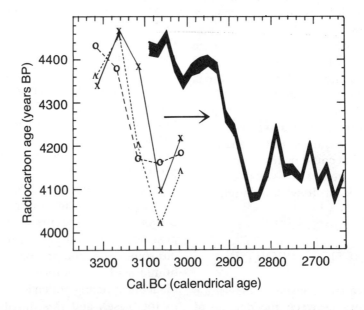

Fig. 3.3 A portion of the radiocarbon calibration curve from Pearson et al. (1986), with three simulations of a wiggle match based on five consecutive samples.

Table 3.2 Five model distributions of radiocarbon dates produced by adding Gaussian noise to the radiocarbon age of five contiguous slices of peat, each of 50 years' duration, centred on a calendar age of 2900 BC

Calendar age BC	Radiocarbon age BP	Models				
		a	b	c	d	e
3000	4360	4274	4456	4442	4344	4366
2950	4400	4382	4419	4387	4473	4445
2900	4260	4345	4276	4173	4391	4207
2850	4080	4070	3929	4167	4102	4030
2800	4180	4128	4182	4189	4221	4146

hypothetical project it would not be, we must turn to a further refinement of radiocarbon dating – the wiggle match (Pearson, 1986; Clymo et al., 1990).

Wiggle matching is a refinement of radiocarbon dating that makes use of the short-term variations in the calibration curve. Some time periods are more successful than others and there is no guarantee of success. The requirements are for about four to seven high-precision samples spaced at about 25 to 50 years. In the present example, we again have to compromise because the 16 g of carbon leading to an ideal precision can only be obtained at the expense of serious stratigraphic errors. Using the 2 cm slice (of 50 years' duration), five contiguous slices could be dated. If we take the total one-standard-deviation errors of each slice to be ±50 years, we can model the results of such a calibration. First we make the assumption of contiguous 50-year samples spanning a calendar age of 2900 BC (roughly the start of the pine decline dates in Ireland). From the calibration curve we can read the radiocarbon content representing this real age, then add Gaussian noise to model what radiocarbon dates one might obtain. Table 3.2 shows five such model distributions. Three of these are plotted in Fig. 3.3 beside the calibration curve to show how successful the wiggle match might be.

The best estimates for the true age of 2900 BC for each of the models are

model a   2900 ± 20 Cal. BC
model b   2880 ± 20 Cal. BC
model c   2900 ± 20 Cal. BC
model d   2890 ± 20 Cal. BC
model e   2920 ± 20 Cal. BC

Thus, this approach, even allowing for less than optimum radiocarbon precision, is capable of defining the true age very much more closely than a conventional approach. In some of the models the date is correctly estimated to within 10 years of the calendrical age and the worst estimate is only out by 20 years.

It is clear from the calibration curve that the example above was chosen to illustrate the technique where the calibration curve showed a major fluctuation. Suppose we take another example. In this case we wish to date a peat layer containing ash of the Hekla-3 eruption. On tree-ring evidence this eruption is thought to date to 1159 BC. Again five models are taken for five hypothetical samples spanning 1159 BC. The best estimates based on wiggle matching of these models are as follows:

model a   1160–1110 Cal. BC
model b   1090–1050 Cal. BC
model c   1190–1150 Cal. BC
model d   1180–1140 Cal. BC
model e   1140–1110 Cal. BC

While clearly not as accurate as the previous example, these dates are all within 100 years and still represent a significant improvement over an individual date.

Now we have a technique that is capable of providing useful answers to phytogeograph-

Now we have a technique that is capable of providing useful answers to phytogeographical questions. Its use is only possible if the line of thinking given above is followed and considered *before the sampling stage*. One of the many implications of this is that sites must be chosen at which sampling by excavation is compatible with conservation interests. Estimates of sample size (given above) suggest that some 40 replicate cores would have to be taken and presumably 40 replicate pollen analyses used to correlate from core to core – not a practical proposition. As the two model examples show, it is also worth considering where on the calibration curve the samples are likely to fall before investing in an expensive wiggle match.

To return to replication, it is clear that the limiting factor is going to be cost. At the present time a single high-precision or AMS date costs about £350-400 and a wiggle match of five high-precision dates would cost about £2000. Some replication is essential when considering phytogeographical problems.

It is to be hoped that the newly discovered tephra layers in Scotland (Dugmore, 1989) and Ireland will provide at least an outline chronology for some of the palynological dating problems and release more radiocarbon resources for the particular problems that only the very best of radiocarbon dating can solve.

4

Great oaks from little acorns...: precision and accuracy in Irish dendrochronology

M.G.L. Baillie

SUMMARY

1. Dendrochronology is providing precise calendrical dates, on a routine basis, in northern Europe. Long oak chronologies provide a chronological backbone for both archaeology and environmental studies.
2. Three dating exercises are used as vehicles for discussion. The first demonstrates the interaction that is possible between precise tree-ring dates and historical information. The second explores some of the limitations on chronological interpretation, even when precise dates are available. The third discusses defoliation in medieval Irish trees and attempts to separate the effects of humans from those of climate.
3. There are now no major problems with the methodology of oak dendrochronology; insertion of precise dates into pre-existing archaeological and environmental records will force a tightening-up of all aspects of chronology.

Key words: dendrochronology, exact dating, environment, defoliation

4.1 INTRODUCTION

Serious dendrochronology in the British Isles started on 1 August 1968 when Alan Smith and Martyn Jope, directors of the then Palaeoecology Laboratory at Queen's University, Belfast, employed a research assistant to investigate the possibility of constructing a long oak tree-ring chronology in northern Ireland. No one, at that time, had any idea of what lay ahead.

What did lie ahead was a remarkably rapid 'filling up of time'. It turned out that it was possible to construct a robust methodology for the tree-ring dating of long oak ring-patterns. Site chronologies were constructed for living trees, building timbers, archaeological sites and naturally-preserved assemblages of bog oaks. These robust chronologies, often many hundreds of years in length, were shuffled into chronological order, cross-matched and extended so that, in the time it took living oaks to put on a mere sixteen growth rings, a year-by-year record of past oak growth had been constructed back to 5289 BC. The story of that 7200-year chronology is already extensively documented (Baillie, 1982, 1983; Baillie *et al.*, 1983; Pilcher *et al.*, 1977, 1984; Baillie & Pilcher, 1987). What turned out to be remarkable were the very close parallels between the tree-ring work at Belfast and that in Germany during the same period. Oak chronologies of comparable length were being completed by separate groups of workers at Stuttgart and Köln (Becker & Schmidt, 1982) and at Göttingen (Leuschner & Delorme, 1984). In retrospect, this was vitally important because it allowed tertiary replication – comparison of completed chronologies from independent workers – to be applied to the European chronology complex (Brown *et al.*, 1986). By the late 1980s, tree-ring studies had extended dramatically back in time, while areal coverage had also increased. German workers could report continuous chronologies running back 9000 years (Leuschner, 1991). A new English chronology was reported back to 5000 BC (Baillie & Brown, 1988) and previous problems were being resolved, for example, with the dating of the Sweet track (Hillam *et al.*, 1990). Dendrochronology now has widespread application and the tree-ring dating of long-lived oak timbers is effectively routine in many areas in northern Europe.

Dendrochronology has engendered a new chronological revolution that will undoubtedly have far-reaching implications for our understanding of prehistory. This applies both to improved chronological control and to the types of environmental information that are beginning to be extracted from the tree-ring records. This chapter will attempt to highlight some of the possibilities offered by the method while demonstrating the ultimate limitations. Because the subject is already so vast it is impossible to be comprehensive; rather it is intended to use some case studies as vehicles for discussion.

4.2 FIRST CASE STUDY

4.2.1 An Ulster crannog

This example demonstrates both the straightforward nature of much tree-ring dating and the mileage that can sometimes be squeezed out of the precise dating of even a single sample. The sample in this case was a large oak tree-trunk, one of several, which had formed part of the fabric of a crannog (man-made island) at Corban Lough, Co. Fermanagh. This sample, Q-3017, was of particular importance because it retained sapwood complete to the underbark surface. This showed that the tree had last grown in AD 1457. Indeed, because the last ring was not completely formed, with growth being truncated immediately after the formation of the spring vessels, it was possible to specify felling in the *summer* of AD 1457. Such dating has the great advantage that it specifies *activity* related to the crannog in the middle of that year. Someone went out and felled a tree. It was then used in the make-up of the crannog. So despite the fact that the tree is not an archaeological artefact in the normal sense, the date has an immediacy that is often lacking in archaeology.

We now know, thanks to R. Warner (pers. com.), that the felling activity most probably relates to hostilities between the Maguires of Fermanagh and the O'Rourkes of Co. Sligo (to the west) in that year. Corban Lough lies only a few miles to the west of Maguiresbridge – one of the seats of the Maguires. So, in addition to the date, the context of the activity suggests strongly that this crannog was 'royal', in the sense of a site associated with the activities of the chieftain. So this tree-ring date, from a single timber, suggests that the Maguires were refurbishing a defensive crannog as part of the ongoing hostilities. The date adds some substance to what was otherwise not a very significant historical reference – indeed it suggests that the reference is based on factual information.

4.3 DISCUSSION

4.3.1 Precise dating

Of course, not every tree-ring date has this immediacy. Many samples are incomplete and the felling date has to be estimated by allowing for missing sapwood. Most sites do not produce wood at all, so the method does not enjoy the widespread application of radiocarbon analysis. However, where there are suitable samples, and when those samples are complete, then the tree-ring method allows the production of dates incomparably better than any other dating method, with the exception of written history. For example, we now know that the oak central post of the ritual 'temple' at Navan Fort, Co. Armagh, last grew in 95 BC and was felled in late 95 BC or early 94 BC (Baillie, 1988). We know that the timbers for Cullyhanna Hunting Lodge, also in Co. Armagh, last grew in 1526 BC (Hillam, 1976; Baillie, 1985). Indeed, one of the earliest precisely dated archaeological features in the British Isles is the Sweet Track in the Somerset Levels, where the timbers last grew in 3807 BC and were felled in late 3807 or early 3806 BC (Hillam et al., 1990). The earliest site may be the ephemeral Post Track, which seems to have marked out the general line of the Sweet Track and for which at least one timber was felled in 3838 BC. It should be noted that we are now talking about dating resolution that can easily separate the dates of structures like the Post and Sweet Tracks – something that would previously have been unthinkable.

4.3.2 Climatic and environmental effects

The cases cited above involved straightforward tree-ring dating. Across Europe, many hundreds, if not thousands, of timbers are being dated every year, adding to an expanding database of precisely dated sites and buildings. In addition to the dates for archaeological features, a whole range of dates are becoming available for environmental effects, deduced from such things as growth initiation phases, die-off phases and extremely narrow growth rings. Oaks could grow in regenerating systems on the surface of peat bogs for many hundreds, sometimes thousands, of years. Both growth initiation and die-off phases imply altered conditions. When such effects are observed synchronously on different bog systems, it can be assumed that climatic change is involved (Baillie, 1979; Leuschner & Delorme, 1988). In Germany, extensive studies of oaks embedded in river gravel deposits have allowed workers to trace the history of valley development during most of the Holocene. In addition, the periods of initiation of such deposits can suggest moves to wetter conditions with increased run-off (Becker & Schirmer, 1977). A surprising observation has been the close similarity between periods of widespread occurrence of narrowest rings in Irish prehistoric bog oaks and large volcanic dust veils implied from the acidity of annual layers in Greenland ice cores (Hammer et al., 1980; Baillie & Munro, 1988). This observation of effects in the tree-ring record, suggestive of climatic downturns associated with dust veils, appears to provide precise dates for what may have been significant events for human populations (Baillie, 1989, 1991a). This exposes a dilemma for archaeologists, volcanologists and others reliant on radiocarbon-based chronologies. For example, while tree-rings specify a significant, apparently volcanic, event in 1627/1628 BC (LaMarche & Hirschboeck, 1984; Baillie, 1990b), the failure of radiocarbon to specify a narrow date range for the important Bronze Age eruption of Santorini, in the Aegean (Housley et al., 1991), has led to a major controversy on the effects of that particular volcano (Pyle, 1989; Manning, 1990).

4.3.3 The 'suck-in' effect

One additional consequence of the exact specification of dates for environmental events in prehistory is what can be called the 'suck-in' effect. Because archaeological chronology, based on radiocarbon analysis, is inherently flexible, there is a likelihood of exactly dated environmental events being used to explain a wide swathe of archaeological phenomena. This may be quite justifiable, in some instances, because it is well known that radiocarbon tends to 'smear' the dates for point events (such as Santorini). However, in other cases such explanations could create completely artificial horizons. These contrary effects, with exactly dated events sucking in evidence while radiocarbon tends to spread evidence through time may necessitate a wholesale refinement of chronological studies (Baillie, 1991b).

4.3.4 Proxy data: quantitative climatic records

These environmental effects, deduced from bog and river gravel oaks, whether they relate to past dry periods or volcanic dust veils, clearly represent proxy data. In addition to such qualitative studies, there has also been a whole suite of studies aimed at the extraction of quantitative climatic records from large grids of tree-ring data. Most of these studies have involved calibration of modern tree-ring data against instrumental records. For example, Briffa *et al.* (1983, 1986, 1987) have demonstrated various levels of success with reconstructions of rainfall, temperature and pressure patterns in northern Europe. Unfortunately, so far, most of these reconstructions have been restricted to the last few centuries. Recently, however, Briffa *et al.* (1990) produced a significant summer-temperature reconstruction for Fennoscandia, back to AD 500. This may pave the way for other long reconstructions exploiting the long oak chronologies.

As noted above, the exact specification of environmental effects in prehistory has some interesting chronological consequences. The next case study exemplifies the types of problems that now have to be faced as attempts are made to date more marginal tree-ring samples.

4.4 SECOND CASE STUDY

4.4.1 An attempt to apply dendrochronology to the dating of the inner ditch at Haughey's Fort (after Baillie & Brown, 1991).

Due to the desirability of establishing an absolute calendrical date for a site, there is a natural tendency to attempt to use dendrochronology to date wood samples whose suitability is inherently marginal. This marginality may be due to an insufficient number of rings or because the sample(s) originate from a source potentially too distant from that represented by an available reference chronology. This is particularly the case where the archaeological site is important and where a precise date is highly desirable or where there is a need to establish the chronological relationship between an important site and some fixed, historical or environmental, event. Burgess (1989) has suggested that later Bronze Age hill-fort building in Britain might be a consequence of one of the apparently volcano-related environmental events identified in Irish trees – the 1159–1141 BC event. It is rare to be able to test such a theory and when the possibility does occur, special efforts have to be made to attempt to date any samples that become available.

Haughey's Fort, Co. Armagh, is an important, triple-ditched, late Bronze Age hill-fort. In 1988, Mallory & Warner presented the results of seven radiocarbon determinations associated with various contexts from the site. The consistency of the radiocarbon dates, ranging from 2833 ± 55 BP to 2923 ± 55 BP, was remarkable, given the fact that they were drawn from pits, hearths and the basal layers of the ditch. Mallory & Warner (1988, p. 39) claimed that an archaeologically satisfactory 'date' for the site would be in 'the later half of the 12th century or during the 11th century BC, in the period subsequent to the "Hekla-3 effect" ' (referring to the 1159–1141 BC event). So the question was whether the construction of Haughey's Fort was part of a response to unsettled climatic conditions in the mid-12th century BC. Although the radiocarbon evidence suggested that this might have been the case, the dating was too blurred to allow any definite conclusion. It was in this context that a further section of the inner ditch was excavated in 1989. One hope was for the discovery of some wood remains that might allow the application of more accurate dendrochronological dating.

While a large quantity of waterlogged wood was retrieved from the inner ditch, only one sample, Q-7971, had sufficient rings to warrant serious consideration. Before going on to discuss the possible dating of this piece, it is worth looking at the theoretical possibilities. In the past, prehistoric archaeologists have seldom found themselves talking in terms of annual precision. Questions such as 'Was Haughey's Fort constructed after 1159 BC?' would have been unimaginable just a few years ago. Partly because of this lack of precedent, there is almost no theoretical framework for dealing with precise calendrical dates in prehistory. So, for this reason alone, it is worth giving some consideration to the possible interpretations of any date for Q-7971.

When an archaeologist is investigating a site that is remote in time, it is not normally possible to assign a close date on the basis of material finds. In such a situation, wood from the bottom of a ditch will certainly supply a date that is, in general terms, close to the digging of the ditch. Seen from a perspective of several millennia, the wood in the ditch 'dates the ditch'. In fact, of course, this is only an archaeological perspective. Moreover, it is the wrong perspective to use when considering any form of precise dating.

To be objective, we must switch perspective and look at the process from the point of view of a Bronze Age ditch digger. What are the relationships likely to be between the date of finishing the ditch and the felling date of any wood ending up in that ditch? We can surmise that there is a range of possible answers (Boyd, 1988). Wood can be deposited fresh or when it has become redundant. If we consider a piece of riven oak, we can conjecture that it was probably used for something before being deposited. So the felling date of a piece of oak from the ditch could bear several relationships to the construction date of the ditch itself:

1. It could be contemporary if it came from an oak structure built on the site at the same time as the digging of the ditch.
2. It could be later than the digging of the ditch and represent subsequent activity.
3. It could be older than the construction of the site, brought in from elsewhere. The recent observation, at Deer Park Farms, Co. Antrim, of reused oak doorposts in a rath context means that it would be quite possible to have oak timbers turning up in a ditch that pre-dated the digging of that ditch (Lynn, 1987).

It is immediately apparent that, even if a felling date can be obtained for pieces of oak from a ditch, the interpretation will not necessarily be straightforward. The very threat of precise dating causes a whole new look to be taken at what is technically possible. It becomes clear that dating a piece of wood from the ditch of Haughey's Fort will not yield a definitive answer regarding the question of the relationship of the ditch to the 1159–1141 BC event.

In fact, the situation is worse than that. If we return to our oak sample from the ditch at Haughey's Fort, one immediate problem is that the sample is not complete; its ring pattern does not run out to the bark. Indeed, the sample retains no sapwood, so at least 32 ± 9 (95% confidence 14–50) rings are missing from the outside of the sample (Baillie, 1982), and it is possible that there may be heartwood rings missing as well. So in the case of Q-7971, the actual sample is spaced an unknown number of years back in time from the felling date of the tree from which the sample came. So the date of the last existing ring on our Q-7971 sample bears a decidedly complex relationship to the date of the construction of the inner ditch. Viewed from the present day, any date obtained may represent a refinement over the range of dates inferred from the radiocarbon evidence; viewed realistically, in the context of the relationship of the construction of Haughey's Fort to the calendar date 1159 BC, the prospects of the sample refining our understanding are slim. All of the above can be deduced before even attempting to date the sample.

4.4.2 Attempts to 'date' Q-7971

One consequence of a small piece of timber with a large number of growth rings is that inevitably those rings are extremely narrow. In order to extract a usable ring pattern from Q-7971, a large number of repeat measurements had to be made. It was eventually possible to extract a total pattern of 161 rings from this sample whose maximum dimension was 10 cm. As a starting point, the sample was compared with the Belfast master chronology (BLC7000). No significant correlations were found at any point in the last seven millennia. However, taking account of the radiocarbon evidence, it was known that, in the first and second millennia BC, the Belfast chronology is heavily biased towards more northerly sites, particularly Garry Bog. Only one constituent chronology, Charlemont/Tullyroan, which spans the whole of the second millennium BC, was from the area south of Lough Neagh, i.e. in the region of Haughey's Fort. When Q-7971 was compared with this southerly chronology, a significant correlation value was obtained at a *possible* end-date of 1166 BC. If this were shown to be correct, for example by future replication, we would be able to assert that, owing to missing rings, the best estimate of the felling date would be 1150–1116 BC *or later*. However (and here we run into that fundamental dilemma for the dendrochronologist), at present it is not possible to confirm this 'date' – insufficient support exists to allow unequivocal dating.

4.4.3 Inferences

So, Haughey's Fort is not dated by dendrochronology. Moreover, it is clear that, irrespective of whether a tree-ring date had been achieved, the date of the piece of wood *could not* have dated the site in any precise sense. Let us imagine that Q-7971 did represent felling in the range 1150–1116 BC or later. If correct, this would certainly suggest activity at or around Haughey's Fort at or after, but fairly certainly not before, the 'Hekla-3' event. It would appear that this is the limit of possible interpretation. This example admirably demonstrates the effects of dendrochronology. We are in a position to begin to ask a whole new generation of questions and to define the limits of chronological refinement.

4.5 THIRD CASE STUDY

4.5.1 Early 16th century defoliations in Ulster

We have already noted some examples of the environmental information that is beginning to come out of the oak chronologies. It would be a mistake, however, to think that everything is straightforward. There are many cases where it is clear that trees have been affected by 'something', but it is by no means easy to separate out the specific cause. In

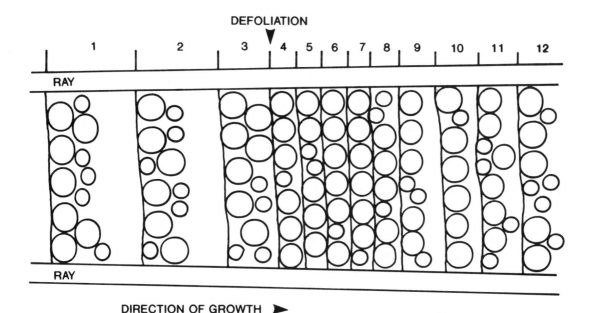

Fig. 4.1 Schematic representation of dramatic growth reduction in oak, presumably associated with defoliation/pollarding.

such cases, hypotheses can be set up for future testing. The 16th century AD defoliations are a case in point.

While the original Belfast chronology was under construction it was noticed that, in timbers from Hillsborough Fort, Co. Down, a number of ring patterns showed evidence of dramatic defoliation. This was of particular interest because it introduced an unwanted measurement difficulty – which was a considerable nuisance at that developmental stage. The effect was illustrated by Baillie (1982) and can be described as follows. A tree that has been growing strongly, often with double lines of spring vessels, suddenly fails to develop any summer wood in one year. In the first year without summer wood, the spring vessels are often still double because the tree is growing using reserves from the previous year. However, the following year it has few resources and puts on a very narrow ring, usually with vessels smaller than normal (see Fig. 4.1). It then takes approximately ten years to recover to normal growth. It seems unlikely that the effect was caused by insect attack as oaks are generally capable of rapid recovery, even following total insect defoliation. The dramatic growth curtailment in the Hillsborough samples appears to be more consistent with severe damage, possibly the effect of branch loss.

Normally in dendrochronology, if some climatic influence has affected the trees, we expect the effects to be observed in many different trees at the same time. It was interesting, therefore, to document the years of dramatic growth reduction (DGRs) in trees of the 15th and 16th centuries in the north of Ireland. If the damage consistently took place in the same year (or years), a climatic factor might have been to blame. If, on the other hand, different trees were damaged in different years, then climate was much less likely to have been directly involved and human influence might be inferred. Of course, climate could still have been an indirect driving force, in that particular conditions could have encouraged certain types of human response.

So, isolating the specific causes of particular effects in tree-ring records can be fraught with difficulty.

To study the effect, first it had to be defined. In order to class as a DGR, there had to be a change from strong normal growth to growth with no summer wood. In addition, the effects had to be obvious in the ring pattern for at least five years. All of the dated ring patterns from the 15th and 16th centuries were then examined and examples of DGRs were identified and dated. It was quickly discovered that the effects were not randomly distributed in time. Table 4.1 lists the years and numbers of occurrences from a survey of all the late-medieval samples from the north of Ireland. Not only was there a series of dates, but, in several instances, it was noticeable that the effects occurred in consecutive years. It was also found that, although the samples were drawn from the whole of the north of Ireland, the effect appeared to be limited to timbers from the east of the province, in north Down and south Antrim. The effects seemed not to occur in the quite wide spread of timbers from the rest of Ulster.

Table 4.1 Years in which oak trees in eastern Ulster suffered severe growth reduction. (Calendar years AD)

Year	Number	Year	Number
1477	1	1518	1
1488	2	1519	3
1502	1	1520	2
1503	1	1528	1
1506	1	1530	1
1507	1	1531	4
1516	2	1541	1

This effect only appears to be common within a relatively short, late-medieval, period. No examples could be found from the century before AD 1477 or from the century after AD 1541. Here, then, is an interval of time when it seems clear that oak woods in eastern Ulster were being heavily interfered with by humans (the occurrences in consecutive years would tend to rule out direct climatic effects). If we assume that the trees that have come down to us represent a reasonable statistical sample of the trees growing across this period, then some 12% of all oaks growing in eastern Ulster between AD 1474 and 1550 were defoliated or 'pollarded' at some stage in their lives (12% of the available specimens show the effect). This would represent an enormous number of original trees affected. Exactly what social conditions this reflects is not clear; whether people were stripping trees for cattle fodder or for fence building or indeed for firewood is not known. The observation can at least be passed back to the historians for further consideration.

This example contrasts with the first case study, where the information was categoric – felling at Corban Lough in the summer of AD 1457. In the defoliation case study, it is not possible to be categoric. We may know the exact years in which the trees were defoliated but we cannot be certain just what produced the effects. This is not untypical of the problems encountered when attempting to infer past conditions from particular effects in oak ring patterns.

4.5.2 Localized effects and lessons therefrom

One aside on this phenomenon is that in building the Belfast chronology we were lucky that the effect was not more widespread. Plenty of timbers crossed this period without showing any dramatic effect. This was important because in the affected timbers the damage tended to obscure the common growth-forcing 'signal' registered in most trees. As a result, these disturbed ring patterns were in some cases difficult to crossdate. Because the dramatic growth curtailments are so visually apparent, the initial tendency is to want to line up the patterns so that the dramatic events coincide (see Fig. 4.2). This of course would produce nonsense.

4.6 CONCLUSIONS

Dendrochronology in northern Ireland has come a long way since 1968. It should be stressed that Irish dendrochronology is merely part of European dendrochronology, so although the examples used have been Irish, similar case studies could have been drawn from a dozen regions across northern Europe. There are essentially no major problems with the methodology of oak dendrochronology. More and more dates are becoming available and more detailed interpretations of building activities and environmental changes are bound to follow. The insertion of precise dates into the pre-existing archaeological and environmental records will inevitably force a tightening-up of all aspects of chronology.

Acknowledgements

I thank David Brown for assistance in attempting to date the timber from Haughey's Fort and Andrew McLaurence for dating the late-medieval DRGs. Alan Smith, along with Martyn Jope and Jon Pilcher, designated a research assistantship in 1968 (occupied by this author until 1970 and by Jennifer Hillam from 1970 to 1976): without their foresight there would be no long oak chronologies in the British Isles.

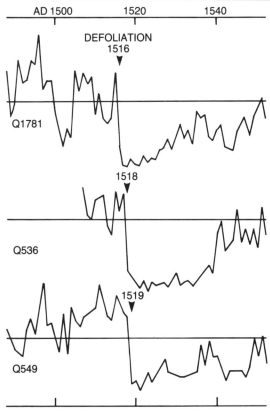

Fig. 4.2 Dramatic growth reductions in different years in oaks from eastern Ulster.

Fortunately, the number of unaffected timbers and the overall power of the method to match uniquely when the valid pattern is taken into account, allowed us to bypass these potential problems.

Part Two

Climatic Change on the Landscape

Introduction

Some of the earliest work on Holocene climatic change was based on studies of the mires of north-west Europe. From this work emerged a climatic division of the postglacial (the Blytt–Sernander scheme). Its terminology is now enshrined in a chronostratigraphic division of the Holocene, employed in the Scandinavian countries. The scheme is frequently used elsewhere, particularly in other parts of Europe, with varying precision and with dubious legitimacy, to indicate climatic or biostratigraphic zones. In Chapter 5, Dr Jeff Blackford examines this scheme in his chronicling of peat stratigraphic work in north-west Europe. He evaluates recent research and suggests a strategy for future peat-based studies of climatic change. Next, in Chapter 6, Dr Richard Bradshaw considers a major debate in Holocene palaeoecology – the relationship between forest trees and their response to climate. Taking three case studies, *Fagus grandifolia* from the USA, *Pinus sylvestris* in Europe, and the boreal–nemoral ecotone in central Sweden, he re-examines the questions of climatic thresholds, migrational lag, and inertia, raised by A.G. Smith in 1965.

Professor J. John Lowe (Chapter 7) considers the relationship between climate and vegetation in Scotland during the early and mid Holocene, including the spread of *Pinus sylvestris* across Scotland, and considers the suitability, or otherwise, of pollen analytical methods to investigate this relationship.

Finally, in Chapter 8, Dr John A. Matthews considers the landscape above the present tree-line in Norway, much modified by recent (Holocene) glacial advance and retreat, and reviews attempts to radiocarbon date buried soils in this arctic-alpine environment.

5
Peat bogs as sources of proxy climatic data: past approaches and future research

Jeff Blackford

SUMMARY

1. Mires have been used as sources of proxy climatic data since the last century. The progress of research into the relationship between peat characteristics and climate is reviewed.
2. It is argued that changes in the methodological approach to the subject, as well as technological advances, have led to progressively more accurate results. Reductions in the scale of palaeoecological reconstruction have coincided with improved understanding.
3. Future research should include improved dating techniques and the analysis of a wide range of variables, enabling dry periods to be detected as well as wet-shifts. The range of sites used could be broadened to include all ombrotrophic mires including watershedding blanket bogs.
4. Researchers should understand each individual site, before results are combined into regional syntheses.

Key words: climatic change, ombrotrophic mires, peat stratigraphy, peat humification, dating

5.1 INTRODUCTION

Sub-fossils obtained from peat deposits have often been used for climatic reconstruction (cf. Birks, 1981; Coope, 1986). Characteristics of bogs themselves, including stratigraphy, have also been used (cf. Barber, 1982, 1985). This paper addresses the latter type of study, and aims to look first at previous approaches and then to make suggestions for future research.

5.2 PAST APPROACHES

5.2.1 The Blytt–Sernander scheme

Peat bogs have been used as sources of proxy climatic data since the 19th century, and until the advent of pollen analysis, studies of mire

Table 5.1 The classic divisions of the post-glacial, with terminology slightly modified (from Sernander, 1908, p. 472)

Climatic period	Climatic condition
Sub-Atlantic	Humid and, especially at the beginning, cold
Sub-Boreal	Dry and warm
Atlantic	Maritime and mild, probably with warm and long autumns
Boreal	Dry and warm
Subarctic (Preboreal)	Climatic conditions undetermined

stratigraphy formed the main body of evidence for post-glacial climatic changes in north-west Europe (Birks & Birks, 1980). Peat-stratigraphic units were regarded as directly resulting from, and therefore being indicative of, periods of different climate. Layers of relatively undecomposed peat, often with *Sphagnum* as a major constituent, were believed to show wetter climatic conditions. Horizons of darker, decomposed (humified) and apparently slower-growing peat, often containing tree stumps, were attributed to drier, warmer periods.

Sernander (1908) used this approach and proposed a series of broad divisions of the post-glacial. He developed parts of Blytt's (1876) work, and correlated mire stratigraphies, lake levels, archaeological periods and plant immigration, concluding that climatic changes linked them all (see Table 5.1). The Blytt–Sernander scheme was by no means universally accepted. To use Sernander's own words, the question was '... a vexed one, and in Sweden, where the problem has been most eagerly debated, the investigators of peat mosses are at present ranged in two opposite camps...' (1908; p. 466). When Samuelsson (1910) linked the forest beds in Scottish peat stratigraphies, recorded by Lewis (1905, 1906, 1907), to the Scandinavian evidence, the theory gained greater acceptance. This was despite the reservations of Lewis (1911), who questioned the contemporaneity of the forest beds in different districts of Scotland, stating:

This is really a most important problem, and upon it hangs the whole question of the stratification of the peat. If Blytt's conception of each forest bed as representing a dry period could be proved, then indeed we have a proof of the contemporaneity of these strata, but at any rate in the case of the lower forest, evidence from this country does not give any support to that theory. (Lewis, 1911, p. 824)

The post-glacial was divided into five phases, whose names then took on a wider meaning in the subsequent expansion of Holocene palaeoecology. Changes in pollen assemblages were correlated with, and attributed to, these broad and apparently synchronous climatic phases. For example, the beginning of the Atlantic period was associated with the rise of *Alnus*, and the decline of *Ulmus* appeared to be a result of changing conditions at the Atlantic–Sub-Boreal transition.

The downfall of the scheme came about with the application of radiocarbon dating. The work of H.H. Birks (1975) is most pertinent, as it is a reappraisal of the forest beds in Scottish peats, one of the original sources of evidence used to establish the scheme. Birks dated the stumps of pine and showed a considerable age range, not consistent with the Boreal and Sub-Boreal chronozones. She concluded: 'little evidence can be drawn from the palaeoecological studies for the hypothesis that large-scale climatic changes are reflected by the occurrence of forest beds.' (Birks, 1975, p. 221)

As well as the original evidence being thrown into doubt, the pollen zones were also found to be diachronous. Of particular note is the paper by Smith & Pilcher (1973). They presented a summary of the radiocarbon

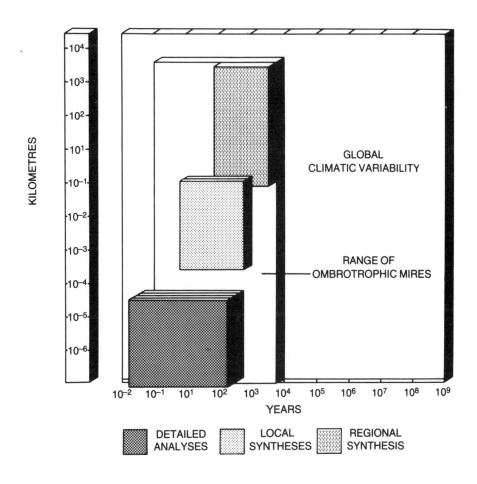

Fig. 5.1 Time–space ranges of different approaches to mire-based palaeoecology. The horizontal scale ranges from months to the age of the earth; the vertical scale ranges from millimetres to the size of the earth. The regional approach of the Blytt–Sernander scheme is illustrated by the upper right-hand block, bounded by the range of ombrotrophic mires. Local studies are represented by the central block. Detailed studies of single profiles at close sampling intervals are represented by the lower, left-hand block. It is argued that the results of a series of detailed studies should be used to move towards a local synthesis, before combining such syntheses for regional modelling.

dates for key features of pollen diagrams and demonstrated a wide range of dates for the establishment of several taxa. The 'rational limit' of *Alnus*, for example, was found to range over 2700 years. The decline of *Ulmus*, first associated with climatic cooling at the end of the Atlantic period, has since been attributed to human activity or disease.

The Blytt–Sernander scheme represented a broad approach in terms of timescale and areal extent, covering long-term changes over the whole of north-west Europe. Fig. 5.1 shows the different approaches of peat–climate studies in terms of the space and time-scales covered. The Blytt–Sernander scheme was a regional model, represented by the 'regional synthesis' block.

5.2.2 Recurrence surfaces

The premise of peat stratigraphic changes being related to climate was continued and refined in what Barber (1982, p. 208) described as a 'search for fixed points'. Layers of less humified peat from raised mires were again correlated with each other, although now more shifts were indentified, the presence of wood layers being considered less important than changes to wetter, *Sphagnum*-rich peat. Recurrence surfaces were first referred to by Weber (1900), and are associated with the general rejuvenation of the bog, caused by increased water availability at the bog surface (Walker & Walker, 1961). The history of recurrence-surface studies has been reviewed by Barber (1982). Granlund (1932) identified five recurrence surfaces (RYI-V), dated by archaeological correlation, including one in the first millennium BC known as the Grenzhorizont (after Weber, 1900). Nilsson (1935), Mitchell (1956) and Lundqvist (1962) added more recurrence surfaces, some of which matched those of Granlund, and some of which did not (cf. Barber, 1982).

Recurrence-surface correlations were later questioned. Van Zeist (1954) and Frenzel (1966) used pollen horizons to show different ages for what was apparently the same humification change. Radiocarbon dates allowed Schneekloth (1965) to show an 1100-year age-range for the Grenzhorizont within a single mire system. This type of study revealed similar asynchroneity elsewhere, including the raised bogs of north Germany (Overbeck *et al.*, 1957) and the Somerset Levels (Tinsley, 1981). Barber (1982, 1985) pointed out that many of the studies showing asynchroneity were based on borehole sampling rather than continuous exposures, and that the horizons dated were not conclusively proved to be the same stratigraphic units. However, the comprehensive survey by Haslam (1987), making full use of radiocarbon dating, confirmed a wide spread of dates for the main humification change, originally thought to be equivalent to the Grenzhorizont, in mires from Ireland to Poland.

Despite the problems in correlating certain stratigraphic horizons, dated shifts to apparently wetter conditions have produced a number of periods favouring the formation of recurrence surfaces. For instance, Dickinson (1975) and Svensson (1988) found changes to wetter conditions close to the dates of Granlund's original recurrence surfaces. Many mires produced proxy evidence in this way for a climatic shift in the period 2800 to 2200 BP. Changes to wetter conditions have also been suggested for the periods around 3950 to 3850 BP, 3500 BP, 2050 BP and 1400 BP (Barber, 1982; Blackford & Chambers, 1991).

In the period during which recurrence surfaces were most widely studied, the approach to peat stratigraphy as a proxy climatic record changed from a regional to a more local scale. Granlund's (1932) work referred to Sweden, and was a more localized model than the Blytt–Sernander sequence. Schneekloth (1965) and Overbeck *et al.* (1957) examined individual mire systems, an approach increasingly favoured by subsequent authors. Recurrence-surface studies also became more concentrated in temporal scale, with little attention being paid to the early Holocene and much concentration on what appeared to be one event, the Grenzhorizont. This approach revealed the complications of mire hydrology and local stratigraphic differences, and showed the possibility of zero peat growth or erosion. Correlations between mires were more carefully judged. More useful information was produced following a change to research in greater detail, concentrating on local syntheses (Fig. 5.1).

Studies of recurrence surfaces did not, however, provide precise or accurate proxy climatic data, being inaccurate in age, imprecise in timescale and unspecific in meteorological implication.

5.2.3 The search for new methods

As a result of the problems of interpreting dated recurrence surfaces, a new emphasis in peat-based climatic work followed. Barber (1982) describes the trend as one towards the reconstruction of climatic curves rather than the identification of fixed points. This was made possible, however, by the continuous and comprehensive nature of the data obtained, and as such the change could be described as a search for new methods to determine a climatic signal from ombrotrophic mires.

An example of a new approach is the work of Aaby & Tauber (1975), who studied the degree of humification, rhizopoda, pollen frequencies and *Sphagnum* remains from 13 profiles from six Danish raised mires. They found a clear relationship between the degree of humification (measured colorimetrically) and surface wetness at the time of peat deposition. This provided evidence supporting the principal assumption of preceding stratigraphic work: that the relative wetness of a peat surface can be detected thousands of years later by its degree of decomposition. This work was supported by intensive radiocarbon dating, including 55 dates from a single profile. Peaks of high humification and *Calluna* pollen percentages alternated with low humification and *Rhynchospora* remains, while records of the rhizopoda *Assulina* and *Amphitrema* spp. peaked in phases of relatively wet conditions. On the basis of this evidence, Aaby (1976) suggested a 260-year cycle in climate.

Another approach was followed by Barber (1981), whose study of a raised bog in Cumbria included macrofossil analysis, in particular the identification of *Sphagnum* remains of known ecological tolerance. This was supplemented by careful stratigraphic observation over a wide area of a raised bog, with pollen work to give a more regional perspective. By identifying 'wet shifts' in former peat surfaces, Barber produced a curve that reflected the pattern of climatic change inferred for the region from other sources (Barber, 1981, p. 212). This study demonstrated the potential of detailed stratigraphic work on a mire where many sections were exposed, and which may have been particularly sensitive to climatic change (see Chapter 19).

An approach to palaeoecology pursued by van Geel (1972, 1978) involves the recording and identification of all microfossils, as well as larger plant remains, at a contiguous 1 cm sample interval. Such studies encompass a wide variety of ecological parameters, improving the overall picture of changing mire-surface conditions. By reconstructing the surface conditions in such detail, the nature and causes of any changes can be more accurately assessed.

Another approach to deciphering a climatic signal from peat deposits is that of isotopic analysis. Brenninkmeijer *et al.* (1982) reported techniques to determine ^{2}H/H and ^{18}O/^{16}O ratios from cellulose. Applied to peat deposits of known or at least estimated age, this method has the potential to show not only climatic change, but also the nature of the change, as discussed by Dupont (1986). Initially, a decline in both isotopic parameters was suggested for the period 3100 to 2400 BP (Brenninkmeijer *et al.*, 1982). Samples from modern material illustrated the differences attributable to species changes, and Dupont & Brenninkmeijer (1984) used a correction technique based on the major component of the peat at a given level. They compared their results to tree-ring densities, ice-core data and fluctuations in *Corylus* pollen influx. Dupont (1986) tentatively concluded that quantitative climatic information could be inferred, and suggested that the area of study (Netherlands) was warmer (by 0.5°C) and drier between 3500 and 4000 BP, and was slightly colder (by 0.5°C) and becoming wetter around 3000 BP. A further warm period (1.5°C above present mean annual temperatures) followed between 2800 and 2500 BP, during which period conditions remained wet. The final change inferred was a shift to cooler (0.5°C below present) conditions after 2000 BP, when humidity was similar to the present.

This was the most specific climatic information directly obtained from peat deposits. Its reliability can only be shown by replication from peat profiles elsewhere and from other sources of proxy climatic data. The potential of isotopic work depends on whether the effects of species composition can be overcome, by using correction factors or by using selected plant material (van Geel & Middeldorp, 1988).

The approach of Wijmstra *et al.* (1984) and Dupont (1986), as well as that of Aaby (1976, 1978) discussed above, was to examine peat-derived proxy data sets for time-dependent variation. Using the results of consecutively sampled, multivariate palaeoecological analyses, Wijmstra *et al.* (1984) found cycles of various lengths in the pollen frequencies of *Corylus* and *Alnus*, and in the material characteristics of the peat. Time-series analysis was made possible by establishing a timescale for the whole profile. This was done by interpolating between radiocarbon-dated samples by estimating pollen densities, a technique developed by Middeldorp (1982). Some of the cycles suggested by time-series analysis corresponded with known cycles in sunspot activity; the strongest signal had a periodicity of just over 200 years. Dupont (1985) produced oxygen isotope data that supported this, with a cycle of 206 years being recorded. The 260-year periodicity apparent in Aaby's (1976, 1978) work remains unreplicated.

5.3 THE PRESENT POSITION

From the first studies of mire stratigraphy onwards, improvements in the quality of the climatic record obtained have come from increasingly detailed reconstruction. This has been aided by technological advances, especially in dating techniques. A parallel move towards multivariate analysis has also enhanced climatic reconstructions. The principal tenet of the Blytt–Sernander scheme – that the degree of decomposition and floral content of different layers of peat reflect past climates – has been shown to be valid. However, the idea that the classic climatic divisions of post-glacial times could be detected by synchronous responses in peat stratigraphy and vegetational change, is not confirmed. This is summed up by A.G. Smith, who concluded: 'The point has come when the simplistic concepts of the Blytt–Sernander scheme must surely be banished for ever.' (Smith, 1981b, p. 143)

The search for new methods has led to significant advances in the value of mires as sources of palaeoclimatic information. Attempts have been made to isolate particular elements of climate, and even possible causes of change. Specific temperatures have been inferred and cyclic tendencies suggested. These advances have coincided with a reduction in the scale of observation, with individual peat profiles being examined in great detail. The style of analysis that has led to the greatest reward was suggested by Godwin (1954), who wrote:

Bog stratigraphy ought to be approached from a knowledge of the ecology and hydrology of existing bogs, and the investigator should be broad enough in his interests to take account of the evidence from all relevant fields of science, whilst retaining his own cautious severity of judgement. (Godwin, 1954, p. 29)

5.4 FUTURE RESEARCH

The quality of proxy climatic records obtained from peat bogs might be improved by continuing the trends in approach followed in recent decades.

5.4.1 Dating techniques

For peat stratigraphic or humification data to be used to their full potential, establishing a precise and accurate timescale for each profile is essential. Great benefits have come from radiocarbon dating, followed by improvements in calibration (Stuiver & Pearson, 1986) and interpolation (Middeldorp, 1982). Relatively precise dates are needed for reliable correlation or time-series analysis. Improvements in precision are particularly important due to the effect of calibration on age ranges (Pilcher, 1990). Table 5.2 demonstrates how the calendar-year age range of a radiocarbon determination can vary depending on the resolution, which can be improved by careful sample selection and pre-treatment, and longer count times. It may be possible to apply AMS dating techniques to very small samples of known integrity, for instance horizontally bedded material that cannot have been moved up or down the profile. Wiggle matching (van Geel & Mook, 1989) of radiocarbon determinations from peat may also be possible (Pilcher, this volume). This would assist their calibration, a particular problem in the time interval 3000 to 2500 radiocarbon years BP (see Table 5.2).

However sophisticated the laboratory determination, radiocarbon dates from peat will always be to some extent inaccurate. Dates from the base of a peat profile have on occasion been rejected because they appeared younger than the samples above, possibly due to roots accumulating at the base (Caseldine & Maguire, 1983). It is likely that roots accumulate at the base of a peat profile as the surface extends away from it, and also that roots from younger levels accumulate and decompose in all horizons, causing all dates to be to some extent 'too young'. Problems may occur especially when the water-table drops, permitting deeper rooting. An alternative, or supplementary, source of age determination for peat horizons is required.

One possibility is the identification of specific layers of volcanic ash in peat deposits (Persson, 1971; Dugmore, 1989). Tephra layers can be identified by analysing the chemical composition of glass shards (Dugmore, 1989). This technique has the potential of locating particular horizons of estimated age, rather than dating the horizons of particular interest. The number of tephra layers available depends upon the location of the site.

Fixed-date points in peat profiles may also come from historical records, especially in the upper part of pollen profiles. For example, O'Connell (1986) showed the palynological effects of the Irish potato famine of AD 1846–1849, and the planting of conifers in recent centuries has been widely recorded in pollen diagrams from the British Isles.

5.4.2 Sites

The production of a continuous climatic curve has been hampered by the apparent inability of peat stratigraphic methods to determine shifts to drier conditions (Barber, 1982, 1985). This has been due to the belief that given stable conditions, continuing deposition of organic matter will raise the surface steadily above the water-table and lead to the deposition of more highly humified peat. Aaby (1976, p. 281) wrote: 'Shifts to drier bog conditions are therefore ignored as indicators of climatic changes'. Drier phases of climate have been inferred, however, from

Table 5.2 Different calibrated age ranges from a single radiocarbon determination, centred on 2630 BP, depending upon standard error. Of particular note is the improved precision in the calibrated age range and apparent increase in accuracy between ± 40 and ± 30 for this particular example. Calibration follows Stuiver & Reimer (1986)

Resolution (1σ)	Age range (Cal. BC)	Age range (years)
± 20	830–793	37
± 30	888–790	98
± 40	892–665	227
± 50	897–663	234
± 60	900–599	301
± 70	969–562	407
± 80	971–451	520

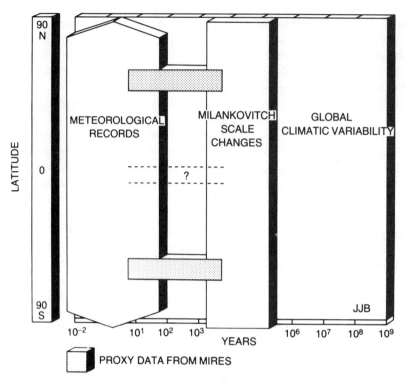

Fig. 5.2 Time–space diagram showing the latitudinal range and timescale of variability of the proxy climatic record available from blanket mire. Ranges of meteorological records, blanket mire growth and 'Milankovitch'-scale variability are nested in a background of climatic variability. Dashed lines indicate that the central African mountain mire type (Fig. 21.3) could be used if shown to be ombrotrophic.

a lack of wet shifts (cf. Barber, 1981), and from layers of highly humified peat. The detection of relatively dry phases of bog growth is possible, then, if appropriate techniques are applied to the right sites.

Most of the aerobic decomposition of organic matter in a peat bog occurs on the surface and in the upper layers, during the first few years after death. Peat will become more humified if the dry season is longer or drier. If this trend is over a long enough time period, the resulting humification record should include a transition to more humified peat. The difference between, for example, four months of aerobic conditions at the surface per year rather than two months, or six dry years in ten rather than two in ten, should be recorded and detected by more humified peat over the time span of a sample. Distinguishing between drier climatic conditions and growth above the water table may be aided by detailed analysis of the onset of change. If a change to more humified peat occurs before any evidence of vegetative succession, then an external cause, possibly climatic change, is most likely.

A mire that regularly has a dry summer season of unwaterlogged surface conditions will produce a record of decomposition containing evidence of both wet and dry phases. This record can be extracted by measuring the degree of humification, either colorimetrically or by other means (Blackford, 1990). Water shedding, and therefore ombrotrophic, sections of blanket mires often fit this description. Blanket peats are a largely untapped source of proxy climatic data (Chambers, 1984). Their distribution and hydrological characteristics, however, suggest that they offer considerable potential (Fig. 5.2; see also Fig. 21.3). Initial research has shown some coincidence in the wet shifts recorded from different areas of the British Isles, especially around 1400 BP (Blackford & Chambers, 1991). Studies of raised bogs and blanket mires might complement each other in establishing proxy climatic records in some areas.

5.4.3 Methods

Studies that have included a range of measured variables have generally produced more meaningful results. The advantage of continuing this approach in future research is clear, especially when climatic inferences are sought, as each variable will respond to a different environmental change. For instance, the degree of humification is linked to effective summer rainfall, and the *Sphagnum* species to mean water-table depth. Isotopic measurements are at least in part controlled by temperature and rainfall.

A further advantage of using a range of variables in studies of peat profiles involves the response rates of the different variables used. Barber (1982) discussed the idea of threshold theory in describing the response of peat growth to climatic change, and suggested that some bogs might be more responsive to change than others. This idea appears to regard a mire as a single entity, either dry or wet, either close to a threshold of change or not. Within a mire system, however, there will always be some component, either a plant species, fungal parasite, amoeba etc., that is close to the extremes of its range. Many of these organisms leave a readily identifiable record in peat profiles. The palaeoecological record, therefore, contains a wide range of interdependent variables, all capable of responding to climatic stimuli. In addition, there is a range of physical characteristics – bulk density, fibre content, degree of decomposition – that can provide a more or less continuous record of past surface conditions.

Each of these 'responder variables' will not only provide a record of a different element of environmental change, but will also have a different response rate. For example, if the *Sphagnum* remains in a raised bog are considered (Barber, 1981; Haslam, 1987) a time-lag is inevitable between climatic change and response (Haslam, 1987). This is especially the case when considering a shift from dry to wet conditions, the change most commonly recorded by peat-based palaeoclimatic work. At the end of a long dry period, a rise in the water table may not reach the surface to cause a species change, and could therefore be missed. After a very short stable or dry period, however, a modest wet shift might be recorded by a change in the surface assemblage. The magnitude of a vegetative response is a function, not only of the size of the climatic change, but also of antecedent conditions. The degree of humification, however, might be expected to respond more quickly, as no species change is required. The lag time is increased if tree development is considered diagnostic of climate, as in the earliest studies.

The concept of threshold theory is much more useful in palaeoecology if all possible 'responder variables' are considered, and the likelihood of detecting a change in the former mire ecosystem is greatly improved when the range of variables considered is increased.

5.5 CONCLUSIONS

Past research into the proxy climatic record directly obtained from ombrotrophic peat profiles has demonstrated the potential of the source and has begun to yield useful information. A movement towards detailed, multiple-variable analysis coupled with radiocarbon dating has improved the record. This methodological approach should be continued into future studies, using detailed analyses of multiple profiles and a range of variables to understand individual mire systems. These can then be put together to allow local and regional modelling, using improved dating techniques for reliable correlation.

Acknowledgements

I thank Dr K.E. Barber for comments and suggestions, and Dr F.M. Chambers for comments on earlier versions of the text and diagrams. The tenure of a NERC studentship, held at Keele University 1986–89, is acknowledged.

6
Forest response to Holocene climatic change: equilibrium or non-equilibrium

Richard Bradshaw

SUMMARY

1. The Holocene dynamics of northern hemisphere temperate forest types have been driven by climatic and human factors. The relative importance of these factors, and the speed of the vegetational response are the subject of intense debate.
2. The Holocene migration of *Fagus grandifolia* in the USA correlates closely with independently reconstructed changes in temperature and rainfall, and Lake Michigan did not present a major barrier to dispersal.
3. The decline of *Pinus sylvestris* populations in western Europe during the late Holocene can be explained by decreased occurrence of wildfire – again in response to climatic change. Human agency played an intermediary, but not a controlling, role.
4. Disturbance by fire also assisted in the recent change from nemoral to boreal forest in central Sweden. Frequent disturbance ensured that the 'climatic inertia' of long-lived deciduous species did not result in extensive forest stands that were out of equilibrium with climate.
5. Smith's concept of 'critical climatic threshold' is valuable in understanding the complexity of vegetational response to climatic change, but 'migrational lag' and 'inertia' of vegetation are not apparent in the examples discussed.

Key words: climatic change, Holocene, temperate forests, disturbance, vegetation dynamics

6.1 INTRODUCTION

The challenge of revealing the underlying causes of vegetational change has exercised the minds of plant ecologists for many years. Smith (1965) discussed problems of inertia and threshold in relation to the post-glacial vegetation history of the British Isles, exploring the nature of vegetation response to particular climatic changes. He argued that simple climatic changes could result in complex vegetational response, and introduced the concept of critical climatic 'thresholds' for vegetational response. He also discussed systems where the inertia of a vegetation type could lead to a 'climatic tension' where vegetation was temporarily out of equilibrium with the prevailing climate. He frequently stressed the inadequate amount and type of data available at the time, and it is therefore appropriate to re-examine this discussion over 25 years later. This period has seen a dramatic improvement in both the quality and quantity of data.

One currently influential school of thought views the vegetation dynamics of the Holocene as primarily a response to climatic change, and therefore a valuable tool in describing past climates (Bartlein et al., 1986; Webb et al., 1987; Prentice, in press). The complex patterns of tree movements recorded from the last 13 000 years reflect the individualistic nature of species response to climate, and the abundance of permutations of temperature, precipitation and wind that have occurred during the Holocene. Properties of vegetation that might thwart this fundamental climatic control, such as migrational lag (i.e. the delay in response of a plant species to a climatic change dictated by the intrinsic rate of spread of the species (Iversen, 1973; Davis, 1984)) or inertia of a vegetation type (i.e. the ability of a community to perpetuate itself under unfavourable conditions and resist the invasion of new taxa), are seen by the climate school as operating over too short time periods, or too small areas to upset the fundamental thesis (Webb, 1986). Support for the overriding importance of climatic control has also come from computerized forest-simulation models, in which complicated dynamics can be modelled by rather few parameters that are themselves climatically controlled (Prentice & Leemans, 1990).

Human influence on vegetation dynamics is potentially a more serious confounding issue, especially in Europe. The arguments in favour of underlying climatic control include some corresponding vegetational developments on both sides of the Atlantic, despite a major contrast in the history and scale of human influence on natural vegetation between the USA and Europe. The recent range changes of *Fagus* and *Picea*, for example, show some parallels in both continents (Huntley & Webb, 1989). One can propose that pre-industrial society lacked the ability to bring about the changes in continental-wide tree distributions that are more easily related to climatic change. Human activity could amplify climatic effects, for example by lowering the tree line, but humans could only burn forests that were already inflammable. Several recent studies have shown that human colonization has not increased the natural fire regime (e.g. Johnson et al., 1990). Yet, human activities have almost halved the area covered by forests in the world over the last 5000 years (Tallis, 1991), and it seems probable that a proportion of the vegetation dynamics of the late Holocene in Europe, for example, are attributable to human activity.

The battle of ideas is not over. If the landscape-scale of vegetation really maintains some type of crude equilibrium with climate, how does one explain the coexistence of two vegetation types over a large area? One example discussed in this paper is the vegetation of the Mälaren valley in Sweden, where mixed deciduous forests with abundant *Tilia cordata* coexist with boreal communities dominated by *Picea abies*. What role is played by the disturbance regime? Can the dynamics of a frequently disturbed community be climatically interpreted? I shall also discuss inertia of vegetation, migrational lag and examine the influence of a geographical barrier to tree migration by reference to a detailed American palaeoecological study.

6.2 CASE STUDIES

6.2.1 Lake Michigan and the migration of *Fagus* in the USA

Davis and colleagues have used pollen data from a grid of sites to map the spread of *Fagus*

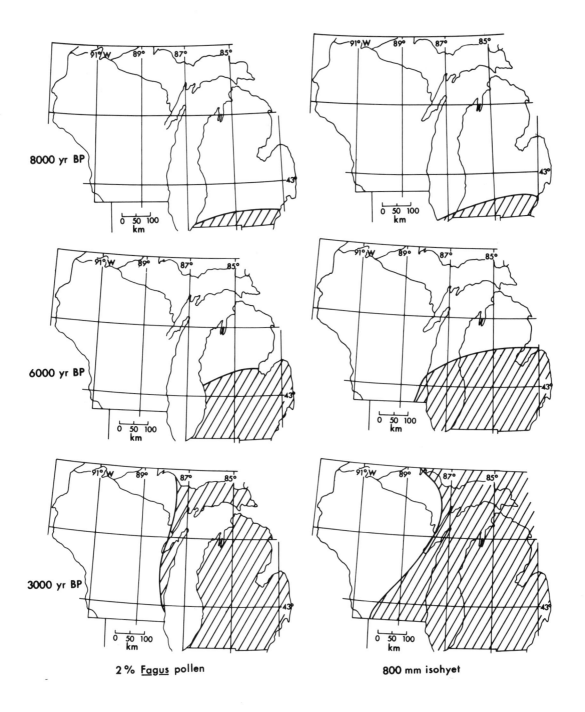

Fig. 6.1 The 2% *Fagus* isopoll and the reconstructed 800 mm isohyet from 8000, 6000, and 3000 BP for Michigan and Wisconsin, USA. Data derived from Woods & Davis (1989) and Prentice *et al.* (1991).

grandifolia from the eastern USA into Michigan and Wisconsin during the Holocene (Woods & Davis, 1989). *Fagus* pollen disperses over relatively short distances (Bradshaw & Webb, 1985), so may be used as a convenient proxy for *Fagus* trees. The western movement of *Fagus* populations is still in progress thousands of years after the last glaciation, and is a potential example of 'migrational lag'. *Fagus* has been, and still is, moving into closed forest communities, so the resistance of these communities to the invader may be studied. How much 'inertia' do the non-*Fagus* communities exhibit? Finally, Lake Michigan forms a potential geographic barrier to this spread, having a present-day width ranging between 100 and 150 km. This example exhibits many problems of dynamic and biogeographical interest.

The immigration of *Fagus* into Michigan and Wisconsin occurred in a step-wise manner, with major range expansions occurring between 8000 and 5000 BP, 3000 and 2500 BP, and during the last millennium (Woods & Davis, 1989). Several separate colonization events took place across Lake Michigan, which did not present a major barrier to dispersal. Woods & Davis (1989) review literature suggesting that low winter temperatures and lack of rainfall are factors limiting the present-day distribution of *Fagus*.

A simplistic model of climate change could not easily account for the complex range dynamics of *Fagus* during the Holocene, but the climatic reconstruction proposed by Prentice *et al.* (1991) provides an elegant explanation for the observed range changes. This climatic reconstruction at 3000-year intervals over the last 18 000 years is based on contemporary pollen–climate relationships for *Pinus, Picea, Betula, Quercus* and prairie forb pollen percentages, so is essentially independent of the Holocene history of *Fagus*. Reconstructed summer temperatures are sufficiently high for *Fagus* growth in the study area by 9000 BP, but January temperatures only became warm enough between 9000 and 6000 BP (Prentice *et al.*, 1991). Precipitation then became the critical limiting factor, both its total amount and distribution throughout the year. The 800-mm isohyet neatly encloses the 2% isopoll for *Fagus* throughout the Holocene (Fig. 6.1). The hypothesis that the range adjustments are under climatic control cannot be rejected, even given the spatial and temporal complexity of these adjustments.

Prentice *et al.* (1991) demonstrate further that the same climatic reconstruction can account for the southward expansion of *Picea* during the same period of time. This example shows that climatic change alone can account for contradictory movements of tree distributions during the Holocene (e.g. the present southward extension of *Picea abies* range and the simultaneous northward movement of *Fagus sylvatica* in Sweden). This is not to say that other factors such as intrinsic dispersal rates and geographic barriers have played no role in vegetation dynamics, but simply that other hypotheses need not be invoked to describe the range changes discussed here.

6.2.2 The Western limits of *Pinus sylvestris* in Europe

Pinus sylvestris has been of considerable commercial significance during the past few centuries within Europe (Björklund, 1984), and is a most likely candidate to show late-Holocene range adjustments purely as a result of human activity. In Denmark, Holland and Ireland it effectively became extinct (some doubt remains) within the historic period, but reintroduced specimens can successfully complete their life cycle today, and populations are becoming naturalized in many parts of Ireland. Its present 'native' western limit (Jalas & Suominen, 1973) (Fig. 6.2) cannot easily be explained by any simple combination of climatic parameters (Prentice, personal communication). *Pinus* has had a dynamic history in Ireland and western Britain during the Holocene (Bennett, 1984; Bradshaw & Browne, 1987), and it

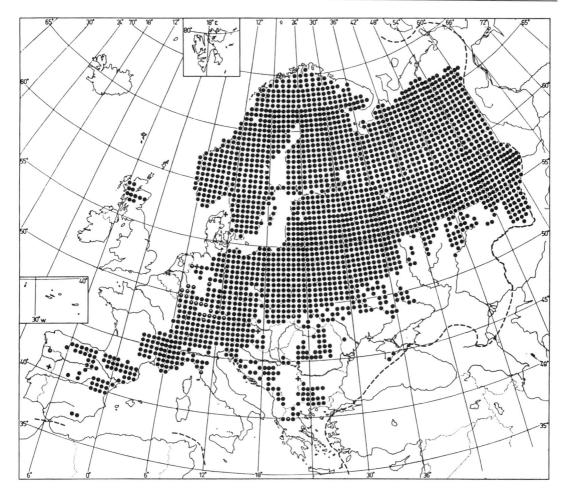

Fig. 6.2 The present natural distribution of *Pinus sylvestris* in Europe (after Jalas & Suominen, 1973).

is instructive to speculate whether its range changes were in response to non-climatic factors.

Pinus trees covered a significant proportion of Ireland by 7500 BP, but were only a minor component of forests on limestone-derived soils in central and northern Ireland, where *Ulmus glabra* and *Quercus petraea* dominated (Bradshaw & Browne, 1987). *Pinus* was ousted from its wetter habitats by the spread of *Alnus glutinosa*, beginning about 8000 BP (Bennett, 1984; Bennett & Birks, 1990), and was confined to western Ireland by 5000 BP (Fig. 6.3). Blanket peat growth further reduced the range of habitat available to *Pinus* and extensive fossil populations lie preserved beneath peat of various ages in western Ireland. The picture emerges of a fast-growing, well-dispersed, light-demanding tree that tolerated a diverse habitat range from acid peats to bare limestone, but was easily ousted by taller, more shade-tolerant tree species or specialist wetland taxa.

How could such a poor competitor have maintained such a large population for thousands of years over much of northern Europe? Furthermore, when a *Pinus* individual dies, the resultant gap is too small to permit

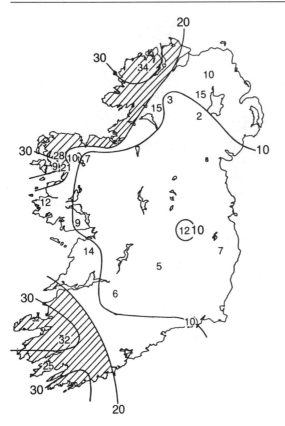

Fig. 6.3 *Pinus* pollen isopolls (with a tree pollen sum) from 5000 BP for Ireland.

replacement by another *Pinus* (Prentice & Leemans, 1990). *Pinus* should have the role of an early successional species such as *Betula pubescens* or a forest fugitive such as *Populus tremula*, and not be a major forest dominant as it is in parts of present-day Scotland and Scandinavia, and undoubtedly was previously over even larger areas.

The paradox is resolved by consideration of the disturbance regime. Many pure stands of *Pinus* in northern Sweden post-date fires that killed its competitors (Zackrisson, 1977). *Pinus* is a fire-adapted species that can survive fire better than *Picea abies*, and regenerates well on burnt ground. The history of *Pinus* in Ireland closely tracks the history of fire. *Pinus* survived far longer at one site out of four studied by Browne (1986) in Co. Mayo. The favoured *Pinus* site had an abundant, almost continuous record of charcoal that was lacking from the other sites, and lay on coarse, free-draining infertile soil, inhospitable to *Pinus*' competitors (Bradshaw & Browne, 1987).

O'Sullivan (1991) shows a close correspondence between *Pinus* pollen values and macroscopic charcoal from a very small bog in Killarney National Park (Fig. 6.4). Peak charcoal values are quickly followed by a rapid expansion in the *Pinus* population. *Pinus* pollen values decline slightly as charcoal values fall to moderate levels, and then fall catastrophically during a 1500-year period without evidence of fire. The last *Pinus* trees, which could have lived for at least 700 years (Engelmark & Hofgaard, 1985), died without replacement at a site that has remained covered by forest throughout the Holocene (O'Sullivan, 1991).

A macroscopic (> 150 μm) charcoal record from beside a small peatland can be interpreted in a different manner to the microscopic charcoal fragments found by Bennett et al. (1990b) beside a large East Anglian lake. The macroscopic fragments are more likely to be *in situ* (Clark, 1988a), and provide a record of forest fire. The fires may have been set by humans, but there is virtually no archaeological evidence to support this (Woodman, 1985). The data are compatible with the hypothesis that forest fire in primarily deciduous forest during the early Holocene created and maintained suitable conditions for *Pinus* dominance. Closer examination of other west European sites will almost certainly reveal similar *Pinus*–fire relationships.

Can we ascribe the decline of *Pinus* in Ireland to the decline in fire frequency? The two processes are closely entwined as *Pinus* forest is the most inflammable forest type in Ireland. I envisage a situation where the *Pinus* populations became smaller and more isolated until they were too small to ignite, and were ousted by competitors. Naturally, the fire regime is strongly influenced by climate

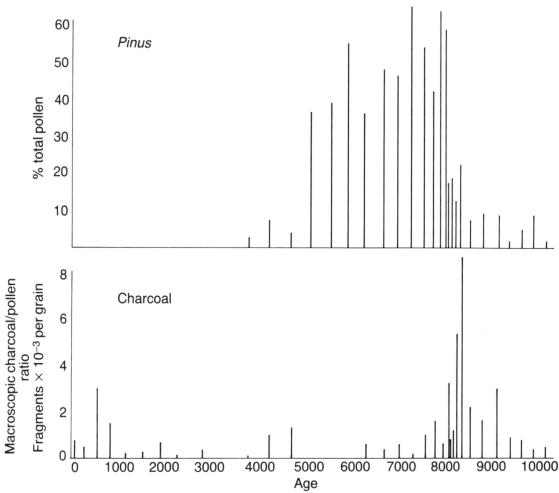

Fig. 6.4 Pinus pollen values and macroscopic charcoal from a small bog in Killarney National Park, Ireland (O'Sullivan, in preparation).

(Clark, 1988b), but in this case we see a species distribution reacting to alteration in the disturbance regime and not directly to climate.

6.2.3 The boreal–nemoral ecotone in central Sweden

How widespread is the influence of the disturbance regime on vegetational change? Can a forest type resist responding to climatic change in the absence of disturbance? Do forest types show some inertial stability as Smith (1965) once suggested? In the Mälaren valley west of Stockholm, Sweden, two forest types lie side by side and rarely mix. Prästholmen is dominated by a rich deciduous forest type that includes *Tilia cordata*, *Ulmus glabra* and *Corylus avellana*. Neighbouring Ytterholmen is dominated by *Picea abies*, *Pinus sylvestris* and *Betula pubescens* – a standard boreal community. Can they both be in equilibrium with prevailing climate? Pollen analysis of a tiny bog (1 m^2) on Ytterholmen shows that it too was once dominated by the deciduous community (Fig. 6.5).

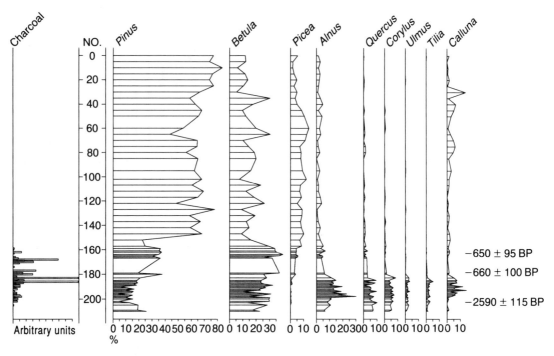

Fig. 6.5 Percentage pollen diagram (with terrestrial pollen sum) from a small bog on Ytterholmen, Mälaren valley, Sweden (analyst: G. Hannon).

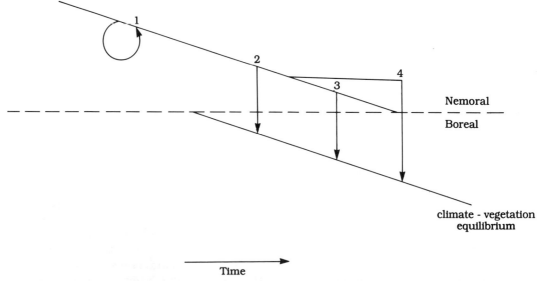

Fig. 6.6 Model of the changing climate–vegetation relationship in the Mälaren valley, Sweden. (1) A secondary succession temporarily displaces the vegetation from the equilibrium position. (2) (3) Catastrophic disturbances cause the vegetation to 'flip' from one stable state to another. (4) Inertia of vegetation permits the community to drift away from the equilibrium position. Disturbance results in return to a new equilibrium point.

The change to boreal forest occurred abruptly after a series of massive fires on the island about 700 years ago. The severity of the burning destroyed part of the deposit itself, which complicates interpretation, but the change of forest types is quite clear. I propose that both forest types are in dynamic equilibrium with the prevailing climate, but catastrophic disturbance tips the system into boreal forest, which is gaining ascendancy with time (Fig. 6.6). Fulton & Prentice (personal communication) suggest that these two stable forest states are maintained by contrasting N-cycling regimes. When the richer deciduous type is catastrophically disturbed, the system changes over to the N-poor boreal forest. We propose to test this hypothesis through soil analyses.

6.3 CONCLUSIONS

These three case studies have demonstrated the overriding importance of climate and climatic change in forcing corresponding vegetational change. Vegetational change may be abrupt after catastrophic disturbance, or sluggish through immigration of a new potential dominant. An existing vegetation type rarely resists change for long. Long-lived stands of *Tilia cordata*, reproducing clonally, show inertia; but in our frequently disturbed northern forest types, such relict populations are exceptional, and never dominate landscapes.

The past 25 years have added fuel to the discussion initiated by Smith (1965). Migrational lag or delayed immigration (Iversen, 1973), and inertia of vegetation to change have not proved to be major factors in vegetational development, but the concept of climatic threshold is valid. It is the variety of critical climatic thresholds for individual species in different parts of their range that leads to the complexity of vegetational response to climatic change.

Acknowledgements

I thank Aileen O'Sullivan and Gina Hannon for providing unpublished data. Colin Prentice initiated stimulating discussion, and Skogs-och Jordbrukets Forskningsråd paid the bills.

7

Isolating the climatic factors in early- and mid-Holocene palaeobotanical records from Scotland

J. John Lowe

SUMMARY

1. The Holocene vegetational history of Scotland is summarized in terms of three broad episodes: the 'pioneer' (10 300–8500 BP), 'afforestation' (8500–5000 BP) and 'deforestation' (5000 BP to present) phases.
2. Problems of using palaeobotanical data to evaluate climatic conditions during the early and middle Holocene are reviewed.
3. Methods are suggested by which the influence of climatic factors can be isolated from other influences on palaeobotanical successions. These require the identification of sites that were sensitive to climatic changes at particular times during the Holocene.

Key words: 'sensitive' v. 'complacent' palaeobotanical records, pollen stratigraphy, pine macrofossils, treelines

7.1 INTRODUCTION

This chapter examines the extent to which climatic variations in Scotland during the Holocene can be inferred from available palaeobotanical data. Whereas the preceding severely cold Younger *Dryas* (Loch Lomond) Stadial and the rapid warming at the start of the Holocene can be fairly well defined from Scottish evidence by using several proxy climatic indicators (see e.g. Sutherland, 1984; Lowe & Walker, 1984), the scale of climatic variations during the Holocene is much more difficult to evaluate. The Holocene appears to have been a relatively stable period in northwest Europe, with only minor climatic shifts that may have had only marginal effects on the prevailing biota. Furthermore, climatic inferences for Scotland during the Holocene have rested almost exclusively on the interpretation of pollen-stratigraphical data; very little palaeoclimatic work has been attempted so far using alternative lines of evidence. This chapter therefore examines the suitability of pollen-stratigraphic evidence for palaeoclimatic reconstruction.

First, a very brief outline of the Holocene vegetation history of Scotland is presented for background. (Fuller coverage of this topic can be found in a review of pollen-stratigraphic data by Walker (1984) and in an overview of the Holocene vegetation history of Scotland by Birks (1977).) This is followed by an examination of the principal limitations of the existing pollen-stratigraphic data as a basis for Holocene palaeoclimatic reconstructions. A third section considers the methods by which improved palaeoclimatic information may be achieved by means of new approaches in palaeobotanical research.

This review is restricted to the early and middle Holocene only (c. 10 000 to 4000 years BP) since it is extremely difficult to separate climatic from anthropogenic effects on the vegetation of Scotland during the late Holocene. The late-Holocene evidence from Scotland is worthy of a separate detailed review.

7.2 POLLEN RECORDS AND THE HOLOCENE VEGETATIONAL HISTORY OF SCOTLAND

In very general terms, the Holocene vegetation history of Scotland can be divided into three broad stages, here termed the 'pioneer', 'afforestation' and 'deforestation' phases.

7.2.1 Pioneer phase (c. 10 300 to 8500 BP)

Most early Holocene pollen records from Scotland show a characteristic succession of pollen spectra from a Gramineae–*Rumex* association at the base followed by spectra dominated successively by *Empetrum, Juniperus, Betula* and then *Corylus* (Figs 7.1, 7.2). Available radiocarbon dates suggest that the onset of this succession may have been as early as 10 600 BP at some sites (e.g. Lowe,

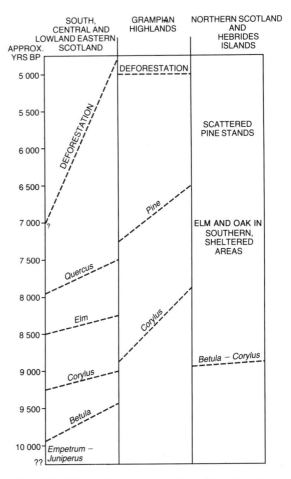

Fig. 7.1 Schematic representation of immigration of main trees and shrubs into Scotland. Inclined dotted lines indicate time-transgressive effects.

1978), though these dates are probably subject to a mineral carbon 'ageing' error (Lowe, 1991). A number of dates also suggest that juniper was widely established in the Grampian Highlands of Scotland (Vasari, 1977; Walker & Lowe, 1979) and the Isle of Skye (Williams, 1977; Walker et al., 1988; Walker & Lowe, 1990) by about 10 000 BP. The expansion of juniper into Speyside and the Island of Mull appears to have been a little later, at around 9700 to 9500 BP.

The colonization of this juniper scrub by birch and then birch–hazel woodland seems

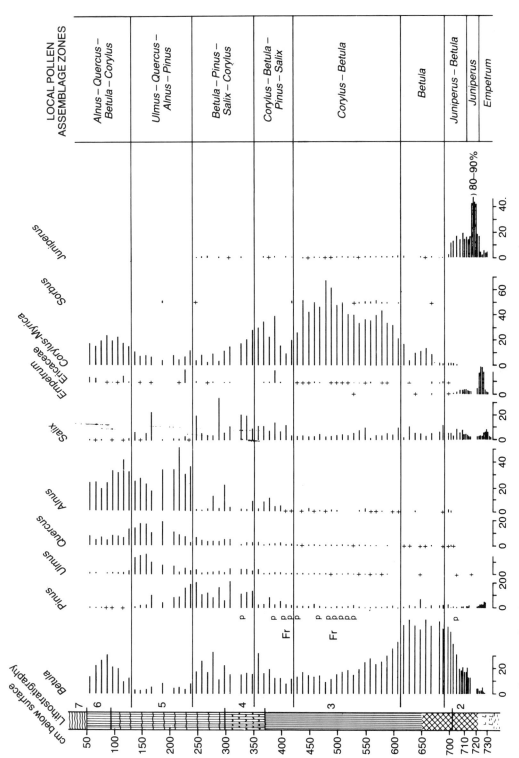

Fig. 7.2 Early- to mid-Holocene pollen stratigraphy at Mollands, near Stirling, Scotland. The site occupies an intermediate position between the central lowlands and the Highlands. The diagram is a composite of two partial Holocene diagrams in Lowe (1978) and Lowe (1982).

to have been rapid across southern and central Scotland. Available radiocarbon dates suggest that birch was established in central and eastern parts of Scotland by as early as 10 000 BP (e.g. Vasari, 1977; Lowe, 1977; Cundill & Whittington, 1983; Whittington *et al.*, 1990, 1991b), and that the spread of birch into the far north of Scotland had been achieved by about 9000 BP (Peglar, 1979).

Whether the succession of *Empetrum–Juniperus* scrub giving way to birch woodland was truly complete in parts of Scotland before or around 10 000 BP is open to question. A number of site-specific and systematic errors affect radiocarbon measurements obtained from materials of late-glacial to early Holocene age, and the true margin of error associated with such dates may be very wide (Pilcher, 1991; Lowe, 1991). However, radiocarbon dates obtained for the levels at which *Corylus* dominates the pollen successions are typically very rich in organic carbon and are less liable to the errors referred to above. They indicate that hazel woodland was established in lowland and eastern Scotland by about 9350 BP (Lowe, 1978; Whittington *et al.*, 1991b) and that it had spread into the Highlands, along the western seaboard and into Wester Ross by about 8800 to 8600 BP (Birks & Mathewes, 1978; Birks, 1972; Williams, 1977; Walker & Lowe, 1982). Since the juniper and birch vegetation phases are nearly always represented in the pollen diagrams by distinct pollen peaks that precede that of *Corylus*, it seems safe to assume that birch was widespread in Scotland by about 9500 BP. However, the true ages of the immigration of birch and of the preceding juniper phase remain problematic.

7.2.2 Afforestation phase (8500 to 5000 BP)

The density and distribution of birch–hazel stands at about 9000–8500 BP are difficult to evaluate, but it is likely that dense woodland had developed locally in favourable, sheltered valleys and on the more suitable substrates. The distinction between 'pioneer' and 'forested' episodes is therefore a blurred one, and varies in age for different parts of Scotland. The distinction is made here, however, on the basis of the development of either a mixed deciduous woodland, involving elm and/or oak, or the replacement of birch–hazel associations by pine.

Elm invaded southern and central parts of Scotland by about 8500 BP (Walker, 1984; Whittington *et al.*, 1991b) followed by oak at around 7800 BP. Both species appear to have thrived in the lowlands of Aberdeenshire (Vasari & Vasari, 1968) and in some of the main valleys penetrating the Highland edge (Donner, 1962; Rymer, 1977; Walker & Lowe, 1981; Lowe, 1982), and they may also have been present in sheltered localities in the Inner Hebrides (e.g. Birks & Williams, 1983; Robinson & Dickson, 1988). On the more exposed Scottish islands and in the far northern mainland, the pollen of elm and oak is usually recorded in such low percentages that it has been assumed that these trees were not established in these areas; however, some recent pollen-stratigraphic research indicates the presence of elm and oak in Holocene woodlands of South Uist (Bennett *et al.*, 1990a) and possibly also in some of the Shetland Isles (Bennett *et al.*, 1992).

Whereas the woodlands of central, southern and eastern Scotland were characterized by mixed deciduous woodland, most of the Grampian Highlands supported pine woodland. Pine may have appeared as early as 8000 BP in Aberdeenshire (O'Sullivan, 1974) but it generally expanded throughout the Highlands between about 7500 and 5000 BP, probably achieving maximum spread and density during that time (Bennett, 1984; Bridge *et al.*, 1990). However, a short-lived expansion of pine into the far north of Scotland seems to have been delayed until about 4000 BP (Gear, 1989; Gear & Huntley, 1991).

Alnus had colonized many parts of Scotland by about 6500 BP. The tree successfully invaded both the lowland deciduous and the Highland pine woodlands. The history of alder is a rather complicated one, however. It may have appeared in coastal localities at

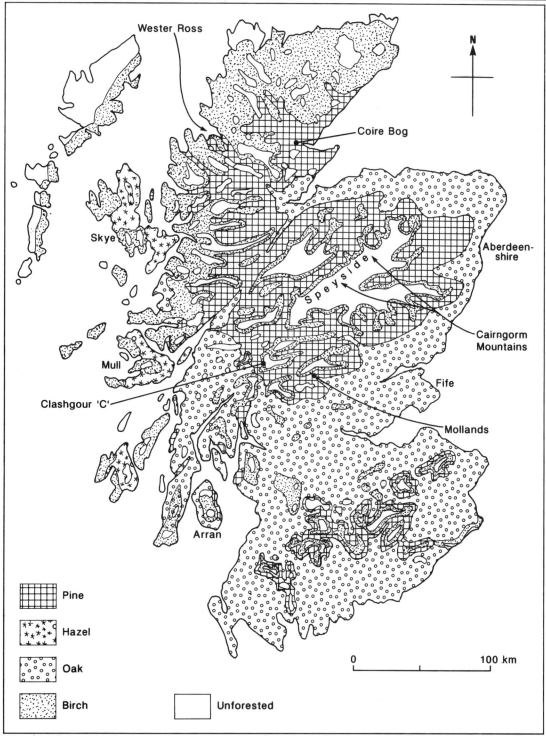

Fig. 7.3 Generalized reconstruction of the dominant woodland cover of Scotland at 5000 BP (from Bennett, 1989).

a much earlier date (e.g. *c.* 7400 BP in Fife (Whittington *et al.*, 1991a)) and there is great intra-regional diversity in the timing and importance of an 'alder rise' (O'Sullivan, 1975; Walker & Lowe, 1981; cf. Chambers & Elliott, 1989).

Maximum woodland expansion in Scotland during the Holocene appears to have been around 6000 to 5000 BP (Bennett, 1989). At that time there was a strong regional differentiation in vegetation cover, which is summarized in Fig. 7.3. This is very much a simplified model as it is not possible at present to assess the variation in density of woodland from pollen data nor the role of several species that are poorly recorded in pollen diagrams, such as *Sorbus* and *Populus*. Plant macrofossil studies are rare (with the exception of the study of pine stumps) so that the presence or absence of trees in a region depends on the subjective assessment of pollen data. This problem becomes acute when interpreting small arboreal pollen percentages recorded for sites of high altitude and latitude and on the exposed western Isles. Also, the map indicates the dominant tree only, whereas each region would probably have supported mixed communities, and there would be much local variation with locally distinct woodland types, controlled, for example, by variations in rock and soil type. However, the overall pattern corresponds closely with the **potential** woodland cover that would develop naturally under the present-day climatic conditions, as deduced by McVean & Ratcliffe (1962; see also Birks, 1977). All of the available evidence indicates that woodland was never established during the Holocene in the far north, most of the islands and the higher mountain regions of Scotland (Fig. 7.3), probably as a result of the shorter growing season and exposure to high winds.

7.2.3 Deforestation phase (5000 BP to present)

The mid-Flandrian forests of Scotland have undergone considerable reduction during the late Holocene. This is usually considered to have occurred during the last 5000 years (Birks, 1977) and there is abundant evidence to show that a dramatic decline in pine occurred throughout the Highlands shortly after 4000 BP (Bennett, 1984; Bridge *et al.*, 1990). The demise of the woodlands heralded an equally dramatic spread of blanket peats throughout the uplands in particular.

The primary cause of deforestation is not known at present. It is possible that increasing climatic wetness prevented tree regeneration in many places through increased soil saturation, eventually tipping the balance towards peat generation. However, there is much circumstantial evidence to indicate human disturbance of deciduous woodland as early as the Mesolithic (*c.* 8500 BP – Robinson & Dickson, 1988). It has also been suggested that there may be a connection between the rather erratic pattern of alder immigration on the one hand and Mesolithic burning on the other (Smith, 1984; Edwards, 1990). The effects of climate and human-induced disturbance are therefore difficult to separate.

Whatever the primary causes, deforestation was certainly under way in some localities by around 5000 BP, the traditional date for the 'elm decline' in this part of Europe (Huntley & Birks, 1983). This appears to have been a selective feature at this time, being more strongly represented in some diagrams than others, and affecting only certain tree species. By about 4000 BP, however, deforestation was widespread and is reflected by marked reductions in total arboreal pollen percentages in most pollen diagrams from Scotland.

7.3 POLLEN STRATIGRAPHY AND PALAEOCLIMATIC INFERENCES

In the early years of pollen-stratigraphic research in Scotland, as was the practice in other parts of Europe, Holocene pollen diagrams were interpreted by reference to a system of pollen zones (the 'Godwin–Jessen' pollen zones) derived from the Blytt–Sernander scheme of stratigraphic subdivision (e.g. Durno, 1958, 1959; Newey, 1968; Vasari & Vasari, 1968; Moar, 1969). The terms 'Boreal', 'Atlantic', 'Sub-Boreal' and 'Sub-Atlantic' were often used to denote significant climatic episodes. The latter were, however, usually described in imprecise terms (e.g. 'cooler and wetter'; 'more humid'). The relationship between vegetational history (pollen zones) and climate was most often assumed rather than examined (see Chapter 5).

The proliferation of pollen-stratigraphic research and especially of radiocarbon dates in Scotland in recent years has provided a fuller picture of the complexity of the country's vegetational history and a better understanding of the limitations of palaeobotanical data. The use of pollen-stratigraphic information to reconstruct climatic history is dependent upon a close and measurable relationship between vegetation and climate. There are, however, a number of factors that complicate this relationship, and that consequently limit the climatic interpretation of pollen data.

7.3.1 Blytt–Sernander scheme

Until very recently, two dominant views were held concerning the Holocene climate of Europe: first, that a thermal maximum occurred during the mid Holocene, and second, that climatic changes were synchronous across large parts of Europe. Both of these conclusions were derived mainly from palaeobotanical information, stemming originally from the Blytt–Sernander scheme of stratigraphic subdivision, and both views have recently been challenged (see Chapter 5).

Birks (1982) has questioned the validity of the Blytt–Sernander scheme even for those regions where it was first applied (e.g. Scandinavia and the British Isles). It appears that the scheme over-generalizes both the spatial and temporal complexity of climatic changes during the Holocene. In addition, the idea of a mid-Holocene thermal maximum does not fit with recent model simulations of Holocene climate based upon the integration of different types of proxy data. The COHMAP (1988) team has recently concluded that temperatures were already warmer and drier than present in most parts of Europe by 9000 BP, and a northern hemisphere radiation maximum would have occurred during the early Holocene according to Milankovitch calculations (Kutzbach & Guetter, 1986). In addition, Birks (1988a) argues that only minor climatic variations have occurred in Europe during the past 8000 years, and these are likely to have affected only those plant communities that were ecologically vulnerable. The precise relationship between climate and vegetation succession during the Holocene in Scotland therefore needs to be established objectively and then compared with appropriate data from other parts of Europe.

7.3.2 Time-lagged responses

A major limitation with climatic interpretations based upon palaeobotanical data is that there can be major time-lags between climatic changes and vegetational responses. Such effects can be measured to a certain extent for the late-glacial period, for which independent, more sensitive palaeoclimatic data exist (see e.g. Coope, 1977). The scale of such time-lags is less well known for the Holocene. Nevertheless, it is clear that most tree species arrived in Scotland some considerable time after they colonized lowland England, and there is evidence for significant time transgression in tree colonization between different parts of Scotland. This has been illustrated to some extent in the previous section and is reflected in Fig. 7.1.

It seems likely therefore that trees invaded Scotland as a response to the warm climatic conditions of the early Holocene, and that the appearance of species in different regions was determined by migration rates. In some areas the establishment of deciduous shrubs and trees was delayed considerably, as, for example, the Isle of Arran, where the sea probably acted as a barrier to fruit dispersal (Boyd & Dickson, 1987; Robinson & Dickson, 1988). The problem as far as palaeoclimatic reconstruction is concerned, is that vegetation succession may have progressed unabated in Scotland during the early to mid Holocene, irrespective of any minor climatic variations. It is entirely possible that gradual climatic cooling took place during the mid Holocene, following the early-Holocene radiation maximum, but that climatic thresholds for flowering and fruiting of trees were not crossed. Tree migration would thus have continued unimpeded, and the maximum development of deciduous woodland in Scotland may not be synchronous with the 'climatic optimum'. It is clear, therefore, that palaeobotanical data must be compared with reconstructions based upon other proxy climatic data.

7.3.3 Indicator-species approach

Sudden changes in the representation of certain trees in pollen diagrams or in the abundance of their macrofossils in sedimentary sequences have been used to infer climatic changes. For example, the sudden appearance of *Alnus* in mid-Holocene pollen records has commonly been interpreted as an indication of increased climatic wetness. This was often used to denote the start of the 'Atlantic' period where the Blytt–Sernander approach was adopted. Several researchers have also interpreted the preservation of pine stumps in blanket peats, which is widespread in the Scottish Highlands, as indicating a change to drier climatic conditions that enabled pine to colonize peat surfaces normally too wet to support the growth of pine seedlings (e.g. Pears, 1972; Birks, 1975; Dubois & Ferguson, 1985). The blanket peat that now buries these stumps is thought to indicate a return to wetter conditions, which accelerated peat accumulation and prevented the regeneration of pine.

While it is likely that climatic factors played a role in such palaeobotanical changes, they may not have played the dominant role. One might expect, for example, the alder rise to be regionally synchronous if it reflects primarily an increase in regional precipitation. The evidence from Scotland so far shows that the immigration of alder was markedly asynchronous in some regions, and probably reacted to local factors as much as to regional climatic changes (O'Sullivan, 1975; Whittington *et al.*, 1991b). The relationships between the abundance of pine stumps in blanket peats, the accumulation of blanket peat and regional climatic factors have been shown to be much more complex than previously thought (Bennett, 1984; Bridge *et al.*, 1990). Indeed, in the view of Bridge *et al.* (1990), it is quite possible that the opposite relationship to that previously assumed can be inferred from records of abundant pine macrofossils: i.e. that pine stumps would tend to be preserved during times of rapid peat accumulation (increased climatic wetness) rather than during dry periods, which would favour wood decay.

7.3.4 Conclusions

This brief survey has emphasized the insensitivity of palaeobotanical records as archives of climatic change. This results mainly from a combination of the wide climatic tolerance of most plants and of the time-lags between climatic change and vegetation response. The available palaeobotanical data from Scotland are also affected by a number of complicating site-specific factors, such as the influence of 1. site characteristics (lake basin size, drainage basin size, drainage control); 2. hydrological changes brought about by local rather than regional changes (paludification, development of perched water tables); and 3. site location with respect to former treelines or ecotones. A combination of many factors will determine whether some sites are more sensitive to regional climatic changes than others, and the extent to which such changes are registered in the resulting palaeobotanical records. During a change to cooler and wetter climatic conditions, very little difference might be registered in a pollen diagram from a site situated in a part of the western Highlands that already experienced high annual precipitation values. On the other hand, critical thresholds may be crossed in the more continental, eastern parts of the country, promoting larger-scale vegetational responses.

There is a need, therefore, for more rigorous testing of palaeoclimatic interpretations based upon palaeobotanical research. Two approaches will help. The first will be an integration of palaeobotanical data with information based upon other independent proxy climatic indicators. The second will be an increased effort to identify those sites that were *sensitive* to climatic changes in the past, where the pollen record derives from vegetation that was ecologically vulnerable during the whole or part of the Holocene. It is possible that many of the available pollen records from Scotland have been 'complacent' to climatic change during the Holocene, and a clearer climatic signal would therefore only be resolved by careful selection of sensitive records. This approach has an obvious parallel in the selection of sites and samples for dendroclimatology.

7.4 ISOLATING THE CLIMATIC FACTORS IN HOLOCENE PALAEOBOTANICAL RECORDS

The question arises as to how sites with 'sensitive' palaeobotanical records might be identified. Some recent research has indicated how this may be possible, and three examples are given in this section.

7.4.1 Early Holocene climatic revertance

Almost all early Holocene pollen profiles published for sites in Scotland indicate a unidirectional vegetation succession from open habitat grass and heathland communities to the establishment of birch and hazel wood and scrubland. The typical succession is exemplified in Fig. 7.1. In some sites with a very high stratigraphic resolution of this period, however, it is possible to detect more subtle shifts in pollen frequency that may, in turn, be indicative of short-lived climatic fluctuations. In two profiles from the Isle of Skye for example (Lowe & Walker, 1991; Benn *et al.*, 1992), a short-lived revertance episode appears to be recorded in the lowermost pollen assemblage zones (Fig. 7.4). A similar minor revertance phase occurs in the early post-glacial succession recorded at Gribun on the nearby Isle of Mull (Walker & Lowe, 1987). All three of these sites contain a high-resolution pollen stratigraphy of the early Holocene, and the data imply some disruption to plant succession during the initial stages of vegetation colonization. This, in turn, suggests a degree of climatic variation along the western seaboard of Scotland during the opening centuries of the Holocene.

A second distinctive and consistent feature of early Holocene pollen records from sites in Skye is a major expansion in *Myriophyllum* (Benn *et al.*, 1992). This tends to occur prior to the early Flandrian *Juniperus* episode and an abrupt decline in *Myriophyllum* values follows the rise in the *Juniperus* curve. The decline in *Myriophyllum* coincides with increasing values for pollen of plants of shallower water (*Potamogeton*) or lake-shore habitats (*Littorella, Equisetum, Menyanthes*). A similar picture is apparent in the early Holocene record from Gribun on the Isle of Mull, where a decline from previously high *Myriophyllum* values coincides with the increase in *Juniperus* in the early Holocene sediments (Dawson *et al.*, 1987; Walker & Lowe, 1987).

This distinctive *Myriophyllum* decline occurs at a consistent biostratigraphic position in several profiles and appears to be independent of both basin depth and sediment type. This suggests that the disappearance of

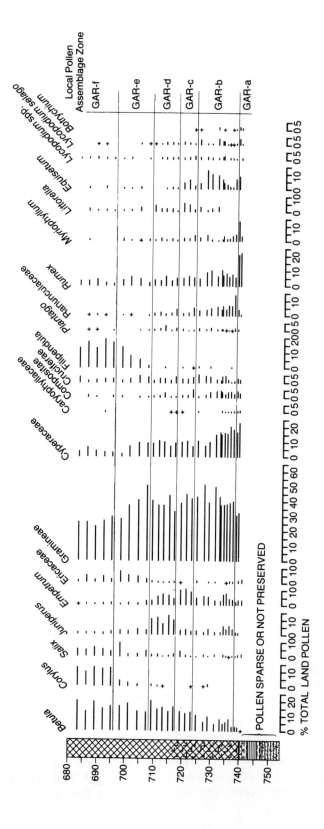

Fig. 7.4 Early Holocene pollen diagram from Glen Arroch, south-east Skye. The reduction and recovery of *Empetrum* and *Juniperus*, and the increased percentages of *Rumex* and *Lycopodium* in zones GAR-b and GAR-c, are interpreted as reflecting a minor climatic revertance phase (from Lowe & Walker, 1991).

Myriophyllum from the pollen records reflects regional forcing rather than the operation of local site-specific factors. One possible explanation is that the decline in *Myriophyllum* during the early Holocene reflects reduced water levels in the lakes of Skye and Mull that may in turn reflect a climatic shift to warmer and drier conditions with a consequent fall in lake water levels.

It therefore seems likely that small, but perhaps locally significant, variations in temperature and precipitation occurred during the early Holocene in the Inner Hebrides, and some of this variation may correlate with climatic oscillations detected elsewhere in north-west Europe (e.g. Behre, 1978; Bortenschlager, 1982). The detection of the effects of these variations is dependent, however, on the availability of early-Holocene stratigraphic successions of high temporal resolution. These are comparatively rare, but do occur in sites situated close to the limits of the former Loch Lomond Stadial ice masses. The evidence from three separate high-resolution records suggests that the pollen-stratigraphic variations are synchronous. However, further research is required to determine their full geographic expression and to examine the hypothesis that they indicate regional climatic effects.

7.4.2 Holocene range limits of pine

Pears (1968, 1970) attempted to reconstruct the upper altitudinal limits of pine growth in Scotland on the basis of mapping the distribution of pine macrofossils preserved in blanket peats. He suggested a maximum Holocene limit of 793 m and a possible 'secondary' limit of 701 m for the Cairngorm Mountains. However, there are two problems with this survey. The first is a lack of dating control and an assumption that the highest pine remains are synchronous at all sites. Bridge *et al.* (1990) have demonstrated considerable variability in age of pine macrofossils found as distinct 'layers' in three sites in the south-west Scottish Highlands. Synchroneity in age cannot therefore be assumed in other areas. The second difficulty is that the highest mapped pine macrofossils may reflect a limit to preservation rather than to the former growth of pine.

Recent research by the author (unpublished) for other parts of Scotland indicates lower maximum limits for pine macrofossils in the south-west Highlands (550 m) and in Wester Ross (500 m) as compared with the Cairngorms. Whether this reflects the former tree-line gradient in Scotland remains to be more firmly established. Current surveys indicate considerable small-scale variations in the altitudinal limits of pine macrofossils, probably reflecting a combination of topography and hill-slope orientation. More comprehensive and systematic surveys are therefore required. If the former maximum Holocene tree-line gradient can be established, however, it would provide valuable, quantifiable palaeoclimatic information.

Pollen-stratigraphic data may also provide some indication of variations in the former pine limits in Scotland. Ward *et al.* (1987) have shown that there is considerable variability in pollen records from sites within a relatively small part of the south-west Scottish Highlands. Pine is consistently recorded in high percentages in sites of low altitude (lower than 250 m) and in low percentages at high altitude sites (higher than 350 m), but shows marked fluctuations at intermediate sites – as, for example, Clashgour C, illustrated in Fig. 7.5. It is possible, therefore, that Clashgour C was located close to the former limits of pine in the region, and was therefore sensitive to variations in pine density. On this hypothesis, the data summarized in Fig. 7.5 would indicate a lowering of the altitudinal limit of pine to below 293 m shortly after 6500 BP and a rise at about 4700 BP.

Gear & Huntley (1991) have provided evidence for short-term expansion and retreat of pine forests into the far north of Scotland at about 4000 BP. Radiocarbon-dated pollen-

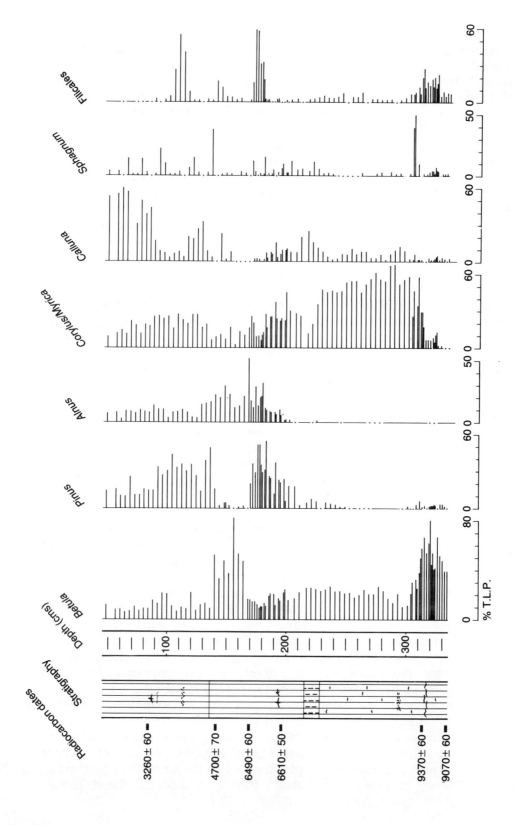

Fig. 7.5 Summary relative pollen diagram (principal taxa only) for the site of Clashgour C (NN2480 4320, 293 m OD), Rannoch Moor, south-west Scottish Highlands. Minimum TLP (total land pollen) pollen sum is 300.

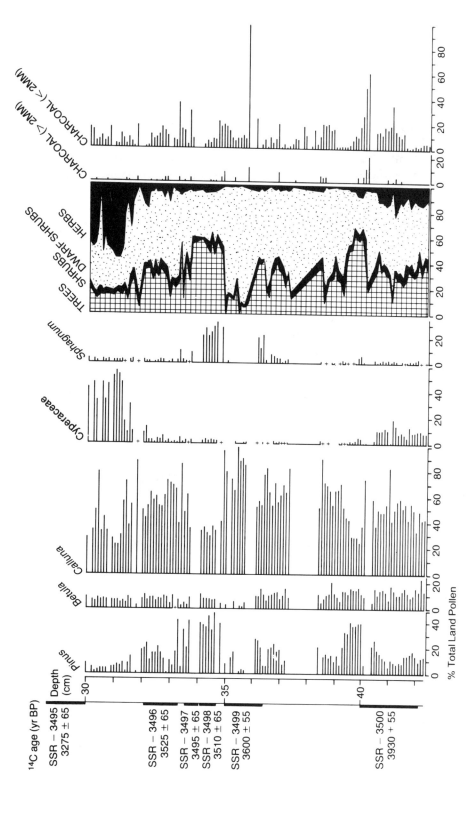

Fig. 7.6 Fine-resolution pollen and charcoal diagram from blanket peat at Lochstrathy, north-east Scotland. The radiocarbon dates on peat shown here are considered to have a 700-year bias. A date of an inter-stratified pine stump from this section gave a radiocarbon date of 4225±60 BP (from Gear & Huntley, 1991). the diagram shows marked variations in pine pollen percentages, considered to reflect migrations of the northern range limit of pine.

stratigraphic data (Fig. 7.6), supported by dendrochronological analysis and radiocarbon dating of pine macrofossils together with stratigraphic charcoal analysis, indicate that pine woodland expanded and retreated some 70 to 80 km within a period of about four centuries. The detailed evidence for changes in the northern range limit of pine is compared with other palaeobotanical data indicating substantial changes in mire surface wetness. The evidence overall indicates significant variations in regional climatic conditions around 4000 BP.

7.4.3 Variations in pine macrofossil abundance

Several authors have interpreted layers of abundant pine macrofossils within blanket peats in Scotland to reflect periods of dry climatic conditions, and layers of peat in which pine macrofossils are rare or absent as indicating a return to wetter conditions (e.g. Pears, 1972; Birks, 1975). Others have suggested that variations in the abundance of pine macrofossils reflect a complex interaction of a number of variables (Bennett, 1984), while Dubois & Ferguson (1985) have concluded that the abundance of pine remains in blanket peat reflects local factors rather than changes in regional climatic conditions.

A study of tree-ring characteristics of subfossil pine stumps preserved in blanket peats in the south-west Scottish Highlands indicates that during the mid-Holocene the pine stands were of exceedingly low density and experienced much difficulty in surviving (Ward et al., 1987; Bridge et al., 1990). The available evidence indicates that the pine woods were ecologically vulnerable, perhaps sensitive to slight changes in local hydrology. Dubois & Ferguson (1985) have published evidence that supports this and suggest a methodology for examining this contention more closely. On the basis of analysis of isotope ratios of deuterium in subfossil pines from the Cairngorm Mountains, they have concluded that particularly wet conditions (what they term 'pluvial' episodes) occurred in the region at c. 7500, 6250 to 5800, 4250 to 3870 and c. 3300 BP.

Bridge et al. (1990) have compared their data with a histogram of all radiocarbon dates obtained from pine macrofossils in Scotland and with the timing of the maximum pine pollen percentages recorded at the site of Clashgour C, discussed in the previous section. Fig. 7.7 illustrates how a wider appreciation of the effects of climatic change may emerge from the integration of several data sets. The peaks in macrofossil radiocarbon dates together with evidence for high pollen percentages from nearby 'sensitive' pollen records may indicate episodes of pine expansion in the Scottish Highlands. The stable-isotope data may reflect the dominant control over such variations – changes in regional precipitation.

Bridge et al. (1990) have suggested that the variations schematically represented in Fig. 7.7 reflect changes from relatively dry (high pine pollen percentages and peaks in pine macrofossil dates) to wet (low deuterium levels) climatic conditions. Future research is planned to test this model further. Clearly, though, comparisons with regional studies of mire stratigraphy (as undertaken by Aaby, 1986, and currently underway in Britain by Barber & Chambers) and with more detailed tree-ring studies (Ward et al., 1987; Gear, 1989) will provide much greater regional context. Integration of these various lines of evidence will help to resolve those effects that are widespread, and related to regional synoptic climatic conditions, from those of only local significance.

7.4.4 Conclusions

The examples provided in this section show the advantage of detailed comparisons of a number of palaeobotanical records from a region and of the integration of palaeobotanical records with independent data (e.g. isotope

Isolating the climatic factors in Holocene palaeobotanical records 81

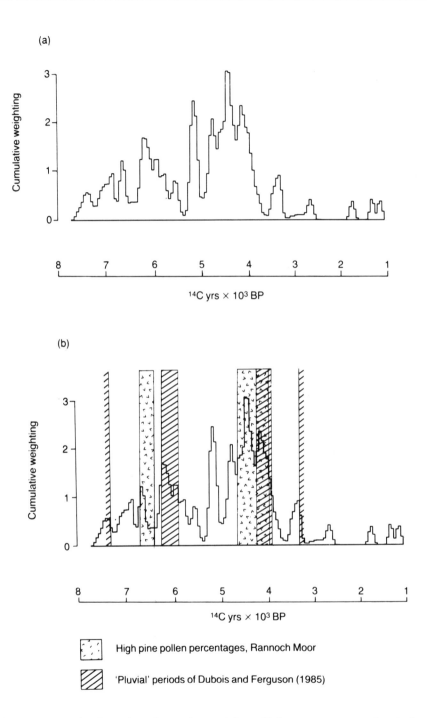

Fig. 7.7 (a) Histogram of radiocarbon dates obtained from Holocene pine macrofossils from Scotland. (b) Comparison of (a) with (i) high pine pollen phases recorded at Clashgour C, Rannoch Moor; and (ii) the 'pluvial' episodes of Dubois & Ferguson (1985). After Bridge *et al.*, 1990.

studies). These approaches enable a clearer focus on regional climatic variations, which should be common to a number of records, and on any bias in particular data sets. However, reconstructions based upon palaeobotanical data remain imprecise; at present they provide an indication of relative climatic changes within the Holocene. Other more precise palaeoecological methods are necessary in order to quantify these changes.

8

Radiocarbon dating of arctic-alpine palaeosols and the reconstruction of Holocene palaeoenvironmental change

John A. Matthews

SUMMARY

1. Radiocarbon dating of soils is reviewed with particular reference to the dating of palaeosols excavated from beneath late-Holocene ('Little Ice Age') end moraines in front of southern Norwegian glaciers.
2. The problems of soil dating are summarized, emphasizing the complex nature of soil organic matter and the importance of sample pretreatment. Details are given of the soil pretreatment procedures used in the Cardiff Radiocarbon Laboratory.
3. Estimates of the time elapsed since burial and the time since the onset of soil formation from buried humo-ferric podsols and brown soils are discussed.
4. The potential of soil age/depth gradients as a chronological framework for palaeoenvironmental studies in the alpine zone is stressed. Applications to the reconstruction of glacier and climatic variations, soil development and vegetation history, are outlined.

Key words: Radiocarbon dating, soil dating, alpine palaeosols, Holocene glacier variations, palaeoenvironmental change

8.1 INTRODUCTION

Research on the dating of soils buried beneath end moraines in front of Norwegian glaciers was begun in the Cardiff Radiocarbon Dating Laboratory in 1978. The immediate aim was to determine the age of Holocene moraines and hence to date glacier variations (Matthews, 1980, 1981, 1982, 1984, 1991a; Matthews & Dresser, 1983; Matthews & Caseldine, 1987). The timing and extent of glacier variations provide a sensitive source of data on Holocene climatic change. There are two major advantages of glacier variations as sources of proxy climatic data. First, they are physical phenomena and are directly related to climate through the balance between winter accumulation of snow and summer ablation (Meier, 1965; Porter, 1981; Oerlemans, 1989). Second, unlike biological phenomena, there are fewer non-climatic relationships that can modify their response to climate (Matthews, 1991b). This is particularly important on the Holocene timescale and in the context of human controls on vegetation and soils. Glacier variations provide, therefore, a means of identifying the natural climatic background to Holocene environmental change.

However, the research is also of significance for insights into the nature of radiocarbon dates from surface and buried soils generally. Despite the considerable uncertainty as to the reliability of dates derived from soil organic matter, there is a wide range of potential applications in many different environments (see Gerasimov & Chichagova, 1971; Scharpenseel, 1971a, b; Scharpenseel & Schiffman, 1977a, b; Goh & Molloy, 1978; Gilet-Blein et al., 1980; Gamper & Oberhaensli, 1982; Geyh et al., 1983, 1985; Matthews, 1985). Here, emphasis is given to the approaches developed at Cardiff for dating soils and for application in the field of palaeoenvironmental reconstruction, including soil development and vegetation history.

8.2 FIELD SITES

Dating of end moraines in the arctic-alpine zone of southern Norway is limited by a sparsity of organic matter above the birch (*Betula pubescens* Ehrh.) tree limit. Palaeosols buried beneath end moraines provide a valuable source of organic material for radiocarbon assay. However, sites where soils have survived burial in a largely undisturbed state are rare. Excavations made in the outermost Neoglacial moraines at 14 glacier forelands in the Jotunheimen/Jostedalsbreen region (Fig. 8.1) have yielded only two sites with apparently undisturbed palaeosols. These palaeosols lie beneath the outermost moraines of Haugabreen and Vestre Memurubreen, and are continuous with the present-day soils 'outside' the moraines.

The two undisturbed palaeosols represent very different soil types (cf. Ellis, 1979, 1980). The humo-ferric podsol at Haugabreen lies at 660 m, immediately above the tree line, and the arctic-alpine brown soil at Vestre Memurubreen lies in the mid-alpine zone at 1480 m. Soil profile descriptions have been published for both the present-day soils outside the moraines and the palaeosols (Matthews, 1980, 1982; Caseldine, 1984; Ellis & Matthews, 1984; Caseldine & Matthews, 1987). Physical and chemical characteristics of the unburied soils have been described in some detail (Mellor, 1984; Matthews & Caseldine, 1987).

8.3 PROBLEMS OF SOIL DATING AND THE IMPORTANCE OF LABORATORY PRETREATMENT

Problems of soil radiocarbon dating are summarized in Table 8.1. The most important problems stem from the complex nature of soil organic matter. Unlike radiocarbon dates obtained from some other types of organic material (e.g. wood, charcoal or bone), those derived from soil organic matter reflect the times of death of many organisms that died at different times. During the development of a soil, the organic matter is continually being rejuvenated from above, but the various organic remains are usually poorly and differentially preserved and/or physically and chemically transformed. Interpretation is aided by the concept of an 'apparent mean residence time' (AMRT) for soil carbon (Campbell et al., 1967), which has also been termed the 'relative age' or 'equivalent age' of the soil (Jenkinson, 1975). This can be envisaged as an average of the individual AMRTs of a wide variety of soil organic components, each with a different age prior to burial. Radiocarbon ages derived from buried palaeosols (whole soil samples or particular

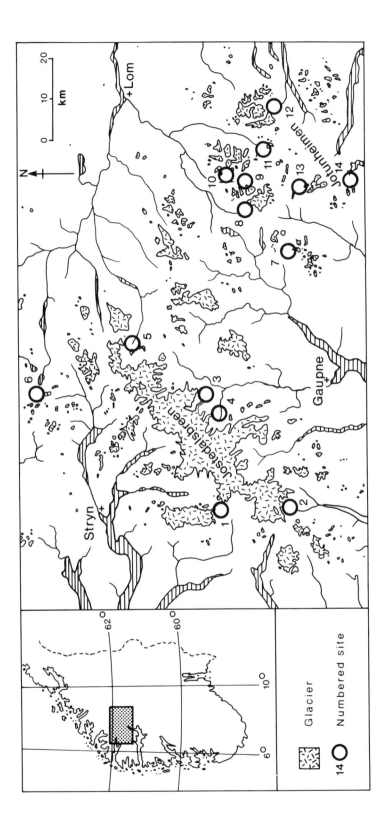

Fig. 8.1 Location of the study sites in southern Norway. Excavations in search of buried palaeosols have been made in the outermost Neoglacial end moraine ridges at the numbered glaciers. Numbered glaciers: (1) Haugabreen; (2) Bøyabreen; (3) Nigardsbreen; (4) Tuftebreen; (5) Kupbreen; (6) Skjerdingdalsbreen; (7) Styggedalsbreen; (8) Leirbreen; (9) Storbreen; (10) Nordre Illåbreen; (11) Visbreen; (12) Vestre Memurubreen; (13) Sagabreen; (14) Koldedalsbreen.

Table 8.1 Summary and check-list of problems associated with the radiocarbon dating of soils

1. Imperfect understanding of the complex nature of soil organic components.
2. Sub-optimal pretreatment procedures for the separation of age-differentiated organic fractions.
3. Age differentiation with depth in soil profiles and single horizons due to the mobility of soil organic constituents and/or accumulation.
4. Non-comparability of different soil types with respect to the age of soil organic constituents.
5. Uncertainties associated with the degree of development of the soil and whether or not soil carbon is in a state of dynamic equilibrium.
6. Sparse information on horizontal (spatial) variability in, and reproducibility of, soil radiocarbon ages.
7. Lack of modern analogues from surface soils unaffected by contamination from 'bomb' carbon (from thermonuclear tests).
8. Susceptibility of shallow-buried soils to contamination from root penetration, 'bomb' carbon and other contaminants from above.
9. Ageing effects from infinitely old carbon present in soil minerals (minerogenic carbon).
10. Surface additions of allochthonous carbon, e.g. from natural aeolian infall or atmospheric pollution.
11. Bioturbation of the soil (e.g. by macrofauna) and physical mixing (e.g. frost disturbances, wetting and drying, or ploughing).
12. Soil erosion prior to or during burial.
13. Continued decomposition and removal of organic matter after burial.
14. The general problems of dating, including limitations on precision and calibration.

fractions) therefore reflect both the time elapsed since burial and the AMRT factor:

Radiocarbon age = Time elapsed since burial + AMRT

Problems arising from imperfect understanding of soil organic matter are compounded by the lack of an optimal laboratory procedure for the separation of different soil fractions that truly reflects the age range of organic material present within the soil. The pretreatment used for soil samples in the Cardiff laboratory (Fig. 8.2) combines some of the features of the classic 'Soviet' approach to the separation of humus substances (Kononova, 1961, 1975; Goh & Molloy, 1978; Schnitzer *et al.*, 1981) with the method of repeated 'acid hydrolysis' (Scharpenseel, 1977, 1979; Stout *et al.*, 1981). The aim is to produce age-differentiated organic fractions from the complex mixture of substances comprising soil organic matter. In practice, the 'Cardiff' procedure retains a degree of comparability with the classical fractions while employing less severe chemical treatments (acid based) before more severe ones (alkali based) (Matthews, 1980, 1985). The more easily extracted 'fulvic acids' (acid-soluble fraction) are thus separated before the 'humic acids' (acid-insoluble, alkali-soluble fraction), leaving the most resistant 'fine residual' fraction (acid-and alkali-insoluble, and < 250 μm). These differ from the classical fulvic acid, humic acid and humin fractions, which are isolated in reverse order by an alkali-first procedure.

The Cardiff pretreatment provides a standardized approach to soil fractionation, but should not be regarded as an optimal laboratory procedure as it is sensitive to such factors as pH, solvent concentration, temperature and centrifugation speeds.

Where positive identification is possible, the shoot fraction is likely to yield the closest approximation to the date of burial of a palaeosol (because the AMRT is likely to be negligible). This fraction is particularly valuable where it can be identified as a distinct layer at the surface of the buried soil and hence where plant death can be assigned directly to burial. The root fraction is not normally dated because, unless deeply buried, it is likely to be contaminated by modern roots penetrating from above. Even if the roots are contemporary with the soil organic matter, a

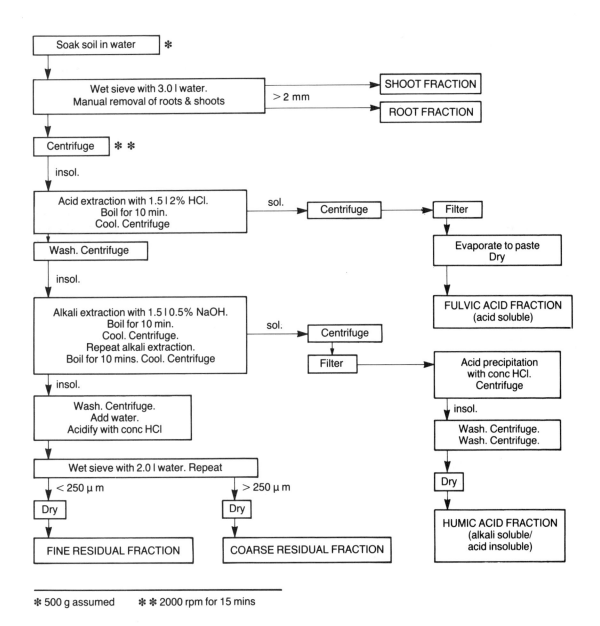

Fig. 8.2 The Cardiff pretreatment procedure for soil samples prior to radiocarbon dating. Defined soil organic fractions are numbered: (1) shoot fraction; (2) root fraction; (3) fulvic acid fraction; (4) humic acid fraction; (5) fine residual fraction; (6) coarse residual fraction.

high proportion may have died in the soil long before burial. Under such circumstances the AMRT of the root fraction could be very large (cf. Griffey & Matthews, 1978; Matthews, 1980).

Being the most mobile fraction, the 'fulvic acids' are likely to contain the youngest soil organic acids. However, as this fraction is also the most likely to contain mobile contaminants, it may not yield reliable results. 'Humic acids' are less mobile and hence are less likely to contain mobile contaminants than the 'fulvic acids'. As the 'humic acid' fraction is isolated after a relatively severe chemical treatment, it is likely to contain relatively recalcitrant (and older) soil organic components. The 'fine residual' fraction is the least mobile and would be expected to contain the oldest soil organic components (with the largest AMRT), including organic substances tightly combined with the mineral matrix and any virtually inert substances surviving from the onset of soil formation. It is less likely than the 'coarse residual' fraction to contain young, partially-decomposed plant fragments, such as small roots. At the Cardiff laboratory, 'shoot', 'fulvic acid', 'humic acid' and 'fine residual' fractions have been dated from different Norwegian soils in various combinations. In practice, the 'fulvic acids' tend to be significantly younger than the 'humic acids' from the same sample, whereas the age of the 'humic acid' fraction tends to be statistically indistinguishable or slightly younger than that of the corresponding 'fine residual' fraction (see below).

8.4 TIME ELAPSED SINCE BURIAL

The time elapsed since burial of a palaeosol may be estimated in several ways. The closest estimate is likely to be derived from the youngest (uncontaminated) organic matter within the buried soil. Detailed research on the Haugabreen palaeopodsol has shown the existence of a steep increase in age with depth within the surface organic horizon (FH) of the soil (Fig. 8.3a), and latest results have demonstrated that three different organic fractions all exhibit a strong age/depth gradient (Matthews, 1991a). It is clear that soil horizons cannot be regarded as homogeneous with respect to age and that the youngest organic matter resides close to the surface of the buried soil.

The age/depth gradients can therefore be utilized to obtain closer estimates of time elapsed since burial than has been possible hitherto. Extrapolation of age/depth gradients to the surface of the buried soil yields predicted dates of burial (with 95% confidence intervals) of Cal. AD 1900 ± 846, 1697 ± 364, and 2123 ± 602 for 'fulvic acid', 'humic acid' and 'fine residual' fractions, respectively. The broad confidence intervals reflect the small sample sizes of five, eight and five dates, respectively. As the three estimates cannot be separated statistically, the best approach would seem to be to combine them, producing a mean estimate for date of burial of Cal. AD 1791 ± 367, or younger than Cal. AD 1424 – in reasonable agreement with soil burial by a 'Little Ice Age' advance in the 18th century AD.

Similar near-linear age/depth gradients have also been established in the buried Brown Soil at Vestre Memurubreen (Fig. 8.3b). In this case, however, there was insufficient 'fine residual' fraction for conventional dating, and predicted dates of burial were somewhat older, although all 'fulvic acid' dates were consistently younger than the corresponding 'humic acids' at the same depth (Matthews & Caseldine, 1987; but see also Matthews, 1981).

Dates on the 'shoot fraction' from similar sites in front of other glaciers in the region also demonstrate very recent burial. At Sagabreen and Storbreen, Jotunheimen, dates as young as 210 ± 65 BP (CAR-749) and 270 ± 60 BP (CAR-839) have been obtained from moss layers within outermost moraines (Matthews, 1991a). At Storbreen, a 2.5-cm thick sample

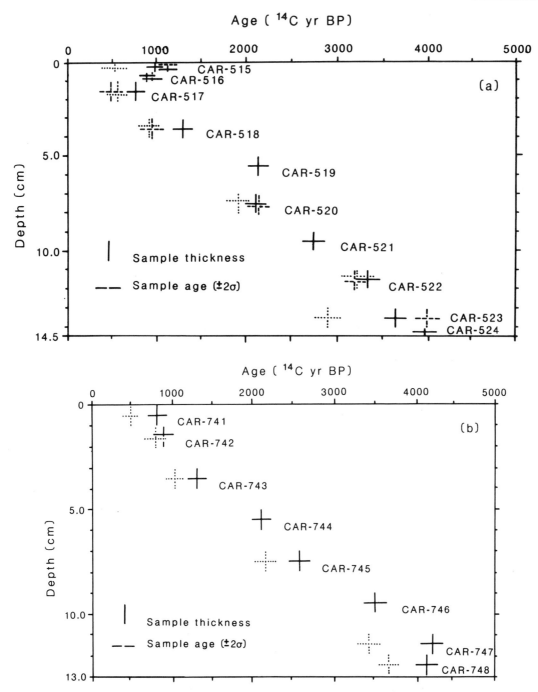

Fig. 8.3 (a) Radiocarbon dates from the buried organic (FH) horizon of the humo-ferric palaeopodsolic soil at Haugabreen. Radiocarbon age is shown in relation to depth for three soil organic fractions: 'fulvic acids' (.....), 'humic acids' (——); and 'fine residual' (– – –). Horizontal bars represent 95% confidence intervals (±2σ) (data from Matthews & Dresser, 1983, and Matthews, 1991a). (b) Radiocarbon dates from the buried Brown Soil at Vestre Memurubreen (data from Matthews & Caseldine, 1987); key as for (a).

Table 8.2 Maximum estimates of time elapsed since burial (and hence moraine age) based on plant remains beneath the historically dated (AD 1750) moraine at Nigardsbreen. Samples were pretreated with a simple acid wash

Sample no.	Age (BP) (±1σ)	$\delta^{13}C$ (‰)	Sample type	Upper age limit (+2σ)	Upper age limit (Cal.)[a] (+2σ)
CAR-783	−30 ± 45	−27.51	Grass	60 BP	AD 1690[b]
CAR-784	85 ± 50	−26.14	Moss	185 BP	AD 1670[b]
CAR-832	−60 ± 60	−26.57	Moss	60 BP	AD 1690
CAR-833	−130 ± 60	−26.36	Moss	−10 BP	Modern

[a] Calibrated according to Pearson et al. (1986); if the calibration of Stuiver & Pearson (1986) is used, CAR-783 and CAR-832 are also modern (further explanation in the text).
[b] Dates from Matthews et al., 1986.

from the uppermost part of a palaeosol from a different site in the outermost moraine had yielded a date of 1070 ± 40 BP (SRR-1085) (Griffey & Matthews, 1978).

Even younger age estimates have been published from Nigardsbreen, an outlet of the Jostedalsbreen ice cap, where Matthews et al. (1986) dated moss and grass remains associated with a disturbed palaeosol beneath the historically dated outermost moraine. This site lies in the sub-alpine zone at an altitude of about 300 m. Historical evidence (Grove, 1985, 1988) indicates that Nigardsbreen attained its 'Little Ice Age' maximum extent, close to AD 1748. The two radiocarbon dates published by Matthews et al. indicated burial after Cal. AD 1670 (CAR-784) and Cal. AD 1690 (CAR-783) (upper confidence limits: +2σ). However, two new dates from further moss samples are given in Table 8.2 together with comparable estimates of time elapsed since burial. The new dates strongly suggest some contamination of the samples by younger organic material as the apparent age of one of them (CAR-833) is so young that it cannot be calibrated. Furthermore, according to the calibration curve of Stuiver & Pearson (1986), two other dates (CAR-783, CAR-832) must also be classified as modern.

As these samples were buried beneath a moraine historically dated to about AD 1750, estimates can be made of the quantity of contaminant necessary to render them modern in terms of radiocarbon years. The oldest of the dates (CAR-784) would require contamination by modern carbon to the extent of 57% of the sample by weight (P.Q. Dresser, personal communication). This level of contamination by modern carbon is considered most unlikely where, as in this case, recognizable plant remains were dated. Contamination is much more likely to be accounted for by a small proportion of 'bomb carbon' that, at an activity level of 150% of modern, could produce these apparent ages with levels of contamination ranging from 2.7 to 7.8%. Contamination by carbon with an activity greater than the modern is in fact required to account for negative radiocarbon ages. Although these samples were buried at a depth of 3–4 m, the unconsolidated and sandy nature of the overlying 'Little Ice Age' till is conducive to the penetration of organic acids from the vegetated moraine surface. Moreover, the moraine is wooded, and Betula pubescens tree roots penetrate to the buried palaeosol with which the dated samples were associated. In the cases of the new dates, the moss specimens were much more decomposed than in the previously dated samples, thus making an explanation involving 'bomb' carbon all the more convincing as decaying roots may not have been recognized.

8.5 TIME SINCE THE ONSET OF SOIL FORMATION

In addition to providing *maximum* estimates of the time elapsed since burial, soil radiocarbon dates also provide *minimum* estimates of the timing of the onset of soil formation at the site (absolute soil age). The oldest organic material surviving within the buried soil provides the closest minimum age estimate. How good such estimates are depends on the extent to which organic components survive from the earliest stage of soil formation. The existence of steep age/depth gradients within the soil (Fig. 8.3) focuses attention on the importance of the oldest organic fraction at depth within the soil.

At the Haugabreen site, minimum estimates ($\pm 2\sigma$) for the initiation of formation of the organic horizon (the predicted dates at the base) are 1882 ± 659, 2438 ± 292 and 2610 ± 573 Cal. BC, for 'fulvic acid', 'humic acid' and 'fine residual' fractions, respectively (Matthews, 1991a). In this case, the peaty nature of the horizon suggests that the estimate from the 'fine residual' fraction at least is close to the true date at which horizon development commenced. However, statistically significant age differences of up to 1000 radiocarbon years exist between fractions of the same sample (CAR-523 at a depth of 13–14 cm below the buried soil surface), which suggests that within-fraction age variation may also be large. Even if the 'fulvic acid' fraction is ignored, some samples exhibit significant differences between the 'humic acid' and 'fine residual' fractions. This suggests that significant within-fraction age differences cannot be ruled out and, hence, that the initiation of horizon formation may have been somewhat earlier.

Although data on the age of organic material in the Bh horizon are more limited, it seems clear that it is no older than the material in the organic-rich (FH) horizon (Ellis & Matthews, 1984). This suggests that podsolic soil formation at the site was no earlier than that of the surface organic horizon (see below), a conclusion in agreement with earlier work on spodosols by Scharpenseel (1972, 1975), which indicated no increase in age with depth in the soil profile as a whole.

The reported unreliability of soil radiocarbon dating from podsols as a measure of soil absolute age (Gilet-Blein *et al*, 1980; Geyh *et al*., 1983) relates to the rapid turnover of carbon in some podsols at lower altitudes (e.g. Tamm & Holmen, 1967; Guillet & Robin, 1972). In the relatively high-altitude Haugabreen podsol, however, it appears that radiocarbon dating has permitted a close approximation to the timing of the onset of soil formation. If so, the discrepancy between the oldest dates from the soil and the age of the land surface on which it is formed (regional deglaciation of the terrain outside the moraine is believed to have occurred about 9000 BP) must be attributed to the initiation of podsol formation as a result of environmental change.

The existence of similar near-linear age/depth gradients in the Brown Soil at Vestre Memurubreen (Fig.8.3b) can be used in a similar way to provide a minimum estimate of the date of soil formation at that site. However, the minerogenic nature of this soil, the lack of distinct horizons, and particularly the likelihood of the existence of a small quantity of older carbon deeper in the soil profile suggest the possibility of a greater underestimate of absolute soil age here than at Haugabreen (cf. Matthews & Caseldine, 1987).

These ideas have been extended to present-day, unburied *surface* soils by Caseldine & Matthews (1987), who investigated the basal ages of 'humic acid' and 'fine residual' fractions from surface organic horizons on the valley-side slopes adjacent to the palaeosol site. Results were fully consistent with those from the palaeosol. Thicker horizons yielded older dates, the oldest being 5265 ± 70 BP (CAR-757C) from a depth of 34–35 cm. The older dates are interpreted as reflecting earlier initiation of podsol development in favourable topographic situations, each site requiring an individual threshold to be crossed before podsol development commenced.

A similar approach has been applied in relation to Brown Soils beyond the outermost

Table 8.3 Minimum estimates of moraine age and the onset of soil formation based on dating of Brown Soils from the crest of a moraine of presumed Preboreal age at Vestre Memurubreen. Samples were pretreated with a simple acid wash

Sample no.	Age (BP) (±1σ)	$\delta^{13}C$ (‰)	Sample type	Depth (cm)	Lower age limit (−2σ)
Site 1					
CAR-970	4440 ± 70	−24.66	Soil	46–50	4300 BP
Site 2					
CAR-971	860 ± 60	−25.34	Soil	15–16	—
CAR-972	3600 ± 70	−25.34	Soil	36–40	3460 BP

Neoglacial moraine of Vestre Memurubreen. Samples were taken from soils up to 50 cm deep on the crest of a moraine ridge of presumed Preboreal (c. 9000 BP) age located about 0.5 km down-valley from the palaeosol site. A simple acid wash was employed for three reasons: first, this is likely to remove the youngest contaminants associated with the 'fulvic acid' fraction; second, experience with dating the Vestre Memurubreen palaeosol indicated that it would not be possible to isolate sufficient quantities of 'fine residual' fraction for dating; and third, the low organic matter content at depth in this soil type precluded the use of the normal pretreatment procedure.

Results from the present-day surface soils at Vestre Memurubreen are given in Table 8.3. Although there is great uncertainty as to the extent to which these soils have been affected by 'bomb carbon' or other contaminants such as plant roots operating over a longer period of time, they nevertheless provide minimum age estimates for the onset of soil formation at the site. They also demonstrate conclusively that the moraine on which they are found could not have been formed from a Neoglacial glacier advance after 4000 BP. The question must remain open as to whether the moraine could have been formed between about 9000 BP and the mid Holocene, although this would seem unlikely given the retracted state of other southern Norwegian glaciers at this time (Nesje et al., 1991; Matthews, 1991a).

8.6 HOLOCENE GLACIER AND CLIMATIC VARIATIONS

The age/depth relationships present in the buried palaeosols establish the existence of late-Neoglacial ('Little Ice Age') glacier maxima at Haugabreen and Vestre Memurubreen. That is, these glaciers were more extensive in the 'Little Ice Age' than at any time since at least the mid Holocene, soil development continuing until the soils were buried during formation of outermost Neoglacial moraines. Use of well-established age/depth relationships based on soil organic fractions (particularly the 'humic acid' and 'fine residual' fractions) permit relatively precise maximum age estimates to be given for the timing of the 'Little Ice Age' maximum. However, these estimates are not as close to the likely true age (about AD 1750 based on historical evidence) as those derived from plant remains (the 'shoot fraction').

8.6.1 Climatic implications

The climatic implications are clear. The climatic deterioration that accompanied the 'Little Ice Age' glacier advance in southern Norway must have been more severe (either in terms of intensity or duration) than any other climatic deterioration of at least the last 4000 years. By comparing the response of southern Norwegian glaciers with that of glaciers in northern Scandinavia, Matthews (1991a) concluded that the relatively large scale of the 'Little Ice Age' glacier expansion episode in the south could be accounted for by a more southerly extension of the north Atlantic oceanic polar front and associated persistent atmospheric weather patterns. Precise climatic parameters, or the relative contributions of reduced summer temperature and increased winter precipitation, cannot as yet be ascertained. However, southern Norwegian glacier variations during this century have tended to be more closely correlated with summer temperature variations (e.g. Bogen *et al.*, 1989) than with winter precipitation, and dendroclimatological evidence suggests that summer temperatures were on average 1–2 °C lower during the 'Little Ice Age' than they are today (Matthews, 1976, 1977).

8.7 SOIL HISTORY AT HAUGABREEN AND VESTRE MEMURUBREEN

At Haugabreen, the nature of the organic surface (FH) horizon, combined with the linear increase in age with depth, indicates a rather simple pattern of development (similar to a peat) in which organic matter accumulation exceeds decomposition. However, a more complex history is indicated for the Bh horizon by the radiocarbon dates shown in Fig. 8.4. The radiocarbon age of 'fulvic acid' and 'humic acid' fractions from the Bh horizon reveal: 1. age/depth gradients; 2. ages no younger than about 2000 BP; 3. ages no older than the oldest organic material in the FH horizon; 4. ages at the *base* of the Bh horizon that do not differ significantly from those of comparable fractions at the *base* of the FH horizon; and 5. relatively large differences in age between different fractions at the same depth (Ellis & Matthews, 1984; Matthews, 1987). These features of the results are explained by the following model of podsol development.

8.7.1 Podsol development

Similarity in age between similar fractions at the base of the two horizons suggests that illuviation of organic matter in the Bh horizon commenced shortly after the FH horizon began to accumulate. Large differences in age between fractions at the same depth in the Bh horizon probably reflect the translocation of material with a wide range of ages from the FH horizon. In the early stages of soil development at the site, illuvial organic material would have originated from the present base of the FH horizon. In the later stages, the addition of younger material from higher in the growing FH horizon would have led to an incremental increase through time in the age of the illuvial material reaching the Bh horizon. As the Bh horizon developed, illuvial organic material appears to have accumulated upwards *within* the Bh horizon, producing the age/depth gradient. The absence of dates younger than 2000 BP in the Bh horizon is therefore accounted for by large AMRTs inherited from the FH horizon.

This pattern of development is consistent with an upward movement of the position of maximum colloidal deposition within the il-

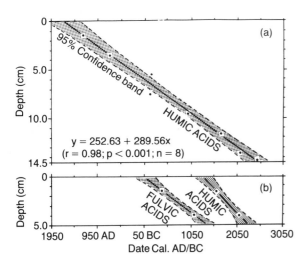

Fig. 8.4 Age–depth gradients (calibrated ages), with 95% confidence intervals, within the Haugabreen palaeopodsol: (a) 'humic acids' within the FH horizon; (b) 'fulvic acids' and 'humic acids' within the illuvial Bh horizon. Depths refer to depth below the top of the appropriate horizon (after Ellis & Matthews, 1984).

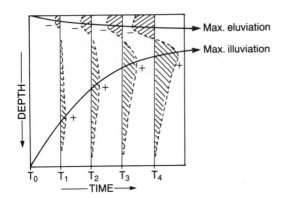

Fig. 8.5 The Scandinavian concept of ascending illuviation (based on Aaltonen, 1939, and others). Shaded areas represent the depth distribution of eluvial removal (−) and illuvial accumulation (+) at different stages (T_0–T_4) in the development of a podsol profile.

luvial zone during soil development (Fig. 8.5), a concept first proposed by Aaltonen (1939) as a result of his studies on podsol development along the emergent coast of the Gulf of Bothnia (cf. Mattson & Lönnemark, 1939; Jenny, 1941).

8.7.2 Brown Soil

Rather different processes must be inferred to account for the age/depth gradient in the Brown Soil at Vestre Memurubreen (Fig. 8.3b). An element of accumulation is necessary to account for this increase in age with depth. Matthews & Caseldine (1987) account for the pattern of ages in terms of a combination of accumulation, downwash and immobilization of organic components. They envisage a net upward accumulation of organic material as a result of 1. aeolian deposition of loess-like fine sand and coarse silt (20–200 µm) at the soil surface; 2. surface or near-surface additions of organic matter predominantly from vegetation growing at the site; and 3. differentiation of a B horizon close to the soil surface. Locally, accumulation of aeolian sediments may be much more extensive, as appears to be the case at the site of the dated profiles up to 50 cm thick (Table 8.3). Such sites provide a rare opportunity for palaeoenvironmental reconstruction in the alpine zone well above the tree line.

8.8 VEGETATION HISTORY

Soil pollen analyses by C.J. Caseldine have been combined with radiocarbon dating at both the Haugabreen and Vestre Memurubreen sites to provide insights into vegetation history (Caseldine, 1983, 1984; Caseldine & Matthews, 1985, 1987; Matthews & Caseldine, 1987). Within the FH horizon at Haugabreen there is good pollen stratification. Furthermore, good agreement between pollen diagrams from buried and unburied soils, and in particular between maxima and minima of pollen concentrations and pollen incorporation rates (p.i.r.) (Caseldine & Matthews, 1985), suggest vegetation changes associated with shifts in the local *Betula pubescens* tree line (Fig. 8.6).

8.8.1 Haugabreen (podsol) sites

As climate deteriorated from at least *c.* 5000 BP, the tree line appears to have fluctuated close to the Haugabreen sites, with a sharp drop to an altitude below the palaeosol site (660 m) between *c.* 3600 and 3300 BP. Between *c.* 3300 and 750 BP, the tree line remained below or close to 660 m, except for two possible upward extensions at *c.* 2600 and 2200–2000 BP. Finally, with the onset of the 'Little Ice Age' after 750 BP, the tree line again fell below 660 m, with little evidence of recovery since (Caseldine & Matthews, 1987). These results from Haugabreen suggest that the organic surface horizons of humo-ferric podsols close to the tree line are sensitive sources of palaeoecological change in the sub-alpine and low-alpine belts. These changes can be relatively well dated using radiocarbon assay because of the peaty nature of the LFH horizons.

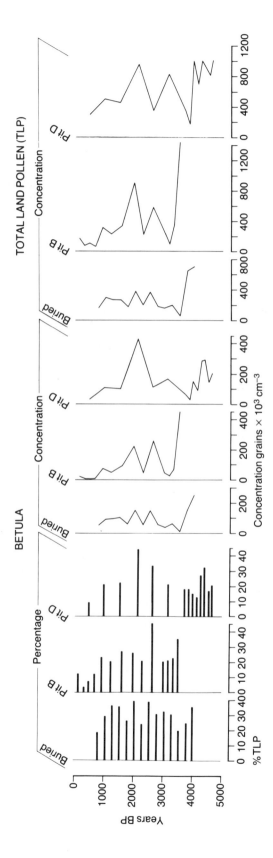

Fig. 8.6 Selected pollen data from buried and unburied (pits B and D) organic surface (LFH) horizons of podsols at Haugabreen. *Betula* and TLP pollen percentages and pollen concentrations are shown in relation to uncalibrated radiocarbon age (after Caseldine & Matthews, 1987).

8.8.2 Vestre Memurubreen (Brown Soil) sites

It is rather more surprising that the Brown Soils from Vestre Memurubreen also exhibit pollen stratification (Caseldine, 1984; Matthews & Caseldine, 1987). This has been interpreted as indicating a single vegetation change at this location during soil development. Assuming the age of the 'humic acid' fraction to be representative of the average age of the pollen at each level, a low-alpine dwarf-shrub heath was gradually replaced by a mid-alpine 'grass' heath over the last c. 5000 years, which in turn suggests a downward shift in vegetation belts and a possible climatic cooling of 2–4 °C (Matthews & Caseldine, 1987).

It appears that several factors combine to preserve organic matter, including pollen, in these Brown Soils: 1. cold or frozen soils for most of the year (reducing soil faunal activity and decomposition); 2. acidic soil pH values (conducive to pollen preservation); 3. minimal down-washing of pollen through the soil profile; 4. deposition of aeolian sedimentary particles (improving the preservation of pollen and enhancing pollen stratification); and 5. an absence of physical soil mixing by frost activity at well-drained sites. The results therefore lend support to the ideas of Dimbleby (1957, 1961a, 1961b, 1965, 1985) and Welten (1958, 1962) concerning the usefulness of soil pollen for palaeobotanical and palaeoenvironmental reconstruction. In view of the dearth of suitable alternative approaches, soil pollen in parallel with radiocarbon dating would seem to offer major opportunities in this field, particularly in arctic-alpine environments.

8.9 CONCLUSIONS

Results have greatly improved our understanding of radiocarbon dates derived from surface and buried soils in the alpine zone, and have made a major contribution to the record of Holocene environmental change from southern Norway. Intensive dating of thin soil slices, which have been physicochemically fractionated, has been particularly important in revealing steep age/depth gradients within single horizons of well-developed soils. The discovery and characterization of such gradients in both humoferric podsols and Brown Soils solves some of the problems associated with soil radiocarbon dating and provides a means of improving environmental reconstructions based on soil dates.

There are also broader implications for the reconstruction of environmental change beyond the Scandinavian mountains. These implications arise partly from the strategic location of southern Norway in relation to the average position of oceanic and atmospheric polar fronts (Lamb, 1979; Matthews, 1991a) and partly from the ubiquity of soils, and hence their potential importance in contexts ranging from the reconstruction of natural environmental change in the alpine zone to human-induced environmental change at archaeological sites in the temperate zone.

Acknowledgements

I thank Professor A.G. Smith for his long-term support of this research, Dr P. Quentin Dresser (radiocarbon dating), Dr C.J. Caseldine (pollen analysis), Dr S. Ellis (pedology and soil micromorphology), Dr N.J. Griffey (geomorphology), Dr A. Mellor (soil descriptions and analysis), and the many individuals and organizations who have taken part in and/or sponsored the field work. This chapter is Jotunheimen Research Expeditions Contribution No. 99.

Part Three

Evidence for Human Impact

Introduction

Human impact upon vegetation has become one of the dominant themes in Holocene palaeoecology, though it is perhaps not often appreciated that significant human impact in many areas may pre-date the Holocene (see Chapter 9). However, it is often difficult to discriminate between evidence for human impact and other environmental change (see Part Four). Furthermore, invoking human impact is not without problems of semantics. A.G. Smith has cautioned us to '... note the view of Kenneth Mellanby on the meaning of the word "anthropogenic". Strictly speaking, he has observed, there are only two anthropogenic activities: evolution and copulation! ... why use fancy words when plain ones will do?' (A.G. Smith, personal communication, 1991).

Part Three commences with a global review by Professor D. Walker and Dr G. Singh of the first palynological evidence for human impact in the major investigated regions of the world. This is followed by chapters reviewing the Mesolithic in the British Isles (Professor I.G. Simmons); fire and the influence of Mesolithic human activity on Dartmoor, England (Dr Chris Caseldine and Jackie Hatton); models of mid-Holocene forest farming (Dr Kevin J. Edwards); and molluscan evidence from the English chalklands in prehistory (Dr J.G. Evans). Three case studies, drawn from Britain and Ireland, illustrate recent practice in palaeoecology and environmental archaeology, and emphasize the impact of human activity on the landscapes of England (Patricia E.J. Wiltshire and Kevin J. Edwards), Wales (Dr M.J.C. Walker) and Ireland (Dr Karen Molloy and Dr Michael O'Connell).

Introduction

9
Earliest palynological records of human impact on the world's vegetation

D. Walker and G. Singh

SUMMARY

1. A broad outline is presented of the chronology of earliest human impact on the world's vegetation, as shown in pollen and charcoal remains.
2. The range of dates, and the diversity in ages reported, both within and between regions, are unlikely to represent the first human impact, by virtue of the inadequacy of palynology as a sensor and because they do not all relate to the same phenomenon.
3. A set of criteria is suggested under which a broad consensus could be found for attributing a palynological change to human impact.
4. The authors report on the huge gaps and discuss the possible reasons behind the lack of palynological information on human impact from regions that are otherwise known to have practised agriculture. These regions are contrasted against those that are devoid of archaeological evidence for agriculture, but have practised slash-and-burn activity for millennia.

Key words: earliest, palynological, world, human impact, vegetation record

9.1 DATA

Fig. 9.1 illustrates the chronology of earliest human impact on the vegetation of many parts of the world as shown by pollen and spore analyses and, in many cases, accompanying counts of microscopic charcoal fragments. Most ages are drawn from radiocarbon dating, either of the pollen diagram in which the disturbance is recorded or from closely related material. Although we have, in some cases, made our own interpolations and extrapolations, we have always accepted an author's dating, however derived, when this has been stated explicitly. A date on the map is sometimes drawn from a single site (e.g. 2 in south China is specifically from Dahaizi Lake on Mt Luoji in Sichuan Province (Li & Liu, 1988)) and sometimes

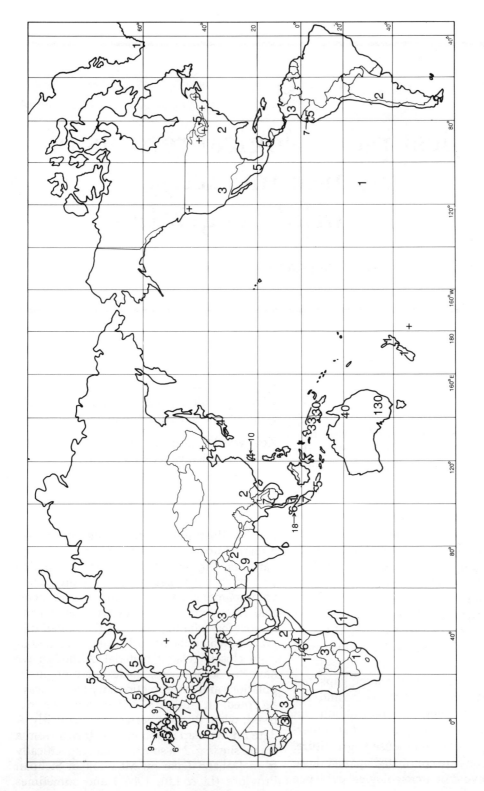

Fig. 9.1 Dates for the earliest palynological evidence of human impact on vegetation. With the exception of one in Canada, rounded to 500 years, all have been rounded to the nearest thousand years BP. Crosses indicate very recent impacts. Larger numbers refer to well-substantiated ages, smaller ones to ages about which there is still some uncertainty. The very early date claimed for Hoxne (England) has been excluded from the map (see text).

Table 9.1 Publications, arranged in geographical groups, that have contributed to the ages shown on Fig. 9.1, or are referred to in the text

British Isles
Bennett *et al.* (1990b)
Bohnke (1988)
Boyd & Dickson (1986)
Bush (1988, 1989)
Dodson & Bradshaw (1987)
Edwards & Hirons (1984)
Molloy & O'Connell (1991)
O'Connell *et al.* (1988)
Peglar *et al.* (1988)
Robinson & Dickson (1988)
Simmons *et al.* (1981)
Smith (1981b)
Tallis & Switsur (1990)
Walker (1984)
West (1956)
West & McBurney (1954)
Whitttington *et al.* (1991a)

Near East
Van Zeist & Bottema (1977)
Van Zeist & Bottema (1982)
Van Zeist & Woldring (1980)
Van Zeist *et al.* (1975)

Europe
Aaby (1988)
Anderson (1988)
Averdiek (1978)
Beug (1966, 1967)
Burga (1988)
Chen (1988)
Clark *et al.* (1989)
Danielsen (1970)
Fredskild (1988)
Hicks (1988)
Indrelid & Moe (1982)
Jeschke & Lange (1987)
Küster (1988)
Nilssen (1988)
Oldfield (1960b)
Reille & de Beaulieu (1988)
Rossignol & Pastouret (1971)
Stevenson (1985)
Stevenson & Moore (1988)
Tolonen (1978)
Turner & Greig (1975)
Van den Brink & Janssen (1985)
Van Zeist (1964)
Van Zeist & Van der Spoel-Walvius (1980)

Africa
Burney (1987)
Hamilton (1982)
Lamb *et al.* (1989)
Livingstone (1967, 1971)

Morrison (1968)
Morrison & Hamilton (1974)
Owen *et al.* (1982)
Scott (1987)
Sowunmi (1981)
Talbot *et al.* (1984)
Van Zinderen Bakker (1989)

Australia and New Zealand
Dodson (1978)
Edney *et al.* (1990)
Kershaw (1986)
Kershaw *et al.* (1991)
McGlone (1983)
Singh & Geissler (1985)
Sutton (1987)

North America
Barnosky (1981)
Betancourt & Davis (1984)
Brugam (1978)
Delcourt & Delcourt (1985)
Maher (1977)
McAndrews & Boyko-Diakonow (1989)
Watts & Bradbury (1982)

India
Jarrige (1981)
Sharma (1985)
Sharma & Chauhan (1988)
Sharma & Singh (1972a, 1972b)
Singh (1971)
Singh *et al.* (1974)

Far East
Fuji (1984)
Li & Liu (1988)
Maloney & McAlister (1990)
Sun *et al.* (1986)
Tsukada (1966, 1967a, 1967b, 1981, 1983)
Tsukada & Stuiver (1966)
Yasuda (1978, 1981–2)

South-East Asia and New Guinea
Flenley (1984, 1988)
Haberle *et al.* (1991)
Hope (1983)
Maloney (1985, 1990)
Morley (1982)
Stuijts *et al.* (1988)
Swadling & Hope (1991)
Walker & Hope (1982)

Central and South America
Bush & Colinvaux (1988)
D'Antoni (1983)
Deevey (1978)

Table 9.1 (continued)

Central and South America (continued)	Van der Hammen & Gonzalez (1985)
Metcalfe et al. (1989)	Wijmstra & Van der Hammen (1966)
Piperino (1990)	
Saldarriaga & West (1986)	*Elsewhere*
Tsukada & Deevey (1967)	Flenley et al. (1991)
Van der Hammen (1962)	Wright & Barnosky (1984)

from a group (e.g. 6 in England). In the latter case, however, the date shown is the earliest recorded in that group and might not refer to all its members. We have avoided trying to define the area to which each age might refer because, in almost all cases, to do so would have been guesswork.

Our literature search has been neither even nor exhaustive and the geographical patchiness of the data reflects to some degree our own interests and the literature available to us. Although we consulted more papers about north-west Europe (including the British Isles) than for any other region represented, as a proportion of the total available this number was relatively very small. We took this line, partly so as not to become involved in complex detail with which we are no longer familiar, but also because several chapters in this volume focus particularly, and with authority, on that region. It seemed inappropriate to identify each number on the map with a specific site or publication; instead, we have provided a geographically grouped list of the publications that have positively informed our decisions (Table 9.1).

In every case we have accepted an author's interpretation of original data and definition of human impact. Naturally, this leads to heterogeneity in the data because, at best, different processes are involved and, at worst, we all have somewhat different imaginations. We return to this matter below.

9.2 COMMENTARY

9.2.1 The nature of human impact

Humans have always had effects on their environment, just as has every other organism. The difference is that humans' cultural development has enabled them to sidestep the homeostatic responses of the ecological systems of which they were and are parts. Indeed, a great deal of that cultural development can be interpreted as a sequence of technological devices for avoiding unwelcome feedback from the exploitation of other organisms and their physical milieu, in order to fuel expanding human populations. In this progression, the degree to which human activities have deflected or prevented changes that would otherwise have taken place, and purposefully instigated entirely new variants, seems to have increased sporadically through time in most parts of the world. At what point, and by what criteria, can pollen analysis detect these effects?

A great deal depends on the sensitivity of an ecosystem to human exploitation and the degree to which this is reflected in the pollen analytical signal. Thus, the burning of forest understorey is more likely to be palynologically recognized if the main beneficiary were the anemophilous *Corylus* with resistant pollen, rather than *Macrozamia* with hardly recognizable and decaying microspores. Forest destruction for agriculture is more likely to be recognized if the crop is a cereal with a characteristic suite of anemophilous weeds than if it is the palynologically cryptic sweet potato. Perhaps, too, we should not expect anything but the most extreme impacts to be as clearly recorded from, say, Mediterranean shrublands as from forests of anemophilous trees.

This basic difficulty almost certainly accounts for some of the diversity in the ages reported here both within and between regions. Not only is none of them likely to represent the first human impact in its region, by virtue of the inadequacy of palynology as a sensor, but they do not all relate to the same phenomenon. Within the British Isles, for example, the low numbers (6000, 5000, 4000 BP) refer almost entirely to supposed forest destruction of the *landnam* type for grazing or agriculture, or both. The entries for 9000 BP represent all those records, pre-dating the above, purporting to represent much more subtle or local human intervention in the affairs of nature, yet arguably distinct enough to justify the imputation of 'impact' or 'disturbance'. Most of the dates on the map do, in fact, relate to the apparent destruction of formerly established vegetation, mostly forests. To this extent the 6000 BP records for, say, the southern Alps and northern Sumatra are comparable but the antecedent processes and the subsequent results, not to mention the crops involved, were quite different in the two places.

At least some of the variation shown on the map and much more of the difference of opinion we have encountered in the literature, reflect the different criteria by which individual analysts attribute a palynological phenomenon to human impact. There is no overriding logic as to whether human activity, or climatic or edaphic change or disease or other environmental or biotic force, should be preferred as the explanation of a palynological change, unless positive contrary reasons can be found. Perhaps climatic causation should be considered first, because the generality of its implications lend it to testing more readily than more localized causes, but all the argument about the *Ulmus* decline in northwest Europe makes us hesitate in this prescription. Most of us would agree that the following together provide sufficient reasons for attributing a palynological change to human impact:

1. It should reflect ecological processes operating at levels and rates that are unprecedented under 'natural' conditions but are readily explicable as resulting from human actions of defined kinds.
2. The necessary human activities should be within the technological capacity of prehistoric peoples of the relevant age and region.
3. There should be some acceptable reason why the humans might have taken the hypothesized action (ideally exemplified by the pollen itself, e.g. crop pollen).
4. There should be strong evidence (ideally artefacts stratified into the pollen-analysed deposits) for human occupation at the appropriate time within the pollen catchment.

Encouragingly, many implications of human activity explicitly meet most, if not all, these criteria. Most of the difficulties arise from the possibility of non-human ecological processes beyond our present comprehension, the probability that some human actions will have been contemporary with, and their effects inextricably mixed with, changes from other causes and the importance attributed to parallel archaeological evidence. A striking example of this last matter comes from a comparison between a site in Ireland (O'Connell *et al.*, 1988) and one in southeast Australia (Singh & Geissler, 1985). In the former, human interference is rejected as a likely explanation of the presence of commonly used indicators from 9000 BP (including a *Plantago lanceolata* peak at about 8600 BP), despite a persuasive charcoal curve, mainly for lack of local Mesolithic remains. In the latter, major vegetation changes and coincident increases in charcoal counts 130 000 years ago are attributed to human activity despite the lack of archaeological evidence for human occupation anywhere on the Australian continent until at least 70 000 years later. This is not to suggest that either interpretation is wrong, but to emphasize that a good deal of the variation between the numbers on the map must come from the degree to which authors are willing to relax one or more of the above criteria in consideration of all the circumstances of a particular site.

Everywhere, it would be desirable to have, for comparison, vegetation sequences that, by virtue of age or location, could confidently be taken to be unaffected by human activity yet controlled by similar boundary conditions of climate and species availability.

9.2.2 Gaps on the map

Some of the large gaps on the map are almost certainly due to our ignorance of the literature, e.g. of the Commonwealth of Soviet Republics and eastern Europe. Others, such as parts of Africa and the swathe from Assam to Vietnam, have not attracted the attention of palynologists, although sites are almost certainly available there. Some parts of the world, however, like much of Australia, have not enjoyed environments suitable for organic sedimentation during the greater part of the Quaternary.

In the northern US and Canada, amply covered by Holocene pollen diagrams coeval with human occupation, it must be that palynologically recorded early disturbance did not occur or has not yet been recognized (except in the case of Crawford Lake, Ontario (McAndrews & Boyko-Diakonow, 1989). The dominant tendency there has been to refine the interpretation of quite small and subtle palynological changes in climatic terms and to seek parallels between these and the changing patterns of human settlement and technology attested by archaeology. Variations in charcoal particle counts, although commonly recorded, have not generally been used there as indices of human activity. This may be wise or unwise, but it contrasts with the habits in some other parts of the world, such as Australia and New Guinea. It would be interesting to know whether this is related to differences in the perception of modern fire-initiation/fire-sensitivity relationships, known differences in the technologies of the prehistoric peoples, or differences in received palynological wisdom.

Whilst the subtlety of the signals of human impacts that were not grossly destructive of the vegetation contributes to their lack of recognition in some parts of the world, this can hardly be the case in China. It is true that most Chinese pollen diagrams are insufficiently detailed to expose human impacts clearly but, even when they have positively been sought in archaeologically rich areas (e.g. Sun et al., 1986), changes attributable to what must have been very active and widespread disturbance have not been recognized. The reason must be that we do not yet know what to look for. Because most of China never had a pastoral history, we are denied its indicators in pollen diagrams, although it would be interesting to seek them in, for example, the north-west, where pastoralism is deeply traditional. For most of China, however, valley agriculture based on cereals (usually rice), pigs and poultry has been at the core of human population expansion. Yet the latter has clearly modified the upland vegetation as well, to the point of virtually destroying most of it. In spite of the difficulty in identifying rice pollen (Maloney, 1990), the effects of agriculture on the Chinese landscape should be able to be documented palynologically.

9.2.3 Time span

The actual destruction of forest, to which the great majority of the dates refer, is a dominantly Holocene event, agriculturally induced. The cluster of older ages – 6000 and 7000 BP – for such major events in western Europe is probably due to the palynological sensitivity of both the trees and the vegetation that replaced them, the numbers of sites investigated and the refinement of the technique used. Certainly, crops were being taken in the Near East before any indications of forest clearing in western Europe, despite the relatively late palynological evidence from the former region. Whether the same is true for Africa is another matter. The rarity of forest-destruction ages greater than 3000 BP might be real, older subsistence having been drawn from root crops, which required less wholesale destruction and did not themselves leave a palynological mark. Or was it that firing rather than felling of trees was practised, a feature that could not be detected easily, especially if the burning were carried out at frequent intervals? Felling of trees might have come much later. Indeed, the use of fire in landscape management may have been prevalent in low and middle latitudes throughout the Africa–Asia–Australia belt.

In the Americas, the age of first clearance seems to parallel the archaeological record of local human population concentrations. It is a matter for conjecture whether the latter was facilitated by the relative ease of cropping, for climatic or other reasons, in one area rather than another. Assuming human immigration to America from Asia in the late Pleistocene, the forest destruction, which began rather later in America than it did in Europe, must be seen as a totally independent phenomenon. By contrast, the still later dates for Easter Island and New Zealand reflect the arrivals there of people with established agricultural traditions.

The mid-Holocene dates for Sumatra and Java are all from the uplands and probably imply earlier, but perhaps less widespread, disturbance in the lowlands. It is striking that, by 7000 or 6000 BP, forest destruction was well under way across the Old World from north-west Europe to tropical south-east Asia.

Outstanding amongst all the numbers on the map (Fig. 9.1) are those for Australia and New Guinea. In New Guinea there is plenty of pollen analytical evidence for horticulturally necessitated forest clearance back to 5000 BP and intimations of it before that. The 30 000 BP date in the east of the island is based on palynological indications of fire-induced forest destruction as evidenced by the first and substantial occurrence of charcoal particles. The slightly earlier date of 33 000 BP in the west, at a site first discussed in this context for carbonized wood dated at 26 000 BP (Haberle et al., 1991), has now been confirmed pollen analytically (Hope, personal communication). The common occurrence of soil erosion between this kind of date and 5000 BP may be partially attributable to human activity.

In Australia, devoid of archaeological evidence for agriculture or husbandry at any time during its 50 000-year, and perhaps longer, span of human occupation, the two very early dates for human disturbance require special justification. That for 40 000 BP in the north-east of the continent depends on association between charcoal-authenticated occurrence of fire, the relative sensitivities to it of different vegetation types, and the comparative absence of charcoal and the supposed fire-induced effects at the comparable stage of an earlier interglacial–glacial cycle. This last fact is thought to substantiate the attribution of fire to human activity, despite the absence of archaeological material.

The 130 000 BP date for south-east Australia marks the replacement of long-established fire-sensitive vegetation by fire-tolerant plants at the same time as a substantial and sustained increase in charcoal particles. No alteration in climatic regime at this time is thought to explain adequately the palynological changes or to account for the increased incidence of fire, leading to burning by humans as the preferred interpretation. This has been challenged on both chronological and ecological grounds (for most recent discussion see Kershaw et al., 1991). The problem is that observations in the ethnographic present establish beyond any doubt the common use of fire by Aborigines as a subsistence tool. The same people, however, pursued a geographically unsettled existence and had few preservable cultural objects, so their presence in a region in the past is not usually directly recorded in pollen-analysable deposits.

There is, therefore, a strong temptation to use the very presence of charcoal peaks in conjunction with corresponding changes in vegetation in a pollen sequence as presumptive evidence of human activity, perhaps as the only evidence that can be expected, even in a general context of a fire-prone and fire-tolerant flora. The argument is, of course, more compelling where charcoal is found in vegetation normally unsupportive of fire. Whether the implications of these early dates for human impact on vegetation are borne out by more direct evidence remains to be seen. In the meantime, the same kinds of argument implicate humans at 20 000 BP at another site in the south-east. To demand more explicit evidence of human involvement would, at this stage, shift the earliest certain dates to the introduction of exotic plants with European settlement two centuries ago.

9.2.4 Oldest disturbance

The oldest unequivocal correlation between human artefacts and pollen analytical evidence for human disturbance comes from the Hoxnian Interglacial at Hoxne, south-east England (West, 1956; West & McBurney, 1954), for which we have not shown a date on the map (Fig. 9.1). It may testify to the technical capacity of Acheulian people to make substantial, if local and temporary, impact on their forest environment. It also encourages the search for more records of its kind and age and for the establishment of ever more percipient palynological arguments for documenting the development of human relationships with other organisms and with landscapes.

9.3 CELEBRATION

Pollen analysis plays a unique role in the study of human–environment relationships. Its most concrete attainment in this field has been the documentation of positive destruction of forests, usually for agriculture and husbandry. This was arguably the single most significant stage in human exploitation of nature or, alternatively, the crucial event in human cultural progress.

Our better understanding of these events through palynology will depend very much on nicely balanced scepticism of the kind that is the hallmark of Alan Smith's work: a disinclination to venerate current dogma or uncritically to espouse the bright new idea, even when it is his own. His evident delight in the weighing of evidence and his uncompromising intellectual fairness in doing so, set the standard for all in this field. This chapter is intended to provide a world context for most of the other contributions to this section and to give Alan, in particular, something to think about.

10

Vegetation change during the Mesolithic in the British Isles: some amplifications

I.G. Simmons

SUMMARY

1. Attention is drawn to a number of themes in the work of A.G. Smith that relate to the interaction of Mesolithic communities in Britain with their natural environments.
2. Evidence is briefly discussed from the period preceding the establishment of deciduous forests, the actions of humans within it, and the transition period from a dominantly hunter–gatherer culture to an agricultural one.
3. The second of these themes is then amplified with work from North Yorkshire, looking in particular at the results of fine-resolution pollen analysis (FRPA) of peats from the later Mesolithic.
4. These findings are then put into the wider context of the likely dynamics of the deciduous woodlands of that time, and one possible avenue for further research is indicated.

Key words: Mesolithic clearance, uplands, forests, mires, British Isles

10.1 INTRODUCTION

The 20 years since Alan Smith's (1970b) synthesis on the Mesolithic period have by no means seen the clearance of all the mists surrounding the relations of humans, climatic change and ecological transformations as deduced from the evidence available. This chapter mostly amplifies evidence rather than re-orders it, but it is hoped that added weight is given to some of Smith's themes concerning the Mesolithic. It reviews the contexts in which his work has contributed to the study of the environment during early prehistory, adds some more evidence from recent work for one particular time and place, and then evaluates both in a wider perspective.

10.2 THE BACKGROUND

10.2.1 The earlier Mesolithic

While the dominance of climate as a forcing factor has never been in dispute, it was some time before the question of human involvement in the rate of species immigration and establishment was raised explicitly by Smith (1970b). Smith suggested that the rise of hazel in many pollen diagrams was not entirely a feature of immigration rates under climatic control, but that early communities of humans may have aided its spread, in particular by firing the landscape.

Further evidence in the British and Irish literature for human-induced vegetation changes during this Flandrian I (Littletonian I) chronozone is patchy. Several authors find charcoal and also pollen evidence of opening or repression of birch and pine woodland; others postulate the burning of heathy vegetation. On the other hand, there are many sites where no such inferences are made. Examples of the first category include northern Dartmoor (Caseldine & Maguire, 1986) before 8785 BP, in a *Juniperus–Pinus–Betula–Empetrum*–Gramineae local pollen assemblage zone (LPAZ). In North Yorkshire (Jones, 1976), silt inwash stripes in deep channel peats are accompanied by the expansion of hazel pollen during the Boreal period, and on the Wolds, Bush (1988) has postulated the keeping open of birch woodland by humans, which resulted in the survival of species-rich chalk grassland. At Waun-Fignen-Felen in upland South Wales (Smith & Cloutman, 1988) before 8000 BP an area of birch forest was kept open apparently by human activity, resulting in heathy vegetation. At several sites in Scotland, fire and vegetation alteration are associated with Mesolithic communities in this period, though there is no necessary association with rises in hazel (e.g. Preece *et al.*, 1986; Robinson, 1983, 1987; Edwards, 1990). This later work is more firm than Edwards & Ralston's earlier (1984) assessment that there were no convincing links between Mesolithic communities and vegetation change in Scotland. Work on the different types of forest present in Holocene Britain raises the possibility (Rackham, 1988a) that the response of hazel to burning may have been different in the various forest provinces. Rackham (1980), however, doubts that *Corylus avellana* is especially fire resistant. Detailed work from the alpine foreland suggests that *Corylus* shrublands were not maintained by fire, but that the transition to *Quercus* forest was in fact assisted by fire (Clark *et al.*, 1989). Bennett *et al.* (1990b) see no relation between lake charcoal frequency in East Anglia and the frequency of any of the woodland dominants, including *Corylus*.

So, Mesolithic settlement from the early Holocene (Flandrian I) is certainly known, there is evidence of fire in several kinds of deposits, and there are sudden sediment influxes into channels. At some sites hazel expands, at others not, and any tree loss seems to be compensated for by expansion of heathy or grassy vegetation. Just as the vegetation seems patchy in much of this period, so is the evidence for deflection of successions and immigration patterns by human activity. It may well have happened but the scale of occurrence (in all senses) seems restricted.

10.2.2 The later Mesolithic

In the Flandrian II chronozone, deciduous forests were at their most extensive. In response to post-glacial climatic changes and the advent of insularity, the major phases in the immigration of tree taxa were completed (Huntley, 1990b). For 6000 BP in the English uplands, this gives a cluster designated as *Corylus–Quercus–Alnus*. That this forest was not absolutely continuous seems accepted. There would be no trees or at best open woodland in circumstances such as climatically-controlled altitudinal tree lines, in areas subject to persistent strong winds (often accompanied by salt burn), on unstable slopes, after freak storms, the presence of browsing and grazing mammals, after natural fires set by lightning, in wetland areas (some of which may have been created by beavers) and in the normal course of forest history as when a tree is lost to senescence or disease, after which it may fall. The 'forest' was then at any one time a mosaic of areas of mature forest, woodland undergoing succession, and unwooded areas that could carry a variety of plant communities or even be bare ground. Within this environment lived an unknown density of hunter–gatherer people who may or may not have been sedentary (the archaeological evidence is too sparse for firm statements) and whose economy may have contained a range of procurement strategies (Edwards & Ralston, 1984). These no doubt were supplemented when agriculture became possible.

Studies on this subject in recent years have sought to answer the following questions:

1. Have human communities significantly affected this vegetation outside the immediate settlement sites? If so, how might these changes be distinguished in the evidence from natural changes?
2. Have such human presences affected not only the physiognomy of the vegetation but also the representation of taxa? Specifically, has alder (*Alnus glutinosa*) benefited in this period in the way postulated for *Corylus* in earlier times (Smith, 1970b; 1984; Bush & Hall, 1987; Chambers & Elliott, 1989)?
3. In the more oceanic climate of the Atlantic period, has human interference with forest allowed the inception and growth of upland peat blankets and other mires in areas climatically and topographically permissive for such growth?

Many studies have endorsed proposition (1), adducing evidence from pollen analytical studies as well as those of other organisms and of peat and lake deposits. Suggestion (2) has more limited support, largely because although the phenomenon of alder rises is well marked and asynchronous, the close association with proven Mesolithic settlement demonstrated at Newferry (Smith, 1984) is not seen elsewhere. Proposition (3) has a number of supporting studies (e.g. Moore *et al.*, 1984; Moore, 1988) although the sequence of events clearly need not be confined to the Mesolithic: provided the terrain is wet enough, it can happen any time.

10.2.3 The Mesolithic–Neolithic transition

Germane to this discussion is the ecology and landscape in which agriculture was first developed in the British Isles. No matter what the cause or causes of any openings, the forests of mid-Holocene times were clearly not an unbroken blanket. There is ample evidence that *c.* 5000 BP there had been a good deal of burning in the landscape, for we find preserved in upland peats many examples of charcoal layers that have large enough pieces of material to be deemed autochthonous. The contribution of domestic fires to the charcoal rain (Bennett *et al.*, 1990b) reminds us that even a low-density population is likely to have used or created open areas.

The result is that agriculture may well have developed in a mosaic of habitats, which would have been equally suitable for any mode of transmission of that cultural practice. That is to say, an indigenous group of hunter–gatherers who had acquired cereals by trade, theft or conquest, would find in their familiar landscapes the types of conditions from which the new food sources had been taken. If they acquired cattle as well (whose dung may well have been an important element in providing nutrients for continued cereal cultivation) then the openings would have contributed to the grazing and browsing resources. On the other hand, if agriculture came as part of an immigration package, with the arrival of new people, new plants and new animals, then the aliens would have found, pre-prepared as it were, a landscape with enough familiar elements in it to encourage them to stay and maybe even to settle alongside indigenous hunters – an inference sometimes made from the archaeological material. Clearly, as many detailed studies as possible of all kinds of evidence for

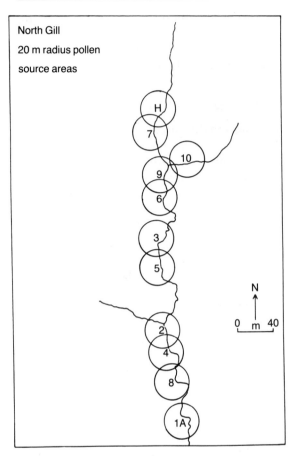

Fig. 10.1 Layout of the sampled profiles at North Gill, showing their spatial relationship and circles of 20 m radius around each.

this transitional period need to be made – as called for by Edwards (1988) and discussed in the context of *Ulmus* declines by Whittington *et al.* (1991a) – without too many pre-set cultural or ecological modes of interpretation.

10.3 TWO MORE-DETAILED EXAMPLES

Two examples of recent palaeoecological investigations are given that fit into the matrix of the above discussion. They throw some light on some of the problems for one place; it is not claimed that interpretations of a universal validity are presented. For reasons of space, not all the data can be presented: they are to be found in Simmons & Innes (1988b) and Innes (1989).

10.3.1 Bonfield Gill and North Gill, North Yorkshire Moors

At Bonfield Gill Head (SE 598 958, 346 m OD), in upland peats of Flandrian II, a stratigraphic sequence is found that gives further detail on some of the questions raised in the previous paragraphs. Although generalized interpretations of this site have been published (Simmons & Innes, 1981, 1988b), this account will focus on specific elements of the site history. The main features of the profile in this context are two layers of charcoal with visible and microscopic sizes, and an upper tree stump that is representative of 14 such remains, some of them trunks of 10 m length, over an area of about 150 × 40 m.

At North Gill (NZ 726 007, 370 m OD), following earlier work (Simmons & Innes, 1988a), 11 profiles have been examined latitudinally down a 400 m stream section (Fig. 10.1) in what at first appear to be blanket peats. Their stratigraphy, viewed collectively, is more complex than at Bonfield Gill, but charcoal (and silt) layers are found, though no 'capping' wood layer is present. These profiles have not yet been published in full though all the data are available in thesis form (Innes, 1989). Close examination suggests the peats are in fact of the upland basin type and the inception of the peat is time-transgressive over the whole length of the stream section. Most of the peats immediately pre-date the *Ulmus* decline and for portions of several profiles fine-resolution pollen analysis (FRPA) was carried out to enhance the understanding of the nature of the ecological processes at that time.

(a) The fate of alder
At Bonfield Gill, and indeed at some other sites nearby, alder is well established by the

time of peat inception. The Bonfield Gill profile, however, is notable for a 'sandwich' of very high alder frequencies. This is contained between two charcoal layers in the actual profile and the peats of this intervening period also show high levels of microscopic charcoal, possibly regional in provenance. Below the sandwich, there are also high micro-charcoal frequencies; above it, very few. Dogmatism on the basis of this kind of coexistence would be unwarranted, but the possibility is hinted that alder is a beneficiary of the conditions then pertaining, whether directly or indirectly. Interestingly, the *Corylus/Myrica* type goes down for nearly all this zone (Bonfield Gill H-8 in Simmons & Innes, 1988b), and, during the course of it, *Calluna* increases overall, while pine and elm seem the main beneficiaries. In this context, therefore, firing of the vegetation at this period does not help hazel, but appears at the very least not to harm alder.

At NG, the rises in alder seen in several profiles are unlikely to be the first of their kind. The radiocarbon dates suggest a regional rise in alder before the onset of deposition there. In these diagrams, therefore, it is an established taxon. That it was well established is seen in several profiles, where the frequency of the pollen is very high indeed, leaving no doubt that it was present in the local vegetation. One example of a rapid rise in alder is seen at NG5B (local PAZ – 3), terminating at 5760 ± 90 BP. Here, the basal 15 cm of the profile are marked by many indicators of open ground and, while few of these vanish during the succeeding 15 cm of alder-dominated peats, several of them fall to lower frequencies. The high alder phase is abruptly terminated by a short phase of very high indications of disturbance of the forest. We might infer that alder had here been a beneficiary of a woodland regeneration phase that was rather suddenly interrupted.

Are we then to suggest that alder is a primary tree-group colonizer after forest disturbance? It seems so, but it is pertinent to ask whether it is alone in being thus. Concentration diagrams are instructive here. At NG5B the period of high alder concentration is entirely paralleled by those of *Quercus* but with the latter running at about 10% of the *Alnus*. Elm and birch also show rises in concentration during this phase, and there are sharp peaks in some indicators of open ground at the time of the highest alder frequencies. At NG4 (PAZ – 2) the phase is deep enough for there to be variations in the alder concentrations, but there is, like NG5B, a movement to a particularly high peak just below the middle of the zone and this once again is paralleled by oak, elm and birch. Some opening-indicators follow these trends, though less sharply than at NG5B. At NG6 (90–70 cm, PAZ NG6 – 2, – 3, – 4), the same overall *Alnus* pattern is followed and is echoed by *Quercus*, *Ulmus*, and *Betula* at its highest concentrations (PAZ NG6-3), though less so in the surrounding subzones. All indicators of opening peak up in this subzone, suggesting it is in fact a period of forest recession rather than regeneration.

The conclusion to be drawn from these movements is that when recolonizing open habitats, alder does so at a time when other trees are also moving into those sites. Thus the cleared area is suitable for all those species and is not confined to very damp ground by the stream, although this locality may be included if we accept that the high values for alder pollen indicate proximity to the sampled profile. So, while alder can be a beneficiary of forest regeneration after disturbance, it is not the only one. Further, the NG6 site suggests that it can also benefit immediately from openings in the woodland, which may suggest changes in the pollen rain due to opening of a canopy. Edwards (1990) notes for Scotland that there is often (but not always) a positive relationship between alder pollen and charcoal occurrence.

(b) The ecology of openings as revealed by FRPA
At five of the North Gill sites, 15 of the Flandrian II disturbance phases have been

analysed at 1 mm intervals using FRPA as described by Simmons *et al.* (1989). These are labelled D2 in Innes's (1989) thesis. Not all such phases have been treated thus, but those that are available reveal fluctuations of pollen curves with a probable resolution of 2–3 years' pollen rain per sample. Each disturbance phase thus examined seems to last about 200 years. Problems of resolution with this technique are discussed in, for example, Moore (1980), Aaby & Tauber (1975), Garbett (1981), Sturludottir & Turner (1985), Turner & Peglar (1988) and Green & Dolman (1988).

Although the ecology of the phases can be discussed in detail, their between-profile chronology cannot be correlated with the data at present available. We have to be content with their location in the latter part of Flandrian II. Another complication in their interpretation is the question of whether each profile has a discrete pollen catchment area, or whether each is receiving pollen (and microscopic charcoal) from a wider area. A circle of 20 m radius around each profile produces no overlapping areas (Fig. 10.1), whereas a 55 m radius brings an overlap segment of about 10% to each of the top and bottom profiles (NG7 and NG1A) and 20–40% to those in between.

One of the unusual features of these phases is that, while each appears on the 1 cm analysis as a single phenomenon, in the FRPA each is tripartite, having three distinct disturbance subzones separated by regeneration periods. These, however, are not of equal magnitude of impact. The earliest records quite a limited impact at profiles 1A, 4, 5B and 7, whereas at NG6 the earliest subzone is the most significant of the three recorded there. By contrast, at NG7 the second of the zones shows the heaviest impact, with 5B and 6 being moderate and 4 and 1A low. The third zone is most severe at 1A, 4 and 5B, relatively large at 6 but a small event at profile 7.

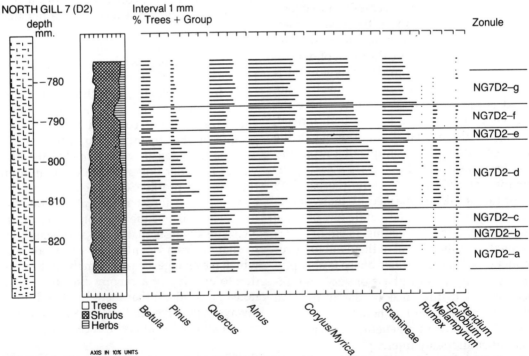

Fig. 10.2 Pollen diagram for part of the North Gill 7 profile, counted at 1 mm intervals and showing the tripartite division into forest recession (d) and regeneration (s) phases. Axis is in 10% units. Source: Innes (1989).

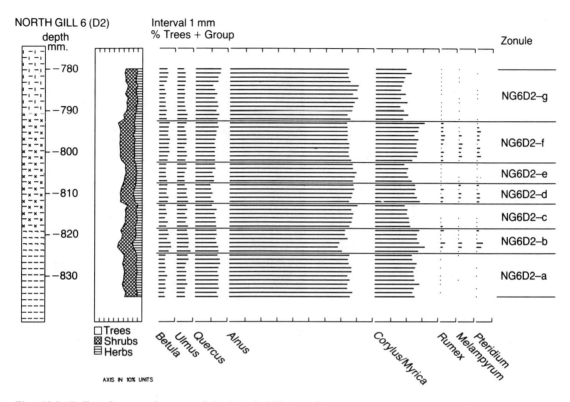

Fig. 10.3 Pollen diagram for part of the North Gill 6 profile, counted at 1 mm intervals. The pollen and spores of open-ground taxa in particular show a strongly three-fold division. Axis is in 10% units.

Close inspection of these zones shows that in a number of cases there is a clear succession of a conventional type in the pollen record. At NG7-B (Fig. 10.2), oak falls sharply and herbs such as *Epilobium* and *Melampyrum* occur, and alder rises. After that, willow and birch rise, followed by *Polypodium*, and finally oak shows a complete recovery of woodland with birch remaining a major constituent. Similar successions are found at NG6-B (Fig. 10.3). In each of these examples, *Corylus* does not seem to be involved as a successional element: in some cases its curve matches that of oak but in others it is high and then falls. This suggests, perhaps, that there is hazel within the woodland as well as on its edges.

However, there are many FRPA zones where the classic successional sequence is not found. The pollen of *Melampyrum* remains high throughout a subzone or even rises towards the end; open-ground weeds are found sporadically throughout the zone rather than strongly represented at the beginning. At NG4 (Fig. 10.4), it is possible to see a tripartite division but it is blurred by the strong presence of *Rumex* through what would conventionally be regarded as a stable, or even regenerating, phase. So, traditionally successional pollen taxa are often homogenized or disordered and not regularly sequential.

How might we explain this? There are several possibilities:

1. that each phase does not represent a single impact but a phase a few decades long in which disturbance was more or less continuous but small-scale;

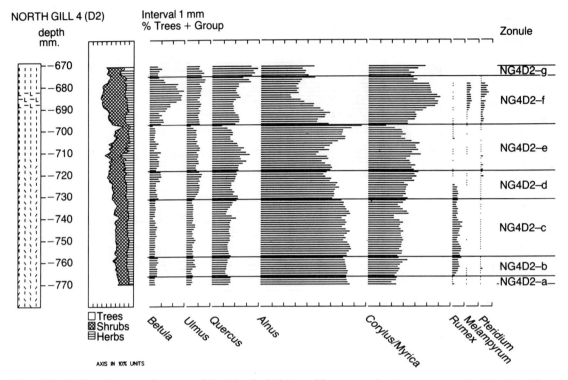

Fig. 10.4 Pollen diagram for part of the North Gill 4 profile, counted at 1 mm intervals. The *Pteridium* curve approximates to a three-fold division, but this sequence is scarcely matched by any of the other non-tree taxa normally involved in such fluctuations. The oak curve, however, mirrors one or other of the open-ground groups. Axis is in 10% units.

2. that there were small-scale, but discrete, impacts;
3. that the taphonomy represents a composite picture of several events within a larger pollen catchment area and that classic successions are taking place at different places within the catchment zone.

(c) The onset of peat growth at North Gill
Above a discontinuous charcoal-rich *Polytrichum* peat lies a deposit of charcoal and charcoal-rich peats that is still largely confined to basins but is spread further than its predecessor. Where the gradient is slightly steeper, then this is the basal biogenic stratum. The next stratum is mostly amorphous peat and it gives extra depth to the areas already covered, plus some lateral spread up the shallow slopes peripheral to the stream and its basins. Paludification on such slopes is most likely to have been preceded by changes within the soil profile, such as the removal of woodland and acidification of soils, along the lines suggested by Taylor & Smith (1980). In the upper layers can be found horizons of charcoal and silt, and in places a high clay fraction as well. Clearly the locality was being affected by fire and soil instability. One such charcoal horizon spreads as far as 25–40 m west of the present stream course.

Thus the burning events of the locality must have had quite strong effects upon the downslope accumulation of the mire, its flora and nutrient status, degree of waterlogging and possibly also upon the sedimentation sequence in actually burned areas, where dry peats may have been consumed by burning. Analysis of profiles away from the stream itself would have to be interpreted with great care. After these deposits of silt and charcoal, the area seems to have been covered with an unconfined blanket peat of the *Eriophorum* type, with the addition of some *Sphagnum* soon after the opening of Flandrian III. At Bonfield Gill, too, there is a similar establishment of true blanket peat after the elm decline.

The conclusion to be drawn from this outline account is that the interpretation of peat profiles in such contexts is complex. The history of sediment accumulation must be considered in its lateral aspects as well as its immediate stratification; the possibility that sequences are interrupted by having been consumed by fire, and the changing pollen catchment area as forest is replaced by mire are variables that have simultaneously to be fed into any interpretation.

10.4 A WIDER CONTEXT

A broader framework is provided by the recent detailed work of Huntley (1990b) on the immigration of tree taxa into western Europe during the Holocene and the pollen assemblage data on the forest provinces of various times. These make clear the dominance of mixed deciduous forest as the regional vegetation type but do not, given their scale, reveal any details of, for example, the upper limits of forest in relatively low uplands like those of England, nor do they deal with the yearly and decadal changes in the physiognomy of the forests that result from its autochthonous processes and that may also be affected by human activities. Into any consideration of the interpretation of pollen, charcoal and macrofossil fluctuations of the types described above, the insights of Rackham (1980, 1988a) and the close palynological work on fire in the forests of North America (e.g. Swain, 1978) must be allied to the broader discussions of the forcing function of climatic change.

That the forests occupied by human communities were not completely closed-canopy ecosystems seems well established. This shows itself in the North Yorkshire work in indications of open ground throughout Flandrian II peats, right from the base of their accumulation. Older Flandrian II peats than those described above also record high frequencies of open-ground indicators, and in this they continue from Flandrian I where sporadic but persistent indications of interruptions in the gathering forest are found. On Dartmoor (Simmons *et al.*, 1983) it seems that burning at high altitudes was initiated in order to prevent the forest encroaching on open ground (see Caseldine & Hatton, this volume).

The coincidence of charcoal with the pollen analytical indicators of forest recession and peat initiation points to the importance of fire throughout Flandrian II. It is less in evidence, though, at the elm decline. Given the spreads of charcoal recorded at North Gill, and the presence of large pieces of this material, it seems unlikely that domestic sources were the main cause of the frequencies observed, as has been postulated for East Anglian lake deposits (Bennett *et al.*, 1990b). The importance of the immediacy of peat deposits containing large pieces of charcoal is emphasized by the conclusions of MacDonald *et al.* (1991) of the inability of evidence from lake sediments to provide detailed histories of past fire activity.

This still leaves us with the question as to whether the fires and the openings with which they are associated are natural or human-induced, a question not open to proof but only to continued reassessment in the light of the accumulation of circumstantial evidence for a particular place and time. The role of soils, for example, as related to altitude and bedrock may be a differentiating one over quite a small area in terms of response to disturbance or indeed the lack of disturbance (Bartley *et al.*, 1990).

Looked at broadly, the probabilities of lightning setting fire to the deciduous woods of upland Britain must be, for any one place, low. Chandler *et al.* (1983) say of European deciduous forests that the (natural) fire season is bimodal, in spring and autumn, and that severe fires only occur in periods of exceptional drought. The average number of fires in the temperate forests of Europe is, they say, 167 per million hectares, with an average fire size of 0.97 ha (a patch roughly equivalent to 100×100 metres), and the return period is over 6000 years. This is contrasted with eastern North America, where they suggest a return frequency of 10–25 years, with a higher frequency whenever conifers are dominant or present. Crown fires were confined to coniferous stands; surface fires are the rule in unmanaged hardwood forests. Where, then, we have evidence of repeated fire at or very near the same locality, as recorded in peat (and remembering that we could have lost some further strata in other fires), it seems reasonable to conclude that the hands of humans were at work.

This might not necessarily entail the belief that Mesolithic communities actually *created* clearings in the woodlands in the ways that are confidently assigned to their Neolithic successors. It might be the case that such communities took it upon themselves to keep open glades that were already present. That such openings in the canopy might occur from windthrow, senescence and disease is not presumably in question. Large mammal herbivores might also play a part in perpetuating such unwooded areas. Given the resource value of these zones, then the motivation to keep them open would have been considerable and fire an obvious tool since other ways would involve the expenditure of a lot of human energy given that the technology of the time appears very restricted. This possibility does not preclude the other prospect, that of deliberate removal of the tree cover. Both possibilities should be seen as hypotheses to be tested against such evidence as is available at a particular site.

Given the temporally discerning nature of FRPA, it should be possible at the right site to discriminate between patterns produced by 'natural' opening, human-assisted, maintenance, and human-produced openings. We need some closer inspection of episodes such as those discussed above, and perhaps some computer-based simulation of the likely courses of vegetation development under the various possible conditions. Recent North American work on the random elements in post-clearance succession in present-day forests may well be helpful here (Oliver, 1981; West *et al.*, 1981).

Simmons & Innes (1987) have postulated that the land cover patterns of later Mesolithic times, by whatever combination of natural and human-induced factors they were produced, would have been an ideal seed bed, so to speak, for early agriculture. Small clearings provided protection from the wind, the remains of Brown Earths not yet highly acidified, and the input of humic materials from nearby leaf-fall would all have been positive factors. If the openings had previous attractions as gathering-grounds for mammals, then their presence would have needed the input of human energy in an interdictory manner to keep away the animals and to replace their succession-stabilizing role. One lesson of all this seems to be that both the creation of mid-Holocene ecologies and their rather later academic study seem to have required effort and continued application as well as imagination and foresight: in this volume we can celebrate both.

Acknowledgements

I thank Dr J.B. Innes for much helpful discussion and for permission to reproduce diagrams from his thesis. Many of the insights were generated in the course of discussion of the wider body of North Gill work with Dr J. Turner.

11

The development of high moorland on Dartmoor: fire and the influence of Mesolithic activity on vegetation change

Chris Caseldine and Jackie Hatton

SUMMARY

1. Results of pollen and charcoal analysis of blanket peat sites from northern Dartmoor demonstrate the influence of burning and grazing on the transition from hazel woodland to blanket peat during the early and mid Holocene.
2. Within a general phase of enhanced burning between 7700 and 6300 BP the site at Pinswell at an altitude of 461 m shows how woodland was transformed over a period of 600–1000 years into blanket peat via a phase of acid grassland, in some ways similar to heath-derived grassland found on Dartmoor today.
3. The results clearly implicate Mesolithic communities, and their use of the high moorland, in the development of the open peat-covered landscape that is so characteristic of the present Dartmoor environment.

Key words: Dartmoor, blanket peat, burning, Mesolithic, grassland

11.1 INTRODUCTION

The question of the origins of the blanket peat that covers much of the uplands of the British Isles has provided a recurrent theme in British vegetation history (see Moore, this volume). Over 20 years ago, in his review of the possible links between the activities of Mesolithic communities and vegetation change, Alan Smith (1970b, p. 85) commented that there had been 'a tendency to overlook the possibility that the deliberate use of fire, or even accidental burning, may have had quite widespread effects, and that 'up to the beginning of the Atlantic period there is often a remarkable coincidence of vegetational change with indicators of the presence of man and the occurrence of fire'. However, not all these comments on the role of fire were directed at blanket peat inception and he expressed doubt, for instance, as to whether there was 'as yet any causal connection' between the development of blanket peat and the activities of Mesolithic human communities in the Pennines (Smith, 1970b, p. 88). In a series of papers, Moore (1975, 1986a, 1988) implicated human activity in the origin of much British upland peat. According to Moore (1988), blanket peat initiation was a function of waterlogging consequent upon woodland removal and the use of fire, the latter allowing charcoal particles to clog pores in the upper soil horizons and exacerbate impermeability.

By 1984, Smith also favoured such a cause of blanket peat initiation, both on the basis of work such as that of Moore, and from a number of his own sites, especially in Northern Ireland (Smith, 1975, 1981a). Work in southern Wales at Waun-Fignen-Felen, using a network of profiles over a small area served to confirm the importance of burning, as reflected in the ubiquitous presence of charcoal at the base of the peat (Smith & Cloutman, 1988). This link between fire and peat inception, especially during the Mesolithic, has now been demonstrated at a wide range of British upland sites from the North York Moors (Simmons & Innes, 1985; Innes & Simmons, 1988), Pennines (Tallis & Switsur, 1973; Tallis, 1975; Jacobi et al., 1976), Lake District (Pennington, 1975) and Wales (Chambers, 1983; Chambers et al., 1988). Discussion of the reasons for the implementation of burning in upland areas has always required something of an inferential approach, as there is rarely any direct evidence for the purpose of the fire. Hence the nature of the exploitation strategies involving fire must remain somewhat hypothetical. Nevertheless, from what is understood about the economy of the Mesolithic in the British Isles, it is likely that fire was used to modify the woodland to encourage and direct game, and that the deleterious effects on the vegetation occurred as a result of persistent burning and/or fires getting out of control, possibly with occasional phases of increased climatic wetness, the latter especially at the end of the Mesolithic, i.e. c. 5000 BP (Tallis & Switsur, 1990).

On Dartmoor, Simmons (1964, p. 169) pointed out the likely influence of Mesolithic activity on forest recession at Blacklane Brook during Zones V and VI, probably through the use of fire. Although he expressed caution over the implication of human communities '... the possibility of an anthropogenic [sic] cause cannot be totally ignored'. Later, more detailed radiocarbon-based study of the site (Simmons et al., 1983) confirmed both the age and the close association of charcoal and reduced arboreal pollen frequencies. In common with emerging ideas elsewhere, this was interpreted as woodland manipulation to ensure the maintenance of 'open ground and scrub in a landscape that was predominantly mature forest, at least at lower altitudes' (1983, p. 666). The work at Blacklane remained the only study to implicate Mesolithic communities overtly in vegetation change on Dartmoor, until preliminary work at Black Ridge Brook on northern Dartmoor (Fig. 11.1) (Caseldine & Maguire, 1986) showed apparently similar results with the presence of varying amounts of charcoal in peats of early- and mid-Holocene age (c. 10 000 to c. 5000 BP).

It should perhaps be emphasized that there is very little archaeological evidence for the presence of Mesolithic communities on Dartmoor, with evidence limited to a number of find spots of flint scatters (Jacobi, 1979; Hatton, 1991), many of them only representing single or a handful of artefacts. From the high northern moor there is only one find spot at Amicombe Hill, 3 km north of Black Ridge Brook (Fig. 11.1), from which Jacobi (1979) has identified a later Mesolithic geometric microlith. Examination of the recent archaeological evidence for Devon suggests that there was very little activity on Dartmoor up

Introduction

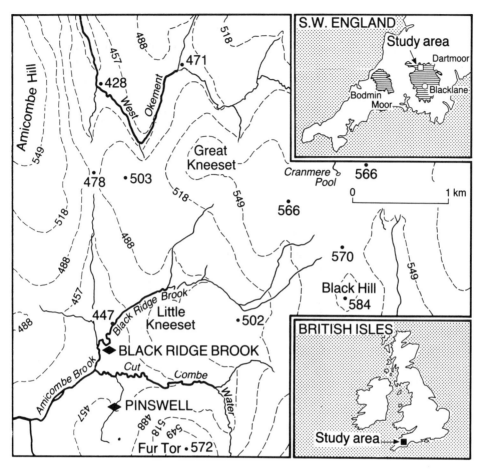

Fig. 11.1 Location map showing Pinswell and Black Ridge Brook sites.

until 8750 BP after which the upland did become more of a focus of activity, although this is reflected more on the peripheries of the higher moor, outside those areas now covered by blanket peat and under which there may be preserved material. It is not possible on the archaeological evidence to be any more specific about the dates involved, although at Postbridge to the south of the Black Ridge Brook area there are several sites with large numbers of probable later Mesolithic flint artefacts between 340 and 400 m OD.

Thus, although there is widespread recognition of the probable impact of human communities on upland vegetation prior to 5000 BP in the British Isles, especially in England and Wales, and there is also strong evidence for linking human use of fire and peat inception (Simmons, this volume), there is only limited evidence for Dartmoor. Furthermore, despite recent detailed work (e.g. Smith & Cloutman, 1988) there is still only a relatively superficial understanding of the vegetation sequences by which upland woodland communities were transformed into blanket peat. The site at Pinswell on northern Dartmoor (Fig. 11.1), close to the basin peat site at Black Ridge Brook, provided an opportunity to examine the nature of vegetation changes associated with the shift from woodland to peat as indicated in the Black Ridge Brook pollen profile for the early/mid Holocene. Results of this study are presented below.

11.2 THE BLACK RIDGE BROOK MODEL

Initial studies at the Black Ridge Brook site (Caseldine & Maguire, 1986), extended by Hatton (1991), have shown a generalized pattern of local vegetation change for the early and mid Holocene that closely parallels that found in other upland areas. The period between 7700 and 6300 BP is characterized by the presence of significant amounts of microscopic charcoal in adjacent peat profiles, associated with pollen changes interpreted as representing woodland recession (Hatton,

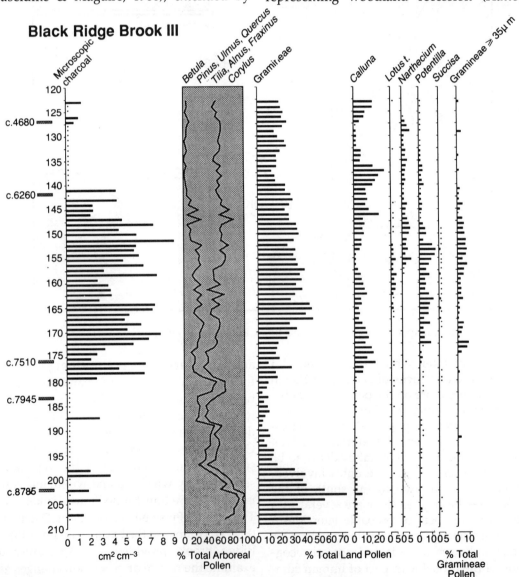

Fig. 11.2 Selected pollen and charcoal data from Black Ridge Brook III (Hatton, 1991). Depths in cm; timescale in years BP.

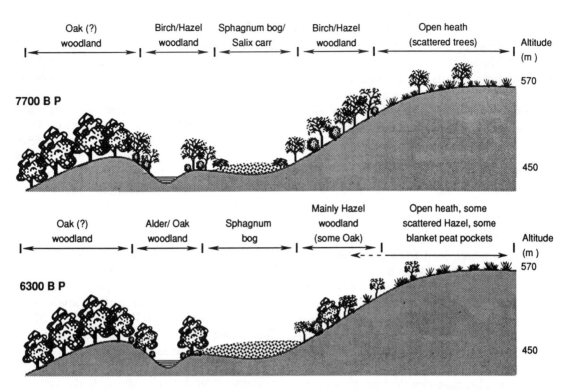

Fig. 11.3 Schematic representations of changes in vegetation on the higher slopes of northern Dartmoor around Black Ridge Brook between 7700 and 6300 BP.

1991). Selected data showing this pattern are presented in Fig. 11.2. From a relatively dense woodland of *Corylus* and *Betula*, with only small areas of open heath covering the surrounding summits at *c*. 7700 BP, the vegetation of the slopes around the bog was transformed by 6300 BP to a hazel-dominated woodland with some oak and local alder, lying at lower altitudes, and partially interspersed with pockets of blanket peat. This change is shown schematically in Fig. 11.3. Apart from the 'snapshot' of the vegetation structure at 6300 BP, the overall results and in particular the later pollen data, show that the changes initiated between 7700 and 6300 BP were essentially a directional change, with a continuous gradual extension of blanket peat from suitable foci, and the expansion of the increasingly open summit areas and higher slopes, largely at the expense of local woodland.

The data from Black Ridge Brook, and especially the microscopic charcoal evidence, demonstrate that the period between 7700 and 6300 BP represents the continuous effects of fire within the upland vegetation communities. The virtual cessation of burning at 6300 BP is a phenomenon seen elsewhere in the later Mesolithic (e.g. Simmons & Innes, 1987, 1988b, c, d) and suggests that after a period of fairly intensive use by human communities there was abandonment of this part of northern Dartmoor, or a change of use of the upper slopes to exploitation not requiring fire, the latter a rather unlikely explanation. Because it lies in the middle of a broad basin, the Black Ridge Brook site produced only a generalized picture of vegetational change for the surrounding area. Herb taxa were quite well represented (Fig. 11.2), but no detailed patterns or short-term events could be resolved. The results are good for demonstrating long-term trends in the vegetation but could not be used to show what was actually happening on specific slopes, i.e. how hazel woodland was transformed into blanket peat.

11.3 PINSWELL

11.3.1 Site location and pollen and charcoal results

The slopes which lead from the edge of the Black Ridge Brook at 440 m to the north-east-facing boulder-strewn margins of Fur Tor (572 m) are covered by blanket peat of varying depth. A series of reconnaissance borings through the peat revealed considerable variation in depth and two skeletal pollen profiles from the lower slopes showed relatively recent, i.e. post-elm decline, peat inception. An eroded peat face in a stream section at an altitude of 461 m provided an ideal location for sampling for both pollen and radiocarbon dating, and provided peat of suitable age for examining the problem of pre-elm decline peat inception. Samples for pollen and charcoal analysis were prepared as for Black Ridge Brook (Caseldine & Maguire, 1986; Hatton, 1991), and at each level a count of 500 total land pollen (TLP) was undertaken. A brief summary of the main pollen changes and associated vegetation history is presented in Table 11.1. For the charcoal analysis the method defined by Clark (1982) was followed. Results of these analyses are presented in Fig. 11.4. The Pinswell diagram has been zoned on a local basis for the purpose of the following discussion. Three samples were sent for radiocarbon analysis at Glasgow University (GU-2046, 2047, 2048). The lowest peat was not used for dating the peat/mineral interface, following the experience at Black Ridge Brook where probable contamination by younger roots decaying *in*

Table 11.1 Summary of pollen characteristics and inferred vegetation changes from Pinswell

Zone	Pollen characteristics	Vegetation
g	Higher *Alnus* and *Quercus*; high *Sphagnum* and *Calluna*, + *Narthecium* and *Potentilla*	Local development of blanket peat and consolidation of valley bottom woodland
f	Increased *Alnus* and *Quercus*; reduced Gramineae, high Cyperaceae and *Rumex*	Increased local wetness with oak/alder woodland in valley bottoms
e	Increased *Betula* and *Corylus* + *Salix*; reduced Gramineae	Persistence of grassland but some woodland recovery
d	Low *Corylus*, *Betula* and *Calluna*; high Gramineae, + *Lotus* and *Rumex*; decline in *Sphagnum*	Opening of woodland, local development of grassland
c	Slight increase in A P; decline in herb taxa; increased *Sphagnum*	Minor woodland regeneration, increasing wetness
b	Reduced *Corylus* and *Quercus*; increased *Calluna*, *Melampyrum* and *Potentilla*	Opening up of local woodland
a	High *Corylus* (arboreal pollen > 55% T L P), with Gramineae; + *Polypodium*, *Pteridium* and *Melampyrum*	Fern-rich, relatively closed hazel woodland

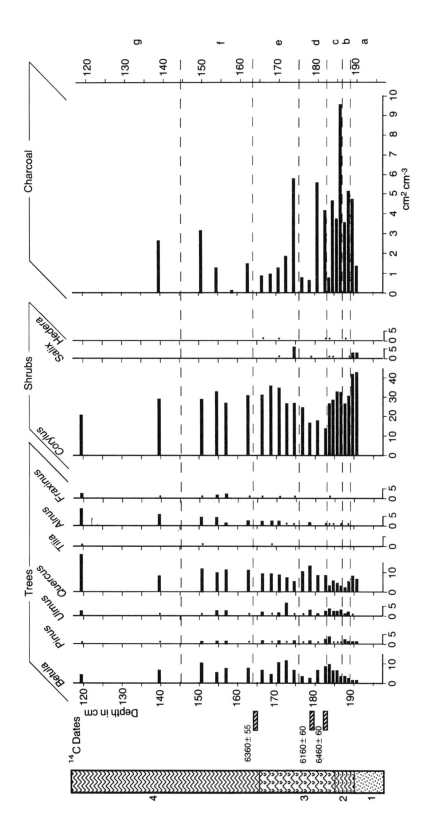

Fig. 11.4 Pollen diagram from Pinswell. Lithostratigraphy: (1) decomposed granite; (2) black, greasy, mor horizon; (3) moderately to well-humified woody peat; (4) well-humified blanket peat.

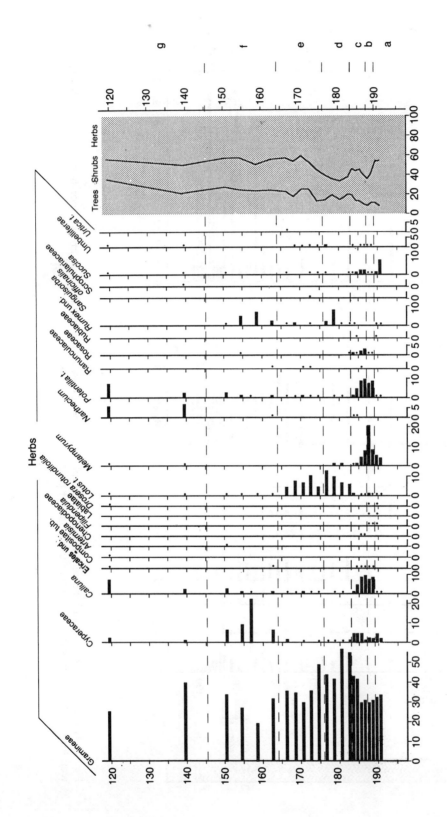

Fig. 11.4 (2).

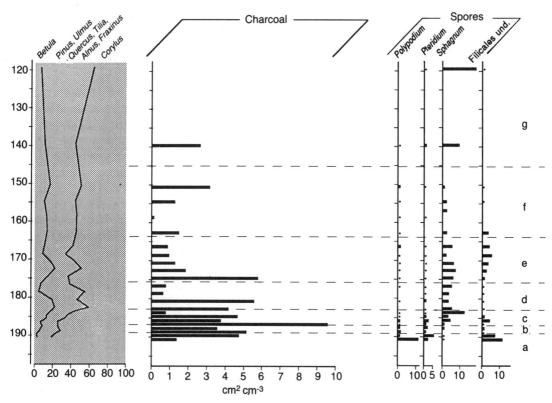

Fig. 11.4 (3).

situ produced an anomalous date. Instead, the samples were taken from above the very base, but in a close series to date the whole of the basal peat section.

11.3.2 Vegetation change and peat initiation

Dating of the peat base can only be estimated by extrapolating from the upper dates, but would appear to be close to 7000 BP. By comparison with the Black Ridge Brook dates, this suggests that mor horizon development at Pinswell occurred within the period of enhanced burning, 700 years after the first indications of vegetation interference were picked up at the lower site, presumably due either to delayed use of the slopes immediately around the site, or due to the soil being able to resist humus accumulation until this later date despite interference. It would appear most likely that the first explanation is more probably correct and that the Black Ridge Brook profile was picking up continuous use of different areas in the overall pollen catchment. As the Pinswell site lies at 461 m this was probably still well within the tree line at 7000 BP, which could have lain close to the upper slopes of Fur Tor above 500 m (Maguire & Caseldine, 1985). It is interesting to note that there was very little birch in the local woodland, despite the altitude, but that there may have been some local oak. In general, however, the local woodland comprised predominantly hazel with a fern-rich understorey also including grasses, willow, *Melampyrum* and *Succisa*. Thus, despite the record from Black Ridge Brook for woodland interference for *c.* 700 years in the general area, the woodland cover at Pinswell was still relatively undisturbed hazel woodland.

The opening of the woodland can be dated to *c.* 6750 BP, parallel to a rise in microscopic charcoal in the profile, and was probably linked to the local use of fire. Direct removal of deciduous woodland by fire has been questioned (Simmons, 1975; Rackham, 1980), so it is most likely that it was the occurrence of fire and increased pressure of grazing animals that together prevented woodland regeneration. This in turn changed the nature of the local community to that characteristic of a woodland edge, with the unusual combination of *Calluna* and *Melampyrum*, a pattern also seen at Soyland Moor in the Pennines (Williams, 1985).

Melampyrum has often been considered as a fire-responsive taxon and an indicator of burnt habitats (Iversen, 1949; Mamakowa, 1968; Wiltshire & Moore, 1983; Moore *et al.*, 1986; Simmons & Innes, 1988b, d; Welinder, 1989). The association at Pinswell with charcoal would initially seem to support this. At Black Ridge Brook the association was not close (Hatton, 1991) and several authors have found high values difficult to interpret, especially as there is no consistent relationship to the appearance of charcoal (e.g. Carter, 1986; Wiltshire & Moore, 1983). The pollen taxon probably represents *M. pratense* L., which occupies a wide range of habitats in woods and heath on acid humus, and does not necessarily imply a direct fire response. It is perhaps significant that *Melampyrum* does not appear later in the profile after its first peak despite later burning, when other conditions and local plant species have changed. The accompanying increase in *Potentilla* supports the idea of increasing acidity as well as openness (cf. Iversen, 1964) as may the reduction in *Pteridium*.

Whilst there is no doubting the impact of fire on the hazel woodland, the expansion of herb species was probably just as much a response to the increased openness and levels of acidification as to the fire itself. Assuming that the firing of the woodland was successful in opening the woodland for game, then added grazing pressure would help to prevent regeneration and in turn benefit the herb flora. In a similar, but much later, context, at Mantingerbos in Denmark (Stockmarr, 1975) the development of an open mor forest of early Medieval age was characterized by the occurrence of *Melampyrum* and an influx of *Calluna* pollen from outside the immediate area. At Pinswell the presence of *Calluna* could indicate either successful post-fire colonization or pollen from summit heath.

There then follows, between *c.* 6750 and 6460 BP, a period of limited woodland regeneration, with eventually hazel, birch and oak all recovering. At Pinswell, peat started to develop over the mor and it is significant to note that this period was not identified at Black Ridge Brook, emphasizing the general nature of that record and that the area was not totally abandoned. With the development of peat and increased soil wetness there is a reduction in the range of taxa represented in the pollen record, especially herb and spore taxa.

Zone d, which begins at 6460 BP, represents further woodland reduction, to the lowest levels indicated anywhere on the diagram, as the local birch/hazel secondary woodland was reduced and prevented from regenerating by a further combination of fire and grazing pressure. This allowed the expansion of grassland characterized by the presence of *Lotus* and, after about 100 years, by *Rumex*. The pollen of *Lotus* was identified to *Lotus uliginosus* on the basis of size criteria (Birks, 1973; Andrew, 1984), and the pollen of *Rumex* included both *R. acetosa* and *R. acetosella*.

The form of grassland indicated in the pollen record has no direct comparison in the Dartmoor of today or of the recent past. The importance of *Lotus uliginosus* suggests a wet grassland, relatively tall and with a pH no lower than 4.5. Keble Martin & Fraser (1939)

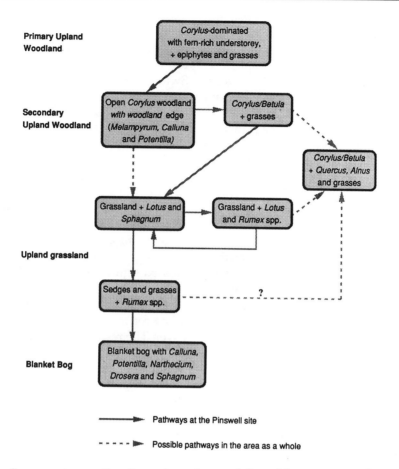

Fig. 11.5 Flow diagram representing the main pathways followed by vegetation change at Pinswell, with possible pathways for other locations in the general area.

record the species to 1600 feet (480 m) on Dartmoor in wet meadows, bogs and damp heath and so the overall impression is of a tall, damp, acid grassland; damp enough to allow the continued presence of *Sphagnum*, and relatively species-poor despite the occurrence of burning at its initiation. The grassland must have been kept open by grazing (*Lotus uliginosus* is tolerant of light grazing (Grime *et al.*, 1988)), but this was not sufficiently intensive to prevent some gradual regeneration of hazel. The clearance of the woodland at the opening of zone d was registered particularly in the birch and hazel records. The higher values for oak in zone d could indicate that this preferentially expanded locally or that the increased openness allowed more oak pollen into the site. After an initial phase lasting about 100 years the grassland became damper with the expansion of *Rumex* spp. The presence of *R. acetosella* would indicate a pH of at least 5.0 and begins to show similarities with grasslands that were extensive on Dartmoor 150 years ago and included 'dwarf dock' (sorrel) (Ward *et al.*, 1972, p. 526) among a range of grass species. This grassland can still be found today and can be derived from heath by either swaling (burning) or more intensive grazing, and tends therefore to have elements of both heath communities, e.g. *Calluna*, and bog communities, e.g. *Narthecium*. Although the pollen record does not have the range represented in the modern and recent flora, there is obviously a similarity both in form and possibly in the origins of the grassland community.

At 6300 BP, at the opening of zone e, there is a further phase of burning, a lessening in the local influence of *Rumex* and a recovery of birch/hazel woodland. There is also pollen evidence for the presence of *Salix*, which was probably the main woody constituent of the peat. The grassland continued to exist but was gradually being reduced in extent over the 140 years covered by the zone as shrubby willows and secondary birch/hazel woodland recovered, possibly with pockets of oak, and increasingly with alder. Despite the original higher level of burning and the continued use of fire, the overall grazing pressure and human use of the slopes must have been reduced. By 6160 BP, the end of zone e, this trend had accelerated and the slopes immediately around Pinswell had become much wetter, returning to a *Rumex*-dominated grassland. By this time the local woodland had recovered to sustainable levels within a general trend to wetter and less-wooded conditions at higher altitudes. The impact of the demise of the higher altitude woodland was partially offset in the pollen record by the expansion of oak and alder into the sheltered lower slopes and valley bottoms. Zone f forms a transition from grassland to true blanket peat. The final zone recognized at Pinswell, zone g, which opened probably *c.* 6000 BP, shows the local development of typical blanket peat with *Calluna*, *Narthecium ossifragum*, *Potentilla*, Gramineae and *Sphagnum* (Proctor, 1989).

Thus at Pinswell the transition from hazel woodland to blanket peat can be seen to have occurred through a complex of communities, including an extended phase of acid grassland. The presence of a grassland phase is relatively rare in records from other upland areas; the transition to blanket peat, especially seen in the presence of *Calluna*, is often more direct. Where there is a grassland phase, for example at Robinson's Moss (Tallis & Switsur, 1990), then there tend to be different species within the grassland, in this example *Succisa* and Ranunculaceae. The pathways from woodland to blanket peat followed at Pinswell, and the possible pathways followed on slopes around Black Ridge Brook are summarized in Fig. 11.5. The transition at the specific site may have taken up to 1000 years from the original woodland reduction, certainly no less than 600–700 years, and was largely a function of the use of fire and grazing, the combined effects of which eventually crossed significant ecological thresholds such that even the abandonment of the area by those human groups responsible did not allow regeneration of the original woodland cover. The results here therefore demonstrate the nature of the irreversible changes that Mesolithic communities initiated on the highest moorland slopes, changes which paved the way for later extension of peat and diminution of woodland as a result of deteriorating climate and later use of the moorland for grazing. The situation on northern Dartmoor was very similar to that described by Tallis & Switsur (1990, p. 875) for the South Pennine uplands where 'substantial elements of the modern mosaic were already in place by the *Ulmus* decline'.

11.4 COMPARISON WITH BLACK RIDGE BROOK AND OTHER DARTMOOR SITES

Comparison with the results from Black Ridge Brook support the idea that the basin results provide a generalized picture of the whole area, but the detail of the short-term changes is missing. Dates from Black Ridge Brook show that the slopes immediately around the Pinswell site were only affected by human activity after woodland reduction had begun at other sites in the area, and there is no signature at Black Ridge Brook that marks the initiation of changes at Pinswell. Although, with the exception of *Rumex*, all the herb taxa found at Pinswell are recorded at Black Ridge Brook, their individual habitats are not obvious and it is not possible to assess their position in the mosaic of communities covering the higher slopes. The abandonment of the Pinswell area coincides with the abandonment of the whole area, at least within the constraints of the errors of radiocarbon dating, and there is no evidence for any secondary pre-elm decline interference. The persistence of charcoal in the record at Pinswell within the blanket peat phase, after the ending of the enhanced charcoal record at Black Ridge Brook, does indicate some continuation of burning in the area, although at Pinswell, charcoal disappears in zone g. Comparison of the two sites reinforces the necessity for searching for as detailed records as possible to elucidate just how upland vegetation responded to the activities of Mesolithic communities.

Elsewhere on Dartmoor the period of Mesolithic interference interpreted for southern Dartmoor from Blacklane, at 457 m (Simmons *et al.*, 1983), is remarkably similar, falling between 7660±90 BP and 6010±90 BP. Although no other sites have been interpreted as having similar evidence, herb taxa are present with reduced arboreal pollen frequencies at Blacka Brook on Shaugh Moor between 6500 and 6200 BP (Beckett, 1981). At Postbridge (Simmons, 1964) the pollen record covers the same period as those from the Black Ridge Brook area, but the samples were analysed at too broad a sampling interval to be confident about identifying indicators of human interference. Increasingly, therefore, the palaeoenvironmental evidence points towards a significant impact by human communities on the higher slopes of both northern and southern Dartmoor, and, as in upland Wales (Chambers *et al.*, 1988), further demonstrates how the virtual absence of an archaeological record for Mesolithic communities need not necessarily imply the absence of the communities themselves.

Acknowledgements

The authors acknowledge grants from The Leverhulme Trust, Dartmoor National Park Authority, The Devonshire Association and the University of Exeter Research Fund. Access was provided by the Duchy of Cornwall. Original sampling was undertaken with Dr D.J. Maguire; preliminary pollen work at Pinswell was by Simon Butler. We thank Art Ames and Terry Bacon for technical assistance in preparation of the paper.

12

Models of mid-Holocene forest farming for north-west Europe

Kevin J. Edwards

SUMMARY

1. Patterns of woodland pollen taxa for forested areas of north-west Europe are considered.
2. There follows a presentation and critique of four models that address the theme of forest farming in mid-Holocene times – the landnam, leaf-foddering, expansion-regression and forest-utilization models.
3. The early conceptualizations of Iversen and Troels-Smith contained elements that continue to be relevant, while the forest-utilization model of Göransson is more radical and ambitious.

Key words: forest farming, landnam, coppicing, palynology, elm decline, models, north-west Europe

12.1 INTRODUCTION

In the pollen curves for forested areas of mid-Holocene north-west Europe, there seems to be a sequence from relatively closed 'Atlantic' woodland to a marked decline in elm pollen and perhaps that of other arboreal taxa, succeeded by woodland regeneration, and followed by a renewed decline in elm and usually other arboreal taxa. There may be variations in this pattern including intermittent reductions in forest pollen types – sometimes associated with the presence of cereal-type pollen – prior to the initial *Ulmus* decline, and multiple reductions in elm. Two examples of the model-type sequence are reproduced in Figs 12.1 and 12.2. Other examples include diagrams from Ireland (Göransson, 1984; Hirons & Edwards, 1986), Britain (Simmons, 1969b; Whittington, *et al.*, 1991a, b), The Netherlands (Janssen & Ten Hove, 1971), Denmark (Aaby, 1988) and Sweden (Göransson, 1987a; Regnéll, 1989).

It is possible to discern deviations from this particular pattern, for instance, where there are no marked regenerations of elm and perhaps other arboreal taxa. These would appear to be associated with poorer soils as at Cannons Lough, Northern Ireland (Smith, 1961b) and the Dommel Valley, The Netherlands (Janssen, 1972), grazing pressures as in Drenthe (Van Zeist, 1959), or for unspecified reasons

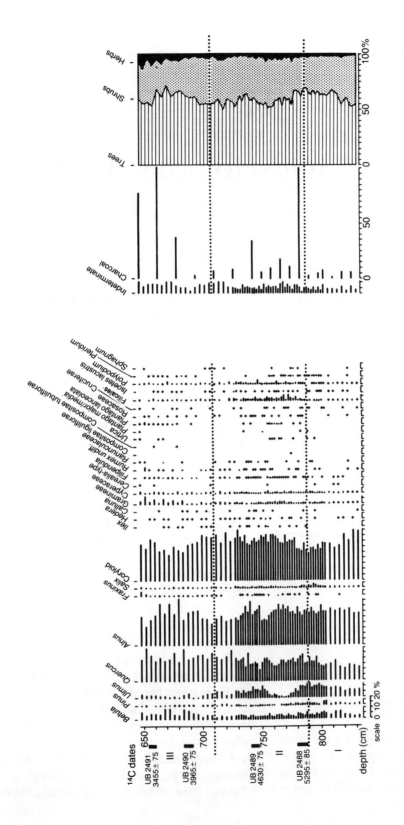

Fig. 12.1 Selected pollen data from Weir's Lough, Northern Ireland (after Hirons & Edwards, 1986).

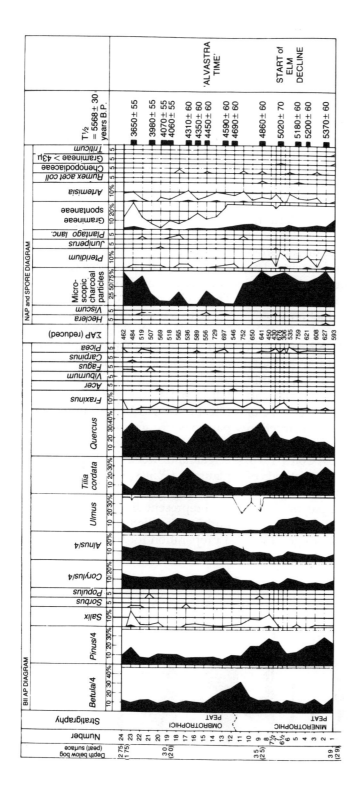

Fig. 12.2 Selected pollen data from Dags Mosse, Sweden (after Göransson, 1986).

as at *Glyceria*-hollow, Denmark (Andersen, 1988).

A further non-conformity with the model appears in areas at or close to the tree line where many broad-leaved arboreal taxa are not well represented, or where the tree pollen component is dominated by a few taxa only, as at Slieve Gallion in Northern Ireland (Pilcher, 1973), Moel y Gerddi in Wales (Chambers *et al.*, 1988), or site 1 Elgsnes in Norway (Vorren, 1986). These sites, by definition, are beyond the remit of a paper on forest farming for the period concerned. The existence of such patterns simply assists the presentation of data, and is not seen as all-embracing. The common factor is the diminution in woodland as the Neolithic advances, usually with a temporary or permanent reduction in elm pollen representation.

The theme is problematical: the first steps toward recognizable agricultural activity must have taken place, as often as not, within woodland, but this is an alien and difficult environment as seen from the perspective of today's open landscapes. The reconstruction of the process is a near-impossible one for archaeology alone (cf. Case, 1969; Coles, 1976; Whittle, 1978; Zvelebil, 1986). The methods of environmental archaeology can assist greatly (Milles *et al.*, 1989), and that of palynology is well equipped to make perhaps the major contribution (Groenman-van Waateringe, 1983; Edwards, 1988, 1989; Edwards & MacDonald, 1991). This chapter examines four palynologically based models that seek to explain the nature of farming in mid-Holocene times in parts of north-west Europe, c. 6000–3800 BP.

12.2 MODELS OF EARLY FARMING ACTIVITY

12.2.1 The landnam model

The first recognizable model of farming activity is that of Johannes Iversen. His 'landnam' (Old Norse: land take) model (Fig. 12.3) presupposed that clearances were made in 'continuous primeval forest' (Iversen, 1941, p. 43). These clearances, or landnam episodes, involved such woodland taxa as oak (*Quercus*), lime (*Tilia*), ash (*Fraxinus*) and elm, and occurred shortly after an initial decline in elm and ivy (*Hedera helix*) that Iversen took to be of climatic origin (Iversen, 1941, 1944). The fall in pollen frequencies for the mixed oak forest trees was followed immediately by expansions in grasses (Gramineae), bracken (*Pteridium aquilinum*) and Compositae, before cereal and weed pollen including ribwort plantain (*Plantago lanceolata*) appeared in a second stage. Charcoal and a rise in birch (*Betula*) representation was taken to indicate the burning of felled trees and the field layer as a preparation for browse creation and cereal cultivation. The plantain was assumed to re-

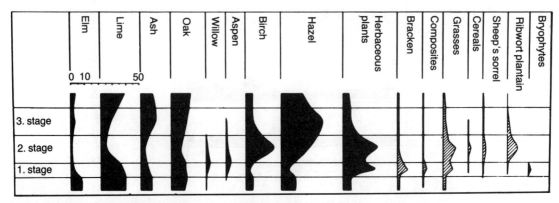

Fig. 12.3 Generalized pollen diagram for a Danish landnam phase (after Iversen, 1973).

flect pasture and the birch was a response to burning on forest mull soils.

A third phase was denoted by the expansion of hazel (*Corylus*) and a reduction in birch and cultural pollen indicators. The *Corylus* rise was that of a woodland pioneer that out-shaded *Betula* and was in turn outcompeted by more slowly regenerating forest trees. Iversen (1956) assumed that the landnam episode took place over a period of perhaps scarcely more than 50 years. In a letter written in 1970, however, and published in part in 1981 (Smith, 1981b, pp. 155–156), Iversen came to regard phase two of his model as lasting at least 100 years. Furthermore, he said that he had

> deliberately avoided using the term 'shifting agriculture' because I thought that agriculture might be understood as 'arable type'. It is the pastoral aspect which in my opinion characterises the landnam phase in Danish diagrams, though pollen grains of cereals indicate some arable farming too. I regard it as essential that much more was burnt than necessary for cereal growing.

This finds an echo in 1941:

> We may assume that a small part of the burnt-off area was sown, whereas the greater part was left to itself. The black surface quickly became green from the emerging herbs, bushes and trees, whose fresh leaves would be welcome fodder for the cattle, which were free to move about. (Iversen, 1941, p. 30)

Iversen certainly regarded his mixed agriculture as having taken place on a shifting basis because he records

> In deposits in large lakes or fjords the curves have a smoother course, and the fall in the curve of the Oak Mixed Forest is not the result of a single 'clearance fire', but more likely an expression of the gradual change in the forest picture of the country round about as a consequence of a whole series of forest clearances, undertaken gradually as the soil-tilling people took possession of new areas. (Iversen, 1941, p. 26)

and

> As a rule the settlements could not have existed long. (Iversen, 1941, p. 48)

In a monograph translated into English after his death, Iversen (1973, p. 85; Danish edition 1967) noted that the initial elm decline was unlikely to be a response everywhere to climatic change and that Dutch elm disease 'is an explanation which can solve the problem in a most elegant way'. He also felt that ring-barking would facilitate the production of leaf fodder beneath the girdled area as observed in the Balkans (Iversen, 1960). Furthermore, he thought that *Corylus* pollen frequencies reached such high levels after the abandonment of cleared areas, that 'hazel groves must have been actively maintained by suppression of the high forest, probably by bark peeling (in lime) or by ring cutting' (Iversen, 1973, p. 92).

Iversen's post-initial elm decline landnam model was formulated at a time when radiocarbon dates were unknown. The extreme longevity of the initial clearance phase, often lasting many centuries, does not support the idea of a single, short-lived occupation episode, but rather the aggregation of a series of clearances (Pilcher *et al.*, 1971; Edwards, 1979) or the process of post-occupation grazing (Buckland & Edwards, 1984). Troels-Smith (1954, p. 52) suggested that 'we must doubtless imagine the so-called occupation phase as a series of individual forest clearances, except that they cannot be distinguished in the pollen diagrams'.

With regard to the pollen curves in large lake and fjord deposits, Iversen (above) remarked that they have a smoother course, and that the fall in the mixed oak forest was not the result of a single 'clearance fire' (for which read 'clearance'), but more likely was an expression of the gradual change in the forest landscape as a consequence of a whole series of forest clearances. While he may be making the correct inference, the smoothed pattern is just as likely to be a response to the mixing action of sediments at the mud–water interface.

Concerning the *Betula–Corylus* sequence occurring within the clearance episodes, Rowley-Conwy (1981) queried whether such a clearly defined pattern would arise from shifting cultivation. I do not find it difficult to accept that similar stages of regeneration within the pollen catchment area of a single pollen site may be attained, or that mismatches in a series of recolonization sequences could still produce predominant palynological patterns that mimic a single area of regenerating forest. The ubiquity of the *Betula–Corylus* sequence is striking, but exceptions may be found outside Denmark and Sweden, for example: a general rise in *Corylus*, but without an expansion in *Betula* (e.g. Ballyscullion – Pilcher *et al.*, 1971; Tregaron and Din Moss – Hibbert & Switsur, 1976); a fall in *Corylus* (Loughs Namackanbeg and Sheeauns – O'Connell *et al.*, 1988; Hockham Mere – Bennett, 1983); and, even in Denmark, an increase in *Corylus* prior to that of *Betula* (Abkær Bog – Aaby, 1988).

Iversen's view of landnam as the provision of grazing in secondary forests with cereal growing as a more minor adjunct, is not the slash-and-burn cultivation of Sweden or Finland (cf. Iversen, 1941, p. 48), but he was sometimes unclear about this: 'Forest clearance by the Stone Age farmer must also have proceeded mainly by fire' (Iversen, 1949, p. 11). However, statements such as 'Iversen argued that this sequence of events marked the arrival of the first Neolithic farmers in Denmark, practising primitive swidden or slash and burn cultivation...' (Barker, 1985, p. 17) are somewhat misleading, as is Rowley-Conwy's observation (1981, p. 85) that 'Iversen's final paper contains a reiteration of the slash and burn arguments'. While using phrases such as 'great clearances by axe and by fire' and 'the forest clearance was accompanied by burning' (Iversen, 1973, pp. 86 and 87), he also said in this popular account of events that 'the landnam phase here was due to local burning... and cattle grazing' and 'there can scarcely be any doubt that the chief purpose of the great clearances was to obtain food for freely roaming cattle' (Iversen, 1973, pp. 89 and 91).

Rowley-Conwy (1982, p. 208) subsequently acknowledged the pastoral aspects of Iversen's landnam paper: 'In Iversen's scheme, not all the cleared and burnt area was for cultivation. Much of it, he suggested, was for grazing...'. Troels-Smith (1954, p. 53) did say that 'it must also be taken for granted that the forest was cleared by fire in order to make room for these pastures'. It is quite clear from his text that he had in mind the burning of felled woodland for pasture creation and not necessarily for cereal cultivation.

12.2.2 The leaf-foddering model

In 1954, Jörgen Troels-Smith proposed that in Denmark a leaf-fodder husbandry occurred at the initial elm decline and prior to Iversen's landnam. Troels-Smith suggested that domesticated animals such as cattle and sheep were stalled in byres or pens and were given twigs and leaves from pollarded trees, while small permanent cultivation plots provided wheat and barley. Hunting, gathering and fishing continued and it was supposed that this first forest farming was undertaken by people of the Ertebølle culture.

Troels-Smith based his theory on four principal observations:

1. The reduction in elm would occur because leaf-foddering inhibits pollen production for 7–9 years after pollarding.
2. Pollen of cereals was found immediately after the elm decline but not before it.
3. No pollen indicators of grazed areas were found at this early stage, e.g. *Plantago lanceolata, Artemisia* (mugwort), *Trifolium repens* (white clover), *Rumex acetosa/acetosella* (sorrel).
4. On the contrary, there was the appearance of pollen types of plants that could not thrive in grazed land, e.g. *Allium ursinum* (ramsons), *Hypericum* (St John's wort).

Table 12.1 The value of different trees for fodder (after Troels-Smith, 1960)

	Norway	Switzerland
Most valuable	Elm, ash	Elm, ash
Valuable	Hazel, lime, oak	Maple, oak, hazel, beech (young leaves), lime
Less valuable	Birch, aspen	Mountain ash, willow, alder, birch
Least valuable	Mountain ash, alder, willow	

Extending his work to include Switzerland, Troels-Smith (1960, 1984; cf. Guyan, 1955) reported palynological, macrofossil, palaeoentomological and archaeological investigations that encouraged him in the belief that the collecting of leaves and twigs from pollarded elm, ash, birch, oak, lime and maple, as well as ivy, for fodder for stalled animals (cf. Table 12.1), was reflected in the behaviour of the pollen curves in the early Neolithic sections of pollen diagrams:

The Neolithic farmer pioneers in Denmark and Switzerland have had to face a continuous primeval forest with hardly any grass vegetation. Still, finds from both countries indicate the presence of oxen and sheep although no pastures existed in such dimensions that they can be traced in the pollen diagrams... Rather extensive pastures... may be demonstrated in Denmark at a slightly later date. (Troels-Smith, 1960, p. 23).

The more extensive pastures and the clearance of the mixed oak forest trees would equate with the subsequent landnam phase of Iversen. Troels-Smith (1954) saw post-landnam agriculture as similar to that of the early Neolithic, and would therefore see leaf foddering recurring after the forest had regenerated in the middle Neolithic.

Thus, for Troels-Smith, the initial elm decline did not have a climatic or disease origin, but arose from the feeding of stalled animals upon leaf fodder. Iversen (1960, 1973), who favoured ring-barking as a means of producing leaf fodder, doubted that the pollarding of elm could explain the sharp decline in the *Ulmus* curve, because this would be a huge task given the area over which the phenomenon has been observed. This view was reiterated by Rackham (1980) and Rowley-Conwy (1982). However, given the ubiquity of the landnam-type episode (wherever it seems to coincide with the first elm decline) in which other tree types are involved and where cereal pollen is so frequently detected, perhaps the power of human-related activities in prehistory is underestimated, even if clearance was aided by disease or climate.

Of great interest also, is that subsequent investigations at Weier in Switzerland (Robinson & Rasmussen, 1989) suggested that over-wintering animals were fed on cereals, in the form of pounded ears or spikelets, as a supplement to leaf hay. This could show that cereal cultivation for the production of animal food may have been as important as that for human consumption. Leaf hay would have been more readily available in a wooded landscape than grass and herb hay, but alone would have probably been insufficient if much of the livestock was to survive the winter. Interestingly, the leaf hay fragments from the manure layers (Table 12.2) were predominantly of ash, lime and willow. Robinson & Rasmussen (1989) commented on the low representation for elm, given its palatability (cf. Table 12.1). It might be suggested that this resulted from the over-exploitation of elm in the vicinity of the lake village site, or perhaps its demise via disease or girdling.

At the *Glyceria*-hollow site (Andersen, 1988), the elm decline corresponded with a fall in most woodland taxa and marked expansions in *Tilia* pollen (from about 50% up to 90% of arboreal pollen) and charcoal 'dust'. This was explained as reflecting the process of woodland clearance, the burning of forest litter and the shredding of lime. Shredding (the removal of adventitious twigs along the

Table 12.2 Species distribution of twig fragments in byre samples from the Neolithic lake village site of Weier, Switzerland (after Robinson & Rasmussen, 1989)

Taxon	Number	Weight (g)	% (wt)
Ash (Fraxinus)	339	114.3	27.1
Lime (Tilia)	285	81.5	19.3
Willow (Salix)	307	71.7	17.0
Alder (Alnus)	215	39.7	9.4
Ivy (Hedera)	186	28.7	6.8
Clematis (Clematis)	69	28.6	6.8
Hazel (Corylus)	60	27.2	6.4
Oak (Quercus)	68	15.1	3.6
Elm (Ulmus)	45	14.5	3.4
Birch/alder (Betula/Alnus)	15	0.6	0.1
Mistletoe (Viscum)	5	0.3	0.1
Total	1594	422.2	100.0

trunk) provides leaf fodder while permitting the tree crown to flower; pollarding may reduce flowering if done frequently. An expansion in *Tilia* also occurred at the start of the elm decline at the Danish sites of Fuglsø Bog (Aaby, 1986) and Holmegaard Bog (Aaby, 1986, 1988) and at Lake Vån, Sweden (Göransson, 1987a).

Within the British Isles, there is supposedly little historical evidence that leaf foddering of trees was common (Rackham, 1988b) unlike the situation elsewhere (Nordhagen, 1954; Heybroek, 1963; Austad, 1988). The medieval use of ivy and mistletoe is known from Britain, and there is the occasional reference to leaves, bark and branches being fed to livestock when other feed was scarce (Rackham, 1988b). Holly (*Ilex aquifolium*) was certainly employed as a winter fodder for sheep in medieval and later England as were other plants for sheep and other animals (Spray, 1981). Indeed, the situation in Britain may be being underestimated. On some farms in the North York Moors to the present-day, ash twigs are fed to cattle as a food supplement, and not only in winter. There is every likelihood that similar practices would have occurred in the wooded prehistoric landscapes of the British Isles.

12.2.3 The expansion–regression model

A long-standing orthodox approach to the interpretation of pollen assemblages is to assume that a reduction in woodland taxa and an expansion in open-land herbaceous pollen types signifies an opening of the landscape, and, by implication, the strong presence of human communities. The subsequent restoration of woodland pollen frequencies has been seen as forest regeneration and often the abandonment of an area. This is implicit in Iversen's landnam model and can be found through to the present (cf. compilations in Simmons & Tooley, 1981; Behre, 1986; Birks *et al.*, 1988).

The approach has been formalized by Björn Berglund (1966, 1969, 1985, 1986a) in his 'expansion–regression hypothesis', embracing the period of an 'early Neolithic landnam' through to 'Modern Time', where he writes of periods of increased human influence (expansion phases to which 'agriculture' is largely restricted), and the converse (regression or stagnation phases with 'closed (forested) stable vegetation... low diversity' and 'weak human impact or restricted to grazing and coppicing') (Berglund, 1986a, p. 82). These phases refer to patterns in the pollen diagram as well as the interpretation of the vegetation dynamics and human impact. For the Neolithic, this involved an early Neolithic landnam phase of expansion followed by a regression period featuring forest regeneration, succeeded by a middle Neolithic expansion phase beginning around 4000 BP.

This model accepts, essentially, that the patterns traced by the pollen curves, are correct. It makes few concessions to further interpretation (or even over-interpretation) of the data and takes a broader temporal view of events than either of the previous two models. The tacit acceptance of pre-elm decline cereal cultivation (Berglund, 1985) and the acknowledgement of possible regression phase grazing and coppicing suggest that there is flexibility in the model, and a realization that there could be something happening behind the woodland pollen screen, but it is not emphasized. The forest-utilization model, discussed below, reflects a much stronger stance that rejects, in many instances, the idea of regression:

> In my opinion, a model that includes only 'anthropogenic pollen indicators' and that excludes the pollen sum of broad-leaved trees during the Stone Age is a useless model. *This early agriculture was based on the presence of broad-leaved trees.* (Göransson, 1987b, p. 44)

In acknowledging the conflict between his model and that of Göransson, Berglund (1986a, p. 83) observed that: 'This emphasises that a correct interpretation of the human impact on the vegetation in a primitive, agrarian society is difficult'.

12.2.4 The forest-utilization model

Hans Göransson (1983, 1986, 1987a) has advanced a model (Fig. 12.4), here termed the forest-utilization model, that portrays his understanding of forest development and, *inter alia*, embraces aspects of forest farming. The schema begins with the Atlantic forest, which Göransson maintains was not an unadulterated primeval climax woodland but was, in part, coppiced woodland (his 'coppice wood phase I'), formed by girdling (ring-barking). The ring-barked tree would sprout below the girdled area, while above it the tree would eventually die. This process would create browse for wild animals but would also pro-

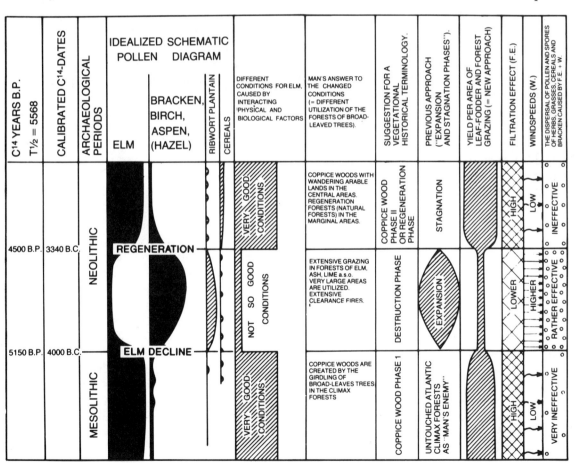

Fig. 12.4 The 'forest-utilization' model (after Göransson, 1983, 1986).

vide stems, leaves and twigs for winter fodder. Girdling would also create a 'manuring' or 'assart' effect, whereby nutrients are released, and would assist such plants as grasses and raspberries to flourish, aided by the light from the opening canopy. The burning of twigs and small branches beneath girdled trees would clear the ground and prepare it for the garden cultivation of cereals. Göransson supposes that the coppiced and girdled woodland mosaic would produce a pollen rain indistinguishable from that of virgin forest. Furthermore, because of the filtration effect (Tauber, 1965; the 'curtain effect' of Göransson) of the coppice woods and high forest, only rare finds of open-land herb and cereal pollen grains would reach deposition sites that were not fortuitously located in the immediate vicinity of farmed forest areas.

The elm decline denotes the beginning of the 'destruction phase', which is seen as resulting from such factors as climatic change (e.g. a series of cold winters) and/or disease combined with browsing. The newly opening landscape meant that farmers and animals had to range further to obtain leaf fodder and browse, thus reducing taxa such as lime, ash and sometimes oak in addition to elm. This is not seen as an intensification of cultivation. The removal of the forest curtain enables open-land pollen indicators, including the Cerealia, to reach more distant pollen-receiving sites. Göransson perceives sequences of arboreal taxa (e.g. *Pinus* → *Betula* → *Corylus* → high forest) during the long (600 radiocarbon years) 'destruction phase' (cf. Fig. 12.4) and these are seen as analogous to the pattern of woodland colonization evident in early post-glacial times. They are also regarded, incidentally, as explaining the stages seen in the Iversen landnam model, while the longevity is such that a series of clearings (cf. Pilcher et al., 1971; Edwards, 1979) is disputed since these would supposedly not allow such distinct successional stages (cf. Rowley-Conwy, 1981).

The restoration of *Ulmus* and other taxa in the pollen spectra is taken to signify the cessation of the processes responsible for the 'destruction phase' and marks the start of the 'coppice wood phase II' or 'regeneration phase'. Once again the coppice woods could be utilized to their maximal extent together with forest farming. At some sites, notably at the Alvastra III and Isberga II sites (Göransson, 1987a), cereal pollen representation actually expands at the start of the 'regeneration phase'. As the phase progresses, 'wandering arable lands' are proposed. A crop is sown in an area of axe-cleared coppiced woodland (the clearance would facilitate nutrient release through natural 'manuring'). The cereal crop would be produced for one year only (which would maximize crop yields) and the land would be left to allow the coppiced woodland to regenerate. This recurrent activity (with areas returning to arable use every 20 years or so) could produce a near-continuous cereal pollen curve.

Combined with the likelihood that extant coppiced woodland could be regularly pruned for foddering purposes, the creation of such rotating arable land would encourage the development of thinner coppice wood with more extensive grass swards (based on a rising Gramineae curve) than was observed in 'coppice wood phase I'. As was the case during the latter, the forest curtain would act as a barrier to open-land herb and cereal pollen, as well as airborne charcoal. The richness of the farmed forest areas was such that cereal pollen still reached suitably located receiving sites. In addition to the use of the coppice wood for the production of leaf-fodder (with grazing on the grass swards), it would also provide resources useful to human inhabitants such as nuts, other foodstuffs and wood for fencing, tools and fires. In areas marginal for farming, the 'regeneration phase' spectra are taken to reflect the regeneration of natural forest.

Göransson repudiates any notion that the early forest was a malevolent obstacle rather than a shelter and resource for humans and animals. Girdling and coppicing would provide utilizable products both before and after the elm decline. Pre-elm decline cereals have a natural place in this farmed forest environment and to Göransson the many Cerealia-type pollen finds in the British Isles (Groenman-van Waateringe, 1983; Edwards & Hirons, 1984; Edwards, 1989; Wiltshire & Edwards, this volume) and Sweden (Göransson, 1987a, 1988; Regnéll, 1989) are not unexpected. Coppicing had been considered likely by Iversen (1941, 1973), of course, and for the post-elm decline period of high arboreal pollen frequencies, coppiced elmwood was found at the Alvastra pile dwelling (Göransson, 1987a), coppiced hazel piles were recovered from the Tibirke trackway in Denmark (Malmros, 1986) and a coppiced hazel hurdle track on Walton Heath in the Somerset Levels of England was dated to c. 4150 BP (Coles & Coles, 1986).

Given that cereal-type pollen has been found at more than one level during the regression–regeneration phase in a number of pollen profiles (e.g. Söborg Sö – Iversen, 1941; Alvastra III and Lake Bjärsjöholmssjön – Göransson 1987a, 1991; Ospelse Peel – Janssen & Ten Hove, 1971; Killymaddy Lough – Hirons & Edwards, 1986; Abbot's Way – Beckett & Hibbert, 1979) then this may provide support for cereal growing within a forest farming system, rather than agricultural stagnation at this time. Many pollen profiles, however, contain no cereal-type pollen during the regeneration phase (where the phase is present). A number of other sites contain agricultural indicators during this phase (e.g. the Sweet Track Factory Site, which also has cereal-type pollen at one level – Beckett & Hibbert, 1979) but were perhaps screened from, or too distant from cultivated areas for the large, short-travel cereal pollen grains to reach the pollen sites.

It is interesting to note that the southern Dutch site of Ospelse Peel, which has Cerealia pollen, is close to the bog edge, whereas that of nearby Griendsveen 2, which does not have Cerealia pollen during the regeneration phase, is a bog centre site and is therefore distant from likely agricultural activity (Janssen & Ten Hove, 1971; cf. Edwards & McIntosh, 1988). If agriculture, of whatever type, was widespread during the woodland recovery phase, then notions of human population decline based in part upon expansion–regression patterns in the pollen record (Whittle, 1978), may have to be reconsidered.

The Göransson model would seem to be plausible. The problem is, of course, to find strong independent evidence to support it. Early Neolithic leaf foddering has been shown from Switzerland, and coppicing is known to have occurred during the regeneration phase. The possibility of coppicing during pre-elm decline times seems likely and, after all, the putative, but probably discredited, link between the elm decline and the beginning of Neolithic farming activity only has relevance in north-west Europe – further south and east, cereal cultivation clearly predated 5100 BP (Ammerman & Cavalli-Sforza, 1984; Edwards, 1989). It also seems reasonable to suppose that there was no proven wholesale population decline for the period 4500–3800 BP. The problem of pollen filtration by woodland screens and distance decay in pollen transport is well documented (Tauber, 1965; Edwards, 1979; Vuorela, 1985), and provides a convincing argument for the non-detection of cultivation for anything other than the most fortuitously located profiles. The truism that absence of evidence is not evidence of absence finds an echo in: 'Is it possible to prove the absence of farming?' (Welinder, 1985, p. 95).

The existence of a 600 radiocarbon year period (c. 5100–4500 BP) of reduction in the pollen of elm and other trees is perplexing. Why should this have persisted so long if disease alone were responsible? Perhaps it was a question of over-exploitation of remaining trees given the depletion of the valued elm. But is there evidence for climatic change that may have repressed elm and other warmth-demanding ecotypes of lime and ash (Nilsson, 1961, quoted in Göransson, 1991, p. 26)? Digerfeldt (1988) has found low lake levels at Lake Bysjön for the period 4900–4600 BP, but there are few other indications of climatic change for this time in southern Sweden. In their study at Waun-Fignen-Felen in south Wales, Smith & Cloutman (1988) found that pollen influx values in a number of profiles was reduced at the time of the first elm decline and climatic change was suggested as a possible explanation (however, see Edwards & MacDonald (1991) for contrary examples).

12.3 CONCLUSIONS

Notwithstanding its deficiencies, Iversen's landnam model has provided most insight into our understanding of the interaction of prehistoric peoples and their vegetational and agricultural environments. Troels-Smith's special contribution has been in giving us an alternative interpretation of pre-landnam events based upon ecological, archaeological and historical observations. Both palynologists participated in the Draved Forest experiments and unwittingly, perhaps, fostered the notion that fire was of greater significance in clearance, as opposed to post-clearance ground preparation, than some commentators would accept. Their papers are, however, models of scholarly observation and perspicacity that still repay study.

The expansion–regression model is much more generalized and reflects a qualified but rather literal acceptance of the paths traced by the pollen curves; simplicity, however, may be the safest, if unexceptionable, route. The forest-utilization model, while taking elements of interpretation already provided by Iversen and Troels-Smith, is more radical and ambitious. Its universality may not be as great as Göransson would imply – and he himself is not fond of the strictures implied by such terms as 'coppice wood phases' I and II, but likes to think of his model as a changing entity (Göransson, personal communication, letter dated 19 July 1991). Yet, in true multiple-working hypothesis fashion, it is surely worthy of addition to the armoury of explanation.

Limitations of space prohibit a serious consideration of other elements of the forest farming debate. Thus, the first elm decline is often the forerunner to three or more reductions in *Ulmus* pollen (Smith & Cloutman, 1988; Whittington *et al.*, 1991a). There may have been elm declines prior to the classical one of *c.* 5100 BP, but it may not be apparent in pollen diagrams because of the swamping effect of other arboreal taxa within the mid-Holocene high forests. The burgeoning number of finds of pre-elm decline Cerealia-type pollen (Edwards, 1989; Göransson, 1991) may force us to look more closely at the nature of the vegetational landscapes of that time, when forest farming involving cereal cultivation was just beginning. We may need to optimize methods for discovering scarce pollen types (Edwards & McIntosh, 1988), while high-resolution studies hold out the promise of peeking inside the clearance episodes in order to approach closer to an understanding of the processes being enacted by the first farmers (Garbett, 1981; Turner & Peglar, 1988).

It is probable that the role of fire will contribute to the debate (Smith, 1970b, 1984; Bennett *et al.*, 1990b; Edwards, 1990). Doubts surrounding the use of 'absolute' pollen studies must temper their uncritical application. Insights into the effects of human interference in woodland areas, however, have been furnished by Aaby (1986, 1988), who showed that tree pollen influx actually increased when disturbance occurred within a heavily forested landscape, presumably because the open canopy enabled prolific flowering either at the woodland edge or of solitary trees. The initial clearance also resulted in decreased herbaceous pollen percentages owing to increased absolute inputs of tree pollen. The wheel turns... it is worth noting that Iversen (1949, p. 21) also said that 'the new open forest must have produced much more treepollen [sic] than the same area before clearing'!

Acknowledgements

I would like to thank Drs J.J. Blackford, P.C. Buckland, F.M. Chambers, H. Göransson and J.P. Sadler for discussing various points contained in the paper.

13

The influence of human communities on the English chalklands from the Mesolithic to the Iron Age: the molluscan evidence

J.G. Evans

SUMMARY

1. Evidence is summarized for the effects of human communities on the English chalklands from the Mesolithic to the Iron Age, as indicated by molluscan analysis.
2. For the Mesolithic period, the evidence is exiguous, but there are a few possibilities of woodland clearance. The natural vegetation was probably woodland, but the continuing absence of data for the high chalklands is noted.
3. Clearance took place from the Neolithic period onwards. There were periods of abandonment when woodland regenerated in some areas, but impoverished grassland is an alternative response to the relaxation of human pressure.
4. Molluscan analysis for later periods is less widely applicable on its own because of increasing cultural and environmental diversity.
5. A research strategy is proposed in which spatial analysis of archaeology, soils and biology at different scales is combined with stratigraphical and chronological studies.

Key words: chalklands, Mesolithic/Neolithic, Mollusca, research strategy

13.1 INTRODUCTION

Viewed ecologically and through time, 'the influence of human communities', 'the chalklands' and 'the molluscan evidence' relate more to the way in which the subject is perceived and studied than to ecological entities. The chalklands (Fig. 13.1), for example, are a unit of study because they have distinctive hydrology, vegetation, soils and land use. These properties, however, were not so distinctive in the past. In the mid Holocene the

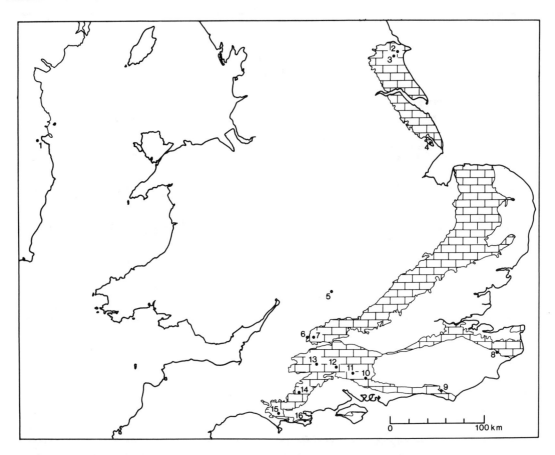

Fig. 13.1 Map showing the solid geological distribution of chalk in southern Britain and sites mentioned in the text. (1) Newlands Cross; (2) Willow Garth; (3) Kilham; (4) Skendleby; (5) Ascott-under-Wychwood; (6) Cherhill; (7) Avebury; (8) Brook; (9) Itford Bottom; (10) Meon valley; (11) Easton Down; (12) Danebury; (13) Stonehenge; (14) Thickthorn Down; (15) Maiden Castle; (16) Blashenwell. Sites in the Avebury area include Millbarrow, South Street, West Overton and Windmill Hill; sites in the Stonehenge area include Coneybury, Durrington Walls, the Wilsford Shaft and Woodhenge; Mount Pleasant is near Maiden Castle.

predominant chalkland vegetation was probably mixed deciduous woodland, and the soils were circum-neutral Brown Earths, similar to adjacent land types such as river terrace gravels and clay vales (Limbrey, 1975; Thorley, 1981).

As with other land types such as wetlands and uplands, the chalklands require particular research techniques. Their aerobic, calcareous and biologically active soils and the lack of peat and deep sediment-filled basins mean molluscs, bones and ostracods, rather than pollen, macroscopic plant remains and insects. However, it only needs a waterlogged site like the Wilsford Shaft (Ashbee *et al.*, 1989), in which there is a greater range of natural history than any other archaeological site in Britain, to emphasize just how restricted individual taxonomic groups like the Mollusca are.

The effects of humans on the environment, and especially the effects of Mesolithic and Neolithic communities on vegetation, have been of interest for half a century. It was one of the main themes of Alan Smith's research from almost the start (Smith, 1961a; Smith & Willis, 1961–62), and remained so throughout (Smith & Cloutman, 1988). Identifying the effects of humans on the environment is, most simply, a device for locating past human activity. Additionally, there are the ecological consequences of interference – the rejuvenation of vegetation and the consequences for animals and humans, as well as resultant soil changes. One must not, however, forget that natural factors have played their part – for example, lightning fires, disease, overgrazing in favoured localities and soil senescence.

I make these points to show how research can become artificially polarized, sometimes without one realizing it.

13.2 MOLLUSCAN ANALYSIS

A recent review of molluscan analysis (Thomas, 1985) allows remarks to be limited to problems specific to the identification and dating of the effects of human activity in molluscan sequences. One of the advantages of molluscan analysis is that in autochthonous contexts, like buried soil surfaces, shells are in their place of life and therefore give information about the sampling point. The reverse of this is that a number of points need to be sampled to build up a wider picture, although in fact a single assemblage reflects a wider area than just the sample point: 'Reconstructions are made with a subtle understanding that the array of micro-environments in a given catchment is conditioned by the nature of the macro-environment' (Thomas, 1989, p. 550). The significance of assemblages is enhanced when they are in a horizontal or vertical sequence, and this allowed Kerney *et al.* (1980) to establish a radiocarbon-dated zonation scheme for the Holocene in Kent, since extended to much of the chalklands, as by Preece (1980) for Dorset and Preece & Robinson (1984) for Lincolnshire. At the local scale, however, modern studies on vegetation islands – for example of grassland in woodland – and of vegetation boundaries (e.g. Boag & Wishart, 1982) are urgently needed to interpret subfossil assemblages in terms of the size of areas of environment and distance from specific types of vegetation. The elegant study of Baur (1987) on the relation of species richness to stone size in an area of karst demonstrates the possibilities.

Detection of human activity is based on particular species, species associations and diversity. For earlier periods, peaks of *Vallonia costata* are important as indicators of local, temporary clearance, although there are also more subtle changes, such as troughs in species like *Discus rotundatus* (e.g. Preece *et al.*, 1986) without the advent of an open-country component. For the Neolithic, the existence of clearance horizons, as evidenced by peaks of *Pomatias elegans* (e.g. Evans *et al.*, 1985), has been questioned by Carter (1990) who has suggested that they are a taphonomic effect of differential preservation. I am not convinced, although there is certainly an element of residuality in the counts of this species; more work is needed.

For archaeological sites, site type – e.g. long barrow, causewayed camp – is important because there is often a relationship between site function and location, and thus to previous land use (Fig. 13.2). Finer detail is related to lithostratigraphical, depositional and archaeological contexts. The absence of a woodland stage in buried soil sequences, for example, is often due to the absence of suitable deposits rather than a real absence. Intersite comparison is only valid in terms of general environment when contexts are equivalent.

In buried soils, ecological resolution is good in the surface few centimetres of the A horizon, but is rapidly reduced lower down by mixing and differential preservation (Carter, 1990). Residuality from earlier deposits, especially of Devensian late-glacial age, can be detected by differences in preservation and

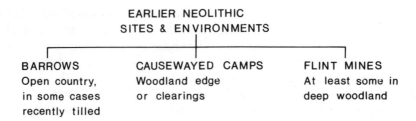

Fig. 13.2 Some different environments of earlier Neolithic sites as related to site type.

by size differences, for example between lateglacial and Holocene shells of *Pupilla muscorum* (Evans et al., 1985).

With ditch and pit assemblages, there is the problem in the primary fill of distinguishing between shells derived from the pre-site soil and those living in the ditch. Additionally, ditch-bottom assemblages often reflect conditions specific to the earliest stages of infilling – rock rubble with a minimum of vegetation (Evans & Jones, 1973). The lower part of the secondary fill is more useful as an indicator of the general environment since assemblages are less biased by local conditions. They also reflect the environment close to the time of site construction, since primary fills often form within a decade (Bell & Jones, 1990; Bell, 1990).

13.3 THE MESOLITHIC PERIOD

Evidence for Mesolithic interference with vegetation on the chalklands occurs in three contexts: tufa, buried soils and organic valley-bottom deposits.

Most convincing is the Blashenwell tufa, Dorset (Preece, 1980), 1.4 km south of the Chalk scarp. Temporary woodland clearance, defined by a maximum of *Vallonia costata* in a standstill horizon, is dated faunally (with reference to Kerney et al., 1980) at c. 7950 BP (6000 bc). There is a loose association with Mesolithic artefacts, and radiocarbon samples of bone in an overlying deposit produced dates of 5750 ± 140 BP and 5425 ± 150 BP (BM-1257, BM-1258) (the middle of the 4th millennium BC). The only other Mesolithic site sealed by tufa is Cherhill, north Wiltshire (Evans & Smith, 1983), where artefacts lay on a palaeosol surface, radiocarbon-dated to 7230 ± 140 BP (BM-447; in the early 6th millennium bc). Molluscan analysis showed no interference with the woodland vegetation, even though the site was substantial – a base or major hunting camp (Whittle, 1990a, p. 106).

At Avebury, north Wiltshire, in the buried soil below the bank of the Neolithic henge (Evans, 1972), an assemblage in the lower part of a treethrow pit, although indicating woodland, was of more open facies than that higher up. The date, on faunal grounds, is prior to 7950 BP (6000 bc). There were no artefacts, but several sites in the vicinity have yielded Mesolithic flints (Holgate, 1988; Whittle, 1990a), the nearest in the valley bottom 0.4 km away, dated to c. 8150 BP by thermoluminescence of burnt flints. In other parts of the chalklands, there are hints of Mesolithic activity, but there are either problems with dating, as with the earliest open-country episode at Brook, Kent (Rifle Butts section; Burleigh & Kerney, 1982), or with taphonomic mixing as in the subsoil hollow at South Street, north Wiltshire (Evans, 1972; Ashbee et al., 1979). The best buried soil site is Ascott-under-Wychwood, Oxfordshire, which, although not on chalk, demonstrates the potential of treethrow pits for preserving evidence of Mesolithic activity (Evans, 1971). Sealed beneath a Neolithic long barrow, Mesolithic artefacts in the lower part of a treethrow pit with a molluscan assemblage of open woodland are separated from an overlying Neolithic occupation and clearance horizon by a closed woodland assemblage and, in some parts of the site, a decalcified soil.

Pollen evidence comes from a carr, Willow Garth, in the Great Wold Valley in north-east Yorkshire (Bush, 1988), where retardation of woodland succession suggests forest disturbance c. 8900 BP (c. 6950 bc). Whether this can be attributed to Mesolithic people is, however, debatable (Thomas, 1989). Furthermore, the implication that grassland lasted into the Neolithic on the basis of an unproven density of Mesolithic sites (Bush & Flenley, 1987) and an alleged desirability of the Great Wold Valley for settlement is controversial, especially as Bush (1988) reports a hiatus in the sequence between c. 7980 and c. 4400 BP (c. 6030–2450 bc). The open-country pollen spectra from the Kilham long barrow (Manby, 1976), cited by Bush & Flenley (1987) in support of their hypothesis, are irrelevant because they are from a cultivated pre-barrow soil profile, not from treethrow pits where woodland assemblages are likely to be preserved.

Palaeoecological analyses can be used to locate Mesolithic sites, which can then be identified precisely by coring and excavation (Bush, 1989). Preece (1980) makes the same point for Blashenwell and another tufa site at Newlands Cross, Co. Dublin (Preece et al., 1986), suggesting that archaeological sites could be located by analysing several profiles on a grid. This has been done in other areas using pollen analysis and surface collecting, for example, Simmons et al. (1989) in the North York Moors. Obvious areas to investigate in the chalklands are river valley bottoms such as the Kennet with its wide age range of sites, and the scarpfoot where tufa deposits are numerous.

For the higher chalklands, a basic problem is that there are no mid-Holocene molluscan assemblages, so the natural fauna that one would use as a basis for comparison is unknown. Nor is the vegetation known, because palynological work (e.g. Thorley, 1981; Waton, 1982, 1986) is from low-lying areas. But in view of the increasing evidence for Mesolithic activity on chalklands (Care, 1982; Holgate, 1988) and the fact that the distinction between chalklands and other areas was probably not so great then as now, evidence for Mesolithic interference with the vegetation similar to that in other regions (e.g. Dimbleby, 1962) should be sought.

13.4 THE EARLIER NEOLITHIC

Although there is support in British archaeology for the earliest Neolithic being in part indigenous (literature reviewed by Whittle, 1990a, 1990b) and only weakly based on cereal agriculture (Entwistle & Grant, 1989), the evidence allows little agreement on the processes involved in, or the chronology of, the Mesolithic–Neolithic transition (Whittle, 1990b; see Edwards, this volume). Certainly in the chalklands there is nothing to indicate that open-country assemblages in the A horizons of buried soils beneath Neolithic sites are other than of Neolithic origin. Sequences are short, relating to immediate pre-monument conditions as shown by gradients of increasing molluscan abundance to the surface turfline. Where Mesolithic interference has been proposed, the events were temporary and separate from later clearance. Of course, there is no reason why there should not have been continuity between Mesolithic clearances and those of the Neolithic, as proposed for the Great Wold Valley, but this has yet to be shown.

Since my summary two decades ago (Evans, 1972), more sites have been investigated that clarify the pattern. Valley bottoms and low-lying slopes and plateaux supported closed woodland (Evans et al., 1985; Evans et al., 1988a). Clearance took place in the earlier Neolithic period. Thus the Giants' Hills 2 long barrow, Skendleby, in the southern part of the Lincolnshire Wolds (Evans & Simpson,

1991), was sited in grassland on the edge of previously cultivated land; prior to that there was deep woodland. Millbarrow in the Avebury area of north Wiltshire, another long barrow, was sited in grassland (Tzavaras, 1991), and the Maiden Castle bank barrow, Dorset, was constructed likewise in open country, although woodland was not far off (Evans et al., 1988b).

In contrast, the causewayed camp at Maiden Castle was constructed at the edge of woodland (Evans et al., 1988b). This site was particularly informative because there were two molluscan sequences from opposite sides that could be related to artefact distribution and soil. On the east side, where molluscs indicated open country and the soils were chalky, surface flints indicated early Neolithic activity. On the west side in contrast, where there was woodland and soil on non-calcareous clays and gravels, no flints were present. Another causewayed camp, Windmill Hill in north Wiltshire, was built close to woodland in land that had been cleared but not intensively cultivated or grazed (M. Fishpool, personal communication). This site is more or less contemporary with the long barrows of South Street, 2 km to the south, and Millbarrow, 0.7 km to the north, built in open country, so a picture of the earlier Neolithic vegetation of the area is emerging. Likewise in Sussex, some causewayed camps were built in woodland that was cleared for the purpose of camp construction or in land that had only recently been cleared (Thomas, 1982).

The different environments of these monuments – open country for long barrows, woodland or woodland edge for causewayed camps – are related to the function of the monuments in the landscape (Evans et al., 1988b). Another type of Neolithic site, the flint mine, relates to flint distribution and might therefore occur in areas otherwise not utilized. Molluscan assemblages from the abandoned shafts of some of these are very rich in woodland snails (Mercer, 1981). Thus it is necessary to examine different kinds of contemporary site to obtain a full picture, and different parts of the same site, as at Maiden Castle, if the detail is to emerge (Fig. 13.2).

13.5 THE LATER NEOLITHIC

For the later Neolithic, it is less easy to generalize with regard to site type. Evidence for the environments of henges (Bell & Jones, 1990), shows that some, such as Durrington Walls and Woodhenge in the Stonehenge area and Mount Pleasant near Maiden Castle, Dorset, were built in extensive grassland but that Coneybury, near Stonehenge, and probably Stonehenge I were built in an area where woodland was not far off.

There is evidence of land abandonment after the earlier Neolithic. This continued for varying periods into the later Neolithic and may be related to a countrywide trend as seen in pollen diagrams and the chronological distribution of archaeology (Whittle, 1978). The nature of the response during abandonment depends on context (Fig. 13.3).

In level areas, grassland may persist, becoming species poor and with concomitant soil decalcification. At Avebury, the surface of the pre-henge soil was non-calcareous in places and weakly calcareous in others (Evans et al., 1985), and the molluscan assemblages indicate poor grassland. This is probably related to a decrease in biological activity brought about by the development of a dense thatch of grass under a regime of relaxed and ultimately non-existent grazing (Evans, 1990). Such vegetation inhibits the establishment of more diverse herbaceous or woody species because the roots of seedlings cannot reach the soil (cf. Smith, 1980).

On slopes, instability persists, leading to soil creep, which allows the soil to remain calcareous. Bare areas of soil and thin vegetation allow a diverse vegetation and ultimately the invasion of scrub and woodland. In terms of vegetation response, this is the extreme to that of flat areas, but the cause – human abandonment – is the same. Such a

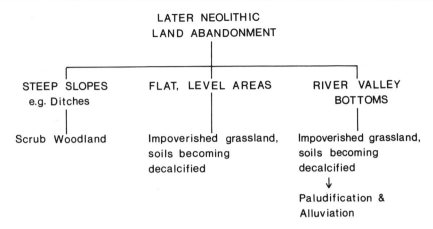

Fig. 13.3 Response to land abandonment in the Avebury area of north Wiltshire in relation to environment during the later Neolithic.

situation obtains in long barrow ditches in the later Neolithic, and, in addition to examples from north Wiltshire, Dorset and Lincolnshire recently discussed (Evans, 1990), we may cite Millbarrow in north Wiltshire (Tzavaras, 1991) and Thickthorn Down in Dorset (Drew & Piggott, 1936; J. Pollard, personal communication), in all of which there was woodland.

In a third type of context, river valley bottoms, palaeosols and molluscan sequences in the upper Kennet have shown a change from closed woodland to dry grassland prior to the later Neolithic (Evans *et al.*, 1988a). Woodland regeneration did not take place, but instead there was decalcification, followed by paludification and alluviation. The earliest radiocarbon dates for alluviation are *c.* 3850 BP (*c.* 1900 bc), which is a few centuries later than the main period of later Neolithic abandonment, but decalcification took place earlier and it is possible that this caused paludification. Hence, the ultimate changes were different in river valley bottoms from those on either slopes or level areas (Fig. 13.3).

13.6 THE BRONZE AGE AND IRON AGE

'By the beginning of the Bronze Age the downs had been extensively cleared...' (Turner, 1970, p. 115). Woodland clearance and cultivation, associated with Beaker pottery (Late Neolithic/Early Bronze Age), is attested in long barrow ditches (Evans, 1990), and at Cherhill, north Wiltshire, there was an associated field system (Evans & Smith, 1983). In the Bronze Age, grassland was widespread, although this may reflect the fact that much of the evidence is from beneath round barrows that were probably sited in areas of upland grazing (Fleming, 1971).

In some areas, the Late Neolithic/Early Bronze Age saw the last episode of clearance that led to the open country of today, with no evidence for subsequent woodland regeneration, *even though suitable contexts exist*. This was the situation around Dorchester, Dorset, where, at Maiden Castle (Evans *et al.*, 1988b) and Mount Pleasant (Wainwright, 1979), molluscan assemblages from Bronze Age and Iron Age contexts indicate open country. Likewise in north Wiltshire, standstill horizons in Bronze Age barrow ditches have open-country assemblages (Evans, 1972). On the other hand, there are some areas where there was more background woodland. In Hampshire, at Easton Down, woodland regenerated in a Bronze Age barrow ditch after earlier clearance (Fasham, 1982), and at the Iron Age hill-fort of Danebury (Cunliffe, 1984; Hewitt, 1990), pit assemblages indicate woodland nearby, although the hillfort itself was built in open country. Because these comparisons are between molluscan assemblages from similar contexts, it can be suggested that they are of regional significance.

For these later periods, molluscan analysis is linked with colluviation and alluviation. There are numerous single sequences from colluvium, many listed by Bell (1983). More useful, however, are studies in which molluscan sequences, colluviation and archaeological sites are linked, both spatially by surface survey and temporally by radiocarbon dating and artefacts. The best example is the work of Bell (1983) on the South Downs, particularly at Itford Bottom.

For alluvium, by contrast, there is only one published molluscan sequence from the whole of the English chalklands, namely that from the upper Kennet valley at Avebury, north Wiltshire (Evans, Limbrey et al., 1988a; Hedges et al., 1987, 1988). Here, the earliest alluviation was c. 3850 BP (c. 1900 bc) with the main deposition in the Bronze Age and earlier Iron Age. It is not certain whether alluviation was massive or incremental, or what the environment of the floodplain was. The molluscan assemblages are difficult to interpret, partly because of taphonomic mixing of land and freshwater species by flood waters, and partly because, under conditions of winter flooding and summer drying, two different molluscan assemblages might have been breeding at the same place but at different times of the year (Robinson, 1988).

13.7 RESEARCH STRATEGY

Research into molluscan analysis in archaeology should take place at different scales (Fig. 13.4).

Overall, there is the distinction between chalklands and non-chalklands – a distinction that becomes increasingly relevant as the Holocene progresses. Within the chalklands, at the regional scale, there are differences such as that proposed for varying degrees of woodland clearance in the Bronze Age. Such differences may relate to the nature of human activity. For example, areas rich in ceremonial and sepulchral monuments like the north Wiltshire downs around Avebury may be ritual landscapes, set aside by people living outside the area, a typical of Neolithic settlement as a whole (A. Whittle, personal communication). Studies of non-monumental areas, like that in the Meon valley of Hampshire (Schofield, 1987), should be made in conjunction if a full picture is to emerge.

Within a region, a variety of site types needs to be studied to obtain the full range of environments, as shown for the earlier Neolithic (Fig. 13.2). It is also useful if regions take in a range of land types – chalk upland, scarp and dip slope, river valley bottom – rather than concentrating on a single valley or upland block, because it is never certain in the initial stages of research (if ever) what area constituted a human annual range. There are also different preservational possibilities. On slopes and plateaux, there are thin soils and localized deposits, with good archaeological visibility. Surface survey is a crucial first stage. Fieldwalking is vital if spatial resolution is to be obtained, but it has only recently been used fully, as in the Stonehenge area (Richards, 1990), largely because it is labour-intensive and costly (see also: Gaffney & Gaffney, 1988; Schofield, 1987). In valley bottoms there are deep deposits where archaeology is hidden but where there are opportunities for sequences, including the preservation of organic material. They have been virtually ignored.

The local scale allows differences between valley bottoms, slopes and level areas to be explored (Fig. 13.3). Information comes from spatial distributions of settlements, field systems and artefacts (as at Itford Bottom), while spatial variation in the environment can be

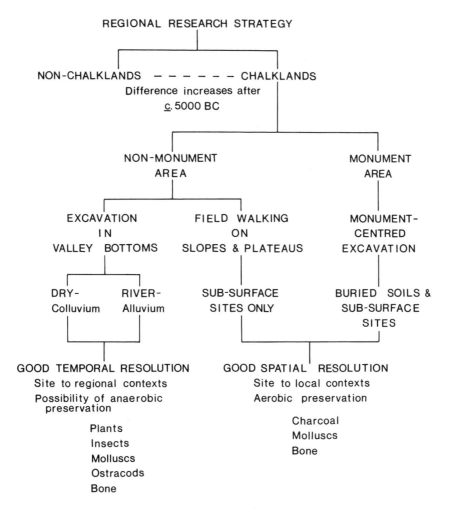

Fig. 13.4 A regional research strategy for molluscan analysis and archaeology on the chalklands.

located by multiple sampling and molluscan analyses of land surfaces and features like ditches and pits.

Equally at the site scale, molluscan analysis and archaeological survey should be used together, as at Maiden Castle. The wider significance of molluscan assemblages from archaeological sites should be supported by biological data that reflect a wider range, such as charcoal collected by people from the surroundings and brought to a site (Smart & Hoffman, 1988) or small vertebrates de-

posited by birds of prey (Evans & Rouse, 1992 for 1990).

It is fundamental for the identification of the effects of humans on the environment that spatial analyses of various kinds and scales – archaeological, biological, pedological, and from the site to the region – are carried out in conjunction with stratigraphy and chronology. But the potentially special nature of any context in relation to its function and modification by humans, whether an individual pit, a buried land surface, or an area the size of a region, should always be taken into account.

Acknowledgements

Stimulating discussion with my colleague Alasdair Whittle enabled this paper to be written. I am grateful to Paul Davies, Mark Fishpool and Joshua Pollard for information on their work in advance of publication.

14

Mesolithic, early Neolithic, and later prehistoric impacts on vegetation at a riverine site in Derbyshire, England

Patricia E.J. Wiltshire and Kevin J. Edwards

SUMMARY

1. Palynological data from a riverine peat in Buxton, Derbyshire, show a record of possible Mesolithic hunter–gatherer vegetational disturbances, continuous arable activity since at least 6000 BP, and extensive woodland clearance from Bronze Age times onward.
2. The fossil record for cultivation may be the earliest yet found for the British Isles. Nearby archaeological excavations have produced dated Neolithic contexts, including remains of wheat and flax from house structures, covering the period 5024 ± 126 BP to 4680 ± 70 BP.
3. The pollen catchment area of the site has always contained locales of open vegetational aspect.

Key words: Mesolithic, Neolithic, early farming, riverine, England

14.1 INTRODUCTION

Many investigations into early human impacts upon vegetation have been based on the palynology of upland sites. These have the advantage of having deposits that have accumulated (nearly) continuously, although they also tend to be located at some distance from areas for which archaeological data are available. Ideally, it would seem desirable to work on pollen sites located at the actual places of past cultural activity. Difficulties include depositional records that are brief and discontinuous, and fossil assemblages that are subject to such taphonomic processes as destruction, mixing, and down-profile movement (Dimbleby, 1985). Some of these problems may be overcome by examining continuously accumulating deposits, such as peats and lake sediments, whenever they survive within or near archaeological sites (Bostwick Bjerck, 1987; Edwards, 1991).

Riverine sites have long attracted human occupation, and pollen sites located in such situations can reveal strong evidence for past human activity (Ewan, 1981; Brown, 1982; Delcourt et al., 1986; Wasylikowa, 1986). Here we present evidence from a site in the town of Buxton in central Britain, where pollen spectra from a river terrace peat, immediately adjacent to an area rich in archaeological finds, contain a strong cultural signal.

14.2 THE SITE

Buxton lies in an area of Carboniferous deposits, on the boundary between the limestone plateau and a surrounding outcrop of shales and thin limestones. These are, themselves, surrounded by outcrops of grits and shales of the Millstone Grit Series and coal measures (Trueman, 1971). Lismore Fields is situated at c. 300 m OD on an interfluve between two streams at the head of the River

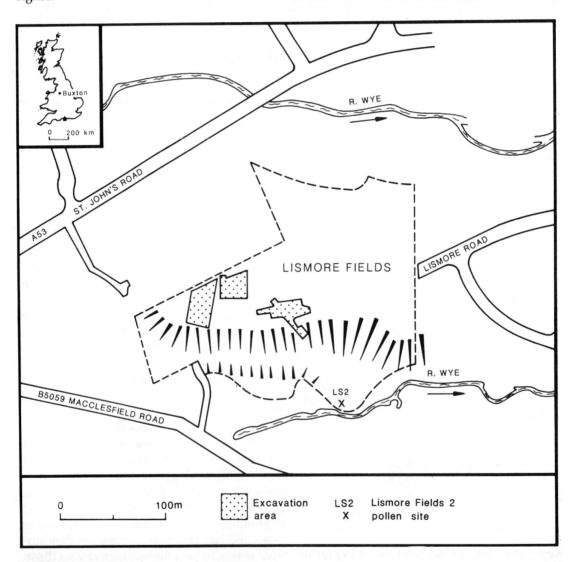

Fig. 14.1 Site location.

Wye (Fig. 14.1). The interfluve is capped by solifluction deposits derived from the Corbar and Kinderscout Grits (Canti, 1987). The southerly edge of the interfluve slopes steeply towards the southern arm of the Wye, and levels off into a narrow river terrace that runs above part of the northern bank of the stream.

Before the construction of a modern housing estate, the site was an area of open, damp meadow. The surviving vegetation consists of flush meadow communities on the open areas, and dense trees and waterside vegetation alongside the stream, including stands of *Filipendula ulmaria*, *Iris pseudacorus*, *Epilobium hirsutum* and *Cirsium palustre*.

Archaeological excavation of a small area of Lismore Fields was carried out between 1984 and 1987 (Garton, 1987). This revealed substantial evidence of Mesolithic and Neolithic settlement, with finds of Grimston/Lyles-Hill type pottery. A radiocarbon date of 7170 ± 80 BP (HAR-6500) was obtained from charcoal within a pit, and dates for *Triticum* grains, *Linum usitatissimum* seeds and charcoal, associated with Neolithic buildings, ranged from 5024 ± 126 BP (UB-3290) to 4680 ± 70 BP (OxA-2435) (Garton, personal communication). There was no indication of exploitation of the area by later peoples.

Previous palynological research from the area has been based on diagrams from sites in extensive areas of upland peat (e.g. Conway, 1947; Tallis, 1964a; Hicks, 1971; Tallis & Switsur, 1990). The nature of the catchment area at Lismore Fields and its apparent reflection of very local events, makes difficult any useful comparison with the other sites.

14.3 METHODS AND PRESENTATION OF RESULTS

14.3.1 Field sampling

Soligenous silty peat deposits of varying depth were found on the river terrace and were sampled for pollen analysis. Three cores were collected and the results from the deepest one, Lismore Fields 2(LF2) (grid reference SK 0510 7306), are discussed here. The core site was located 70 m south-east of the nearest excavated area and 12 m from the break of slope of the occupation area. The 1.4 m core of silty peat was recovered with a Russian sampler (Jowsey, 1966).

14.3.2 Pollen analysis

The uppermost 30 cm of deposit were very friable and were not sampled for pollen analysis. Below this, samples were taken at intervals and subjected to standard treatments (Faegri & Iversen, 1989). The pollen preparations were suspended in glycerol jelly after staining with safranine. The numbers of pollen and spore taxa counted ranged between 350 and 850 with an average of about 450 grains in each sample. Palynomorph nomenclature follows that of Moore & Webb (1978).

Gramineae grains are presented in size classes of $<40\,\mu m$, $40-49\,\mu m$, $50-59\,\mu m$ and $60-69\,\mu m$. Given the use of glycerol jelly as an embedding medium, it is likely that swelling of pollen grains occurred. While only those grains in the $>50\,\mu m$ category may include cereal pollen, the correspondence between grass pollen frequencies in the $40-49\,\mu m$ and $50-59\,\mu m$ size classes is marked, and grains in the $40\,\mu m$ and greater size classes possessed other morphological characteristics of cereal pollen (cf. Beug, 1961; Andersen, 1979; Edwards, 1989). Fragments of microscopic charcoal $>10\,\mu m$ in their longest axis were also tallied.

Two major pollen taxa, *Alnus* and Gramineae, dominate the fossil assemblages. *Alnus* was excluded from the calculation sum because of its variability through the profile and the likelihood that an alder carr dominated the immediate vicinity. Gramineae pollen could have been derived from both local and extra-local sources and, in view of its relative lack of variation through the profile, it was

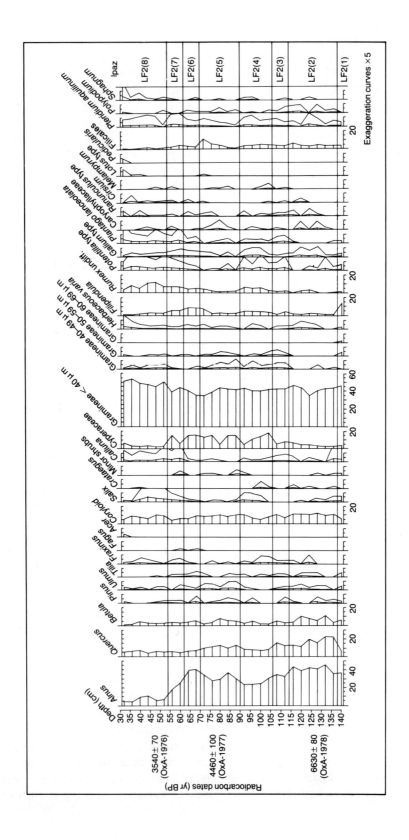

Fig. 14.2 Pollen diagram of selected taxa from Lismore Fields 2. All taxa are exressed as a percentage of total pollen and spores (excluding *Alnus*). *Alnus* is expressed as a percentage of total pollen and spores.

Methods and presentation of results

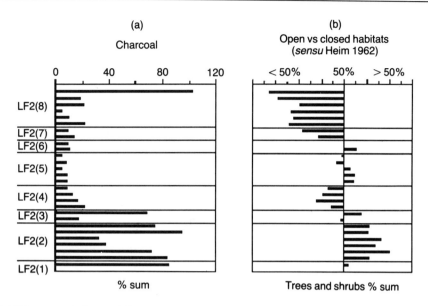

Fig. 14.3 (a) Diagram of charcoal frequency. (b) Diagram of total tree and shrub taxa in relation to 50% of total pollen and spores (*sensu* Heim, 1962).

kept as part of the sum. Since the spore curves did not exhibit unduly erratic behaviour, the non-flowering plants were considered to represent valid components of the plant communities. The sum was, therefore, based on all pollen and spores excluding *Alnus*, which itself was expressed as a percentage of total pollen and spores. Charcoal was expressed as a percentage of total pollen and spores.

The diagram of selected pollen and spores (Fig. 14.2) was produced using the programs TILIA and TILIA*GRAPH (Grimm, 1991). The zones on the pollen and related diagram (Figs 14.2, 14.3) were determined with the aid of the computer program ZONATION (Birks & Gordon, 1985). Each zone represents a local pollen assemblage zone and it is characterized by the prefix LF2 followed by a zone

Table 14.1 Summary data for Lismore Fields 2 pollen profile

Local pollen assemblage zone	Depth range (cm)	Estimated dates (BP)	Estimated duration of zone (radiocarbon years)
8	54–32	3745–2375	1370[a]
		3745–3050	695[b]
7	62–54	4000–3745	255
6	70–62	4255–4000	255
5	90–70	5045–4255	790
4	106–90	5740–5045	695
3	114–106	6090–5740	350
2	138–114	7130–6090	1040
1	140–138	7215–7130	85

[a] Using the mire surface as the upper datum
[b] Based on extrapolation through the upper two radiocarbon dates

Table 14.2 Minor taxa for Lismore Fields 2

Pollen type	32	36	40	44	48	52	56	60	64	68	72	76	80	84	88	92	96	100	104	108	112	116	120	124	128	132	136	140
Shrubs and climbers																												
Genista type								0.5																				
Hedera																+								+				+
Lonicera																+											+	
Prunus type											+				+													
Ulex type												+			0.5													
Viburnum								+																				
Dwarf shrub																												
Vaccinium			+																									
Pteridophyta																							+					
Thelypteris palustris																												
Herbs																												
Anthemis type	+													+							+							
Aster type		+	+	+	+																							
Bidens type									+																			
Centaurea nigra type	+																											
Chenopodiaceae													+											+				
Geum						+																						
Hypericum perforatum type			+								+																	
Lamium type	0.9	0.4																										
Liguliflorae						0.6				+										+					+			
Plantago media/major																+												
Polemonium																		+										
Sinapis type										+	0.5	+																
Stachys type		0.7																+			+							
Succisa			+	+					+					+														
Teucrium	+												+															
Trifolium type	0.6	+	+	0.4	0.5		+	+									+			+		+	+	+	+			
Umbelliferae	0.6	+	+				+	+												+			+					
Urtica type		+	+														0.9											
Valeriana																						+						

+ = Single grain; values = %total sum − *Alnus*

number (1–8). A summary of zone depths and estimated dates for zone boundaries is shown in Table 14.1. Minor pollen and spore taxa are presented in Table 14.2.

14.3.3 Radiocarbon dating

Three accelerator mass spectrometry (AMS) dates were obtained on 1-cm thick samples of peat by the Radiocarbon Accelerator Unit at Oxford University. Straight-line extrapolation from the lowest pair of dates was used for estimating a chronology for the section of the profile below 126.5 cm. Above 47.5 cm, the construction of a dating framework is problematical and at least two strategies are possible: either the curve may be extended to the mire surface, in which case the profile may be truncated at 2375 BP, or the depth–time curve may simply be extended from the uppermost pair of dates, which would suggest a date of 3050 BP for the surface spectrum. Both possibilities are indicated (Fig. 14.4). Within the text, estimated dates are rounded to the nearest five years.

14.3.4 Statistical analysis

The pollen data were subjected to detrended correspondence analysis using DECORANA (Hill, 1979; Hill & Gauch, 1980). Data were not transformed and there was no downweighting of rare species (cf. Turner, 1986). Results are presented in Figs 14.5 and 14.6.

14.4 VEGETATION HISTORY

14.4.1 General

The surface of the peat-covered terrace at the point of sampling is about 5 m below the level of the known Lismore Fields occupation area. Surface run-off from the soils of the area would have provided a probable source for the minerogenic inputs to the peat; there was no indication from the peat stratigraphy that fluvial activity was responsible for the silt content of the deposits. Although the pollen stratigraphy seemed to be ecologically plausible, there always remains the possibility that the microfossil content includes an allochthonous component derived from the downslope transport of material.

Dating evidence (Fig. 14.4) suggests that the profile covers at least the latter half of the Mesolithic and possibly extends to the middle Iron Age. The pollen diagram (Fig. 14.2) shows that the surrounding area supported woodland for much of its history, but consistently high values for non-arboreal pollen (NAP) indicate that at various times Lismore Fields had open ground. Figure 14.3b portrays this changing pattern, based on Heim's (1962) postulate that anything less than 50% of arboreal pollen indicates open conditions. The diagram is, of course, a gross simplification of events and must be treated cautiously.

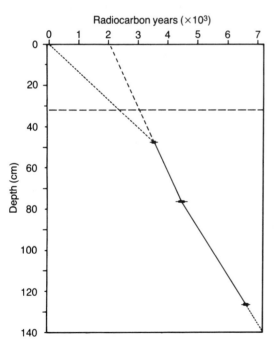

Fig. 14.4 Depth–time curve, showing alternative projections for the age of the top (32 cm depth) of the analysed core. Both 1σ and 2σ limits are shown.

A notable feature of the pollen diagram is the consistent early presence of Cereal-type pollen dating from 6000 BP. The record for microscopic charcoal (Fig. 14.3a) shows abundant frequencies before c. 5740 BP, much reduced levels after this, and substantial amounts in the uppermost sample. There does not seem to be a precise correspondence between the charcoal curve and those for any particular pollen taxon. Although it is conceivable that the high quantities of charcoal in the lower part of the profile are due to the intentional use of fire by hunter–gatherers, this is impossible to prove (Edwards, 1988).

Figure 14.5 is a DECORANA plot of the pollen taxa scores for the first two axes. Axis 1 differentiates between relatively wooded and unwooded conditions. Trees and other taxa frequently associated with woodland, woodland edge or open canopy habitats, such as *Polypodium, Filicales, Melampyrum, Filipendula* and Cyperaceae, plot out towards the left.

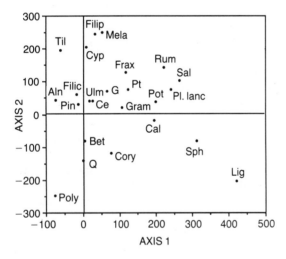

Fig. 14.5 DECORANA plots of taxa scores for axes 1 and 2. *Alnus* (Aln), *Betula* (Bet), *Calluna* (Cal), Cereal-type (Ce), Coryloid (Cory), Cyperaceae (Cyp), Filicales (Filic), *Fraxinus* (Frax), *Galium*-type (G), Gramineae (Gram), Liguliflorae (Lig), *Melampyrum* (Mela), *Pinus* (Pin), *Plantago lanceolata* (Pl. lanc), *Polypodium* (Poly), *Potentilla*-type (Pot), *Pteridium* (Pt), *Quercus* (Q), *Rumex* (Rum), *Salix* (Sal), *Sphagnum* (Sph), *Tilia* (Til), *Ulmus* (Ulm).

Taxa indicative of open habitats such as *Rumex, Plantago lanceolata* and Liguliflorae plot out towards the right. The proximity of Cereal-type to the woodland taxa might indicate that cereal growing and/or processing might have been occurring in clearings (Göransson, 1986). The diagram also indicates that *Quercus, Betula* and Coryloid (assumed to be *Corylus avellana* as hazelnut shells were found during the excavations) may have formed a mixed woodland beyond the alder carr and that *Polypodium* may have been part of that community.

Axis 2 is more difficult to interpret and seems to indicate a complex of environmental variables including hydrology and base status. Taxa associated with positive values (e.g. *Tilia, Filipendula* and *Fraxinus*) may be reflecting wetter habitats and/or those of higher base status, whilst taxa associated with negative values (e.g. *Quercus, Betula, Calluna* and Liguliflorae) may be reflecting relatively dry and/or base-poor habitats.

Figure 14.6 shows the means of sample scores for the first two axes for each pollen zone (cf. Birks & Berglund, 1979) derived from DECORANA. The ranges of scores of the samples in each zone are shown, and the zones are joined in stratigraphic order. The distance between each plot indicates the similarity between the zones in terms of pollen composition. It is immediately apparent that zones with a substantial proportion of woodland taxa (LF2(2), LF2(3), LF2(5) and LF2(6)) are similar in terms of their total pollen composition, and very different from zones LF2(4), LF2(7) and LF2(8), which are dominated by pollen taxa indicative of open habitats. It is also clear that major vegetational changes occur between zones LF2(6) and LF2(7) at c. 4000 BP.

14.4.2 Reconstruction

(a) Zone LF2(1) (c. 7215–7130 BP)

This single spectrum is dominated by *Alnus* and Gramineae pollen with a substantial con-

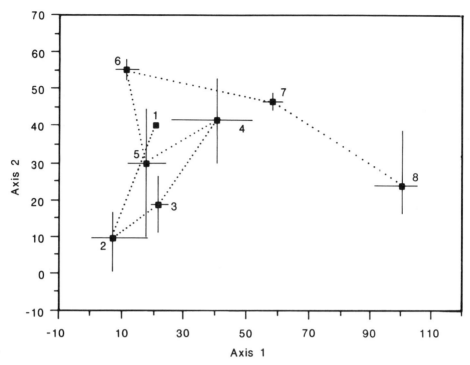

Fig. 14.6 DECORANA plots of mean values of sample scores for pollen assemblage zone LF2(1) to LF2(8) for axes 1 and 2. The range of scores within each zone for each axis is shown.

tribution from *Filipendula*. Other NAP taxa present include *Calluna*, *Rumex* undifferentiated and *Potentilla*-type with some *Plantago lanceolata*. The immediate area seems to have been dominated by alder carr but open ground also existed (cf. Fig. 14.3b). Two large grass grains were found in this zone, but no special claim is being made for their being cereal pollen. It is likely that in view of the proximity of the river, they were of grasses known to produce large pollen grains such as *Glyceria fluitans*. In so far as many of the 'early' Cereal-type grains appearing later in the profile (zone LF2(3)) are taken to represent cereals, the argument is not entirely consistent, especially considering the open-land taxa in LF2(1), which might imply clearance for agriculture. The high level of charcoal may reflect domestic burning by Mesolithic hunter–gatherers, or even attempts at browse creation (Mellars, 1976). As is often the case, it is very difficult to determine whether woodland removal or repression by fire was occurring (Edwards, 1988).

(b) Zone LF2(2) (c. 7130–6090 BP)

At the beginning of the zone there was an overall increase in woodland, involving most of the trees and shrubs. *Alnus* and *Betula* increased quite markedly and *Quercus* maintained its previous level. *Fraxinus* made its first appearance, *Ulmus* and *Tilia* were represented consistently, and *Salix* seemed to decline. Cyperaceae increased progressively and Gramineae declined but recovered towards the end of the zone. Plants indicative of open conditions such as *Rumex*, *Pteridium* and *Galium* were well represented, their maxima being associated with a decline in Gramineae and the appearance of *Fraxinus*. This mid-zone reduction in Gramineae might also indicate heavy grazing pressure from wild, or even managed, herbivores on grassland areas of Lismore Fields, allowing the spread of ruderals. The expansion in Gramineae frequencies in the latter part of the zone may be depressing the values for *Quercus* and *Betula*. The evidence points to the spread of alder carr, with the depression of woodland-edge plants such as *Filipendula* and *Salix*. The consistent representation of *Ulmus* and *Tilia*, the appearance of *Fraxinus* and the presence of open-habitat plants, also indicate that the canopy was rather open. This is supported by pollen of *Polemonium caeruleum* (Table 14.2) found at the end of the zone. This rare plant is now confined to limestone areas of Derbyshire and a few locations in northern England and is characteristic of tall herb communities, especially along river banks (Pigott, 1958).

(c) Zone LF2(3) (c. 6090–5740 BP)
This zone is characterized by a reduction in *Alnus*, the reappearance of *Rumex* and *Potentilla* and an expansion of *Calluna*. However, of greatest interest is the consistent presence, from the basal sample (6000 BP), of Cereal-type pollen in both the 40–49 µm and 50–59 µm size classes. Taken together with the expansions in *Plantago lanceolata* and *Rumex* pollen, this strongly suggests the beginning of cereal cultivation in the pollen catchment area, an activity that appears to be evident to the surface of the profile. Pollen evidence for early agriculture has been claimed for southern Germany (Küster, 1989), and is an increasingly common feature of pollen diagrams from the British Isles (Edwards, 1989). If the conjecture of cultivation is true, then the palynological data suggest cereal growing almost 1000 years before the earliest dating evidence for the Neolithic buildings at Lismore Fields.

(d) Zone LF2(4) (c. 5740–5045 BP)
Great changes appear to have occurred early in this zone (c. 5650 BP) with *Alnus*, *Betula*, *Corylus* and *Quercus* being reduced. *Fraxinus* appears to have benefited from the decline of the major woodland species as did *Salix* and *Crataegus*-type. A marked reduction in local woodland allowed shrubs to flower more prolifically and more distant woodland species to spread into the open areas, or at least disseminate their pollen more freely.

There seems to have been an unselective removal of woodland. The taller trees appear to have been removed, not only from the carr but also from extra-local woodland. Intense grazing pressure is possibly suggested by the relatively high levels of *Potentilla*-type and *Rumex*. Cereal cultivation would appear to have continued. The zone probably reflects an intensification of land use by early farmers.

The charcoal values begin a sustained decline compared to levels in previous zones. Such a pattern found in diagrams from elsewhere has, however, been typical of events around the classical elm decline of c. 5100 BP (Simmons & Innes, 1981; Bennett et al., 1990b; Edwards, 1990) and may reflect, in part, the transition from predominantly hunting–gathering to agricultural activities. No clear elm decline was discernible in the Lismore Fields profile. A fall in elm pollen is seen at 106 cm (c. 5740 BP), but this is based on only a few pollen grains.

(e) Zone LF2(5) (c. 5045–4255 BP)
Pollen values for the major woodland trees increased at the start of this zone and those of open-habitat taxa, especially *Rumex* and *Filipendula*, decreased. *Betula* and *Corylus* might have started to encroach into open areas some time before this. The woodland cover would appear to have extended. Although woodland might have been spreading, Cereal-type pollen grains were more abundant, and cereal plots probably continued to be created. Interestingly, charred remains of *Triticum* spp. and *Linum usitatissimum*, recovered from a Neolithic building at Lismore Fields, were dated to 4930 ± 70 BP (0xA-2434) and 4970 ± 70 BP (0xA-2436), respectively.

The rise in Gramineae frequencies may reflect either an increase in open grassland or a reduced grazing intensity that would have had the effect of allowing more prolific flowering of grasses. In the latter part of the zone, *Alnus* recovered and there was a reciprocal fall in *Betula* and *Quercus*. A decline in grass and cereal pollen could be due to the filtering effect of tall herb vegetation at the edge of the carr.

(f) Zone LF2(6) (c. 4255–4000 BP)
In this zone, *Alnus*, *Betula* and Gramineae percentages rise, *Quercus* and Filicales are reduced and *Filipendula*, Cyperaceae and *Rumex* values are sustained. Open-land indicators were still able to reach the pollen site. In spite of the apparent expansion of woodland taxa, the effect may be illusory and may simply be reflecting more prolific flowering of these taxa in a more open canopy (cf. Aaby, 1986).

(g) Zone LF2(7) (c. 4000–3745 BP)
This zone may be transitional between the Late Neolithic and Early Bronze Age. At c. 4000 BP there were some significant changes in the vegetation at Lismore Fields. *Alnus* declined sharply, *Betula*, *Quercus* and *Corylus* continued to fall, *Filipendula* was reduced, while, at the same time, the open-habitat weeds and *Calluna* expanded along with *Salix*. This clearly indicates extensive opening up of the alder carr and perhaps an unselective reduction of extra-local woodland. Cereal cultivation continued and the higher levels of *Calluna* and *Pteridium* from this point onward in the profile are strongly suggestive of soil acidification, perhaps as a result of long-term land use.

(h) Zone LF2(8) (c. 3745–2375 BP; or c. 3745–3050 BP)
The effects evident in the preceding zone were intensified c. 3700 BP, when *Alnus* was reduced to its lowest levels in the profile, *Filipendula* and Cyperaceae representation was considerably diminished, and *Corylus* and *Salix* expanded. Cereals continued to be well represented. It seems that the carr vegetation was being systematically removed in Bronze Age times, allowing a regional element of the pollen rain as well as the extra-local component to be represented. The increase in hazel and willow could indicate extensive use of these shrubs by coppicing, which enhances flowering. From perhaps 3225 BP, or, depending on the depth–time curve selected, possibly as late as 2375 BP, the locality seems to have been much more open with a relatively species-rich ground flora. The uppermost pollen spectrum may be indicative of conditions in the Iron Age.

14.5 DISCUSSION

For that part of the pollen diagram considered to be wholly within the Mesolithic (i.e. before 6000 BP) and to the period around traditional elm decline times, the landscape was at its most wooded, but was still relatively open when compared with sites from the nearby uplands (Conway, 1947; Tallis, 1964a; Hicks, 1971, 1972; Tallis & Switsur, 1990). The charcoal record at Lismore Fields may indicate that burning was a feature of Mesolithic and earliest Neolithic times, but it is impossible to demonstrate that human-induced fire played a direct role in the patterning of local or extra-local vegetation. There is no firm evidence of hunter–gatherer disturbance, but the mid-zone LF2(2) opening of the canopy could be of human origin. During excavation, several lithic scatters of Mesolithic aspect, but few associated features, were recovered. However, as noted earlier, charcoal from a charcoal-rich layer at the base of a pit yielded a date of 7170 ± 80 BP.

Notwithstanding the difficulties associated with the unequivocal identification of cereal pollen (O'Connell, 1987; Edwards, 1989), the circumstantial evidence from Lismore Fields is strongly supportive of cultivation from an early date. The estimated beginning of farming at 6000 BP places the site as one of the earliest, if not the earliest, yet discovered in the British Isles. The earliest cereal pollen from Moorlands, Machrie Moor, Arran, was dated to 5880 ± 70 BP (Edwards & McIntosh, 1988) and that from Cashelkeelty I in southern Ireland was dated to 5845 ± 100 BP (Lynch, 1981). Earlier 'Cereal-type' finds are known (Bush & Flenley, 1987; O'Connell, 1987; and compare with the finds of large grass pollen in zone LF2(1)).

The records of inferred cultivation persist from 6000 BP onwards, even at times when the woodland was apparently recovering from the extensive period of Neolithic clearance evident in zones LF2(3) and LF2(4). The prolonged phase of possible 'forest farming' (cf. Coles, 1976; Göransson, 1987a; Edwards, this volume) within zones LF2(4) and LF2(5) is supported by the radiocarbon dates on wheat, flax and charcoal found within the Neolithic buildings and spanning the period 5024 ± 126 BP to 4680 ± 70 BP. It is not possible to tell from a single profile whether all of the cultural evidence from the pollen diagram derives from activity in the excavated area of Lismore Fields. There is no reason why farming should not have been taking place to the south of the River Wye where any activity would also affect the pollen record at LF2. It is conceivable that woodland was fairly dense in one part of the pollen catchment area while cultivation was occurring in cleared areas elsewhere. Clearly, in such circumstances, the notion of forest farming would be somewhat spurious.

The length of time represented by most zones (Table 14.1) is considerable. In the main, the phases of woodland disturbance (LF2(3)–LF2(4), LF2(7)–LF2(8)) or farming (e.g. from 6000 BP through to the possible end of the Neolithic – a total of more than 2200 radiocarbon years) are unlikely to be reflecting episodes of sustained activity at any one place. The effect of continuous farming on soil fertility would mean that any location would probably not be exploited for more than a few years or, at most, decades. Widespread soil impoverishment at Lismore Fields was evident from c. 4000 BP. The pattern seen in the pollen diagram probably indicates the cumulative effects of many small, spatially and temporally discrete events (cf. Pilcher et al., 1971; Edwards, 1979; Buckland & Edwards, 1984).

In terms of human influences, the fortuitous or judicious selection of pollen sites in the British Isles has revealed a picture of Mesolithic hunter–gatherer impacts on vegetation and ensuing early farming activity. Most of these sites are found in areas that today are peripheral to settlement and distant from high-grade agricultural land (Pilcher & Smith, 1979; Edwards & Hirons, 1984; Simmons & Innes, 1987; Edwards & McIntosh, 1988). The site discussed in this paper has produced evidence for both possible pre-agricultural disturbance and subsequent cereal cultivation dating from 6000 BP. Palynological evidence for a Mesolithic and a Neolithic presence accords with archaeological evidence for these groups. There were no artefactual finds for the Bronze Age and Iron Age at the site, although the pollen data suggest extensive activity at those times.

Alan Smith has written of 'ecological feasibility' and 'credibility' in the interpretation of the palaeoecological record (Smith, 1970b, p. 93). He might not agree fully with our interpretations, but in the absence of a 'Time Machine' (Smith, 1981b, p. 127) we feel sure that he would sympathize with our attempts.

Acknowledgements

We thank Ms Daryl Garton for valuable information, Dr R. Scaife for core collection, and English Heritage for funding.

15

Holocene (Flandrian) vegetation change and human activity in the Carneddau area of upland mid-Wales

M.J.C. Walker

SUMMARY

1. Five pollen diagrams from the Carneddau area of northern Powys provide evidence for vegetational change in the uplands of mid-Wales throughout the Holocene (Flandrian) and for possible human impact during the last four millennia.
2. The area formerly supported an extensive mixed woodland, but successive clearance episodes in the middle to Late Bronze Age (c. 3370 BP), at the Bronze Age/Iron Age transition (c. 2500 BP) and during the Romano-British period (1800–1900 BP) reduced this previously wooded landscape to open grassland and upland heath, with only isolated stands of birch, hazel, oak and alder.
3. Pastoralism appears to have been the principal farming type throughout both the historic and prehistoric periods.
4. These data from the Carneddau region are discussed in the context of other evidence from upland Wales for human activity during the mid and late Holocene.

Key words: pollen analysis, forest clearance, human impact, late Holocene

15.1 INTRODUCTION

Over the past 20 years, large areas of upland Wales have been afforested. This has led not only to a major change in the character of the landscape, but the ploughing and planting of former open moorland poses an increasing threat to the archaeological heritage of the Principality. In 1988 and 1989, the Clwyd–Powys Archaeological Trust conducted a survey and selective excavation of around 112 ha of moorland in the Carneddau region of northern Powys, an area found to be rich in antiquities dating from the Late Neolithic/Early Bronze Age to the 19th century, and which was subject to a forestry application. The archaeological fieldwork was accompanied by an investigation of peat

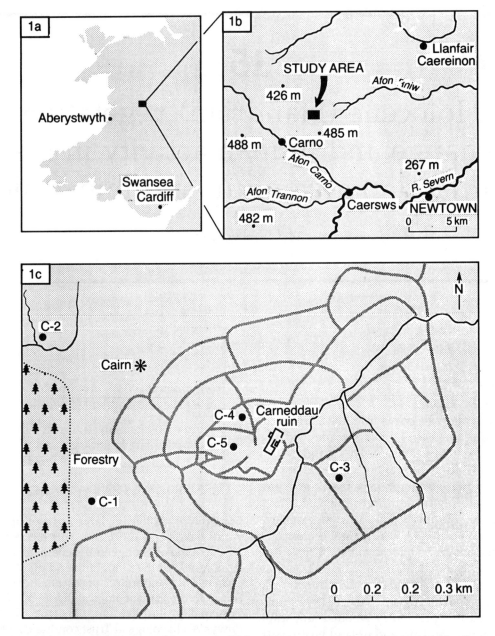

Fig. 15.1 (a) Location of the Carneddau area in Wales. (b) Location of the study area. (c) The Carneddau area showing the field boundaries, antiquities and location of the pollen sites.

profiles, the aim of which was to produce a palynological record of vegetation change and possible anthropogenic impact over the last five millennia. The results of this aspect of the research programme are discussed here.

15.2 STUDY AREA

The study area (Grid Ref. SN 99 99) is located at around 400 m OD in the rolling hills 3–4 km to the north-east of the village of Carno

(Fig. 15.1b) in the old county of Montgomeryshire (northern Powys). The area ranges from heather moorland to improved pasture and is drained to the north-east by one of the tributaries of the Afon Rhiw, which eventually joins the River Severn near Welshpool. The research area was centred on the abandoned farmstead of Carneddau, which is surrounded by large enclosures or fields that form a roughly concentric layout at the head of the valley (Fig. 15.1c). The earliest documentary evidence relating to the farmstead dates from AD 1744, although much of the present fabric of the building may be no earlier than 18th or 19th century. However, there are indications that an earlier rectilinear field system may be Medieval in age (Silvester, 1989). On the ridge to the north-west of the Carneddau ruin is a large Bronze Age cairn, excavated during the summer of 1989, which proved to have a complex sequence of phases of sepulchral and ritual activity (Gibson, 1989). Other archaeological discoveries in the area include a stone circle and several possible prehistoric cairns, one longhouse and numerous stone structures of more recent origin, post-Medieval field banks and evidence of peat cutting and drying on the high moors (Silvester, 1989). Finds to date include a flint scatter, an Early Iron Age bead, and pottery fragments and a spindle whorl of Medieval age (Silvester, personal communication). The Carneddau area, therefore, clearly contains a rich cultural palimpsest, reflecting human occupation from Bronze Age times (and maybe from even before) until the middle of the present century.

15.3 RESEARCH STRATEGY

The recognition of human impact in pollen diagrams has been the subject of a considerable amount of discussion in recent years (e.g. Edwards, 1982; Berglund, 1985; Hicks, 1985), and has been highlighted in two substantial symposium volumes (Behre, 1986; Birks et al., 1988). These studies have emphasized the difficulties that are frequently experienced in detecting and quantifying human impact in pollen data. They include, *inter alia*, problems of site location in relation to former areas of human activity (e.g. farming or woodland clearance), difficulties in establishing whether woodland disturbance was a localized or regional event, the correct identification of so-called 'anthropogenic indicators' in pollen diagrams, and problems that arise from variations in pollen dissemination and recruitment in different types of site. In addition, difficulties have been experienced in the development of a radiocarbon timescale for clearance events, either because of contamination by younger carbon residues, or because insufficient material has been available from sediment cores to allow the precise dating of critical biostratigraphic horizons.

In the present study, three imperatives guided the selection of sites for pollen analysis. First, it was considered vital to investigate a number of profiles from different depositional contexts in order to reduce the influence of local site factors in subsequent vegetational reconstruction (cf. Edwards, 1983). Hence, by selecting both infilled basin sites and blanket peat sites, problems arising from differences in pollen recruitment, pollen preservation and local pollen overproduction (by, for example, Cyperaceae) would at least be partially overcome. Second, in so far as the detection of human interference in the vegetation pattern, in particular, the identification of former agricultural activity, was a primary objective, it was important to find some pollen sites within the area of enclosed fields where both historic and prehistoric farming were assumed to have been practised. Third, it was necessary to locate sites that could be excavated to reveal open sections. In this way, sufficient material could be obtained for radiocarbon dating from narrowly defined stratigraphic horizons. Each of these initial requirements was realized in the sites selected for pollen analysis.

Table 15.1 Radiocarbon dates from Carneddau

Central depth (cm)	Lab. ref. no.	Date (BP)	Central date (bc/ad)	Estimated age for central date	Reference below
Carneddau 1					
22.5	CAR-1244	1790 ± 70	ad 160	Cal. AD 230	1
				Cal. AD 225	3
Carneddau 3					
9.0	CAR-1237	510 ± 60	ad 1440	Cal. AD 1420	1
				Cal. AD 1410	3
22.5	CAR-1238	1880 ± 70	ad 70	Cal. AD 110	1
				Cal. AD 120	3
75.0	CAR-1239	3120 ± 70	1170 bc	1410 Cal. BC	2/3
115.0	CAR-1240	4710 ± 80	2760 bc	3510 Cal. BC	3
Carneddau 5					
12.5	CAR-1241	350 ± 60	ad 1600	Cal. AD 1490	1
				Cal. AD 1510	3
				Cal. AD 1580	3
				Cal. AD 1630	3
27.5	CAR-1242	1960 ± 70	10 bc	Cal. AD 30	1
				Cal. AD 50	1
				Cal. AD 80	3
65.0	CAR-1243	3450 ± 70	1500 bc	1750 Cal. BC	2
				1760 Cal. BC	3
				1855 Cal. BC	3

1–Stuiver & Pearson (1986); 2–Pearson & Stuiver (1986); 3–Pearson et al. (1986)

15.4 METHODS

Five peat profiles were investigated (Fig. 15.1c). Three of these were basin sites (Carneddau 1, 2 and 4) from which 1 m cores were taken using a Russian peat sampler. Of the remaining sites, Carneddau 3 was a blanket peat profile, while Carneddau 5 was a shallow peat accumulation within the area of enclosed fields adjacent to the abandoned farmstead. At both of these sites, pits were excavated down to bedrock and samples taken from the exposed peat faces in monolith boxes 15 cm wide by 20 to 50 cm long.

In the laboratory, samples for pollen analysis were abstracted from the cores or monoliths at 5 cm or 10 cm intervals and treated using conventional methods for organic sediments (Moore & Webb, 1978). A sum of 300 total land pollen (TLP) was achieved at all levels. The pollen diagrams were divided into local pollen assemblage zones (LPAZs) based on fluctuations in the principal terrestrial taxa.

Samples for radiocarbon dating measuring 1.5–2.0 cm in thickness were cut from the monoliths from Carneddau 3 and 5, and also from a monolith obtained from the upper 30 cm of the Carneddau 1 profile. In each case, the levels to be dated were determined on the basis of the pollen stratigraphy. The dates are shown in Table 15.1 in conventional radiocarbon years and in calibrated years, based on the curves in Stuiver & Pearson (1986), Pearson & Stuiver (1986) and Pearson et al. (1986).

15.5 REGIONAL POLLEN ASSEMBLAGE ZONES

The LPAZs from each profile were integrated into a sequence of regional pollen assemblage zones (RPAZ; Table 15.2) as follows:

Table 15.2 Correlation between local pollen assemblage zones in the five Carneddau profiles

Regional pollen assemblage zones	Carneddau 1	Carneddau 2	Carneddau 3	Carneddau 4	Carneddau 5	Radiocarbon dates BP	Interpolated dates
	C-1h: Calluna–Gramineae				C-5f: Gramineae–Cyperaceae–Calluna		
C-6: Gramineae–Cyperaceae	C-1g: Alnus–Corylus–Gramineae–Betula	C-2f: Gramineae–Cyperaceae–Corylus	C-3e: Gramineae–Cyperaceae–Corylus	C-4e: Gramineae–Cyperaceae	C-5e: Alnus–Betula–Gramineae		
	C-1e: Gramineae–Cyperaceae–Calluna				C-5d: Gramineae–Corylus	c. 1870	
C-5: Betula–Alnus–Corylus–Gramineae	C-1e: Alnus–Quercus–Betula–Corylus–Cyperaceae	C-2e: Betula–Corylus–Alnus–Gramineae–Gramineae	C-3d: Betula–Cyperaceae–Alnus–Corylus–Gramineae	C-4d: Alnus–Gramineae–Corylus	C-5c: Betula–Corylus–Alnus		c. 2500 BP
C-4: Alnus–Corylus–Quercus	C-1d: Alnus–Corylus–Quercus–Betula	C-2d: Alnus–Corylus–Betula–Quercus–Gramineae–Cyperaceae	C-3c: Cyperaceae–Alnus–Betula–Corylus–Gramineae	C-4c: Alnus–Quercus–Corylus	C-5b: Alnus–Gramineae–Betula–Corylus		
C-3: Alnus	C-1c: Alnus–Corylus	C-2c: Alnus–Corylus–Quercus	C-3b: Alnus–Corylus–Quercus–Betula	C-4b: Alnus–Corylus–Quercus	C-5a: Alnus–Corylus–Quercus	c. 3120	
C-2: Corylus–Pinus–Quercus–Betula	C-1b: Corylus–Quercus–Betula–Pinus	C-2b: Pinus–Corylus–Quercus–Betula	C-3a?: Corylus–Pinus–Quercus–Alnus–Betula	C-4a: Pinus–Corylus			
C-1: Betula–Corylus	C-1a: Betula–Corylus	C-2a: Betula–Salix–Corylus					

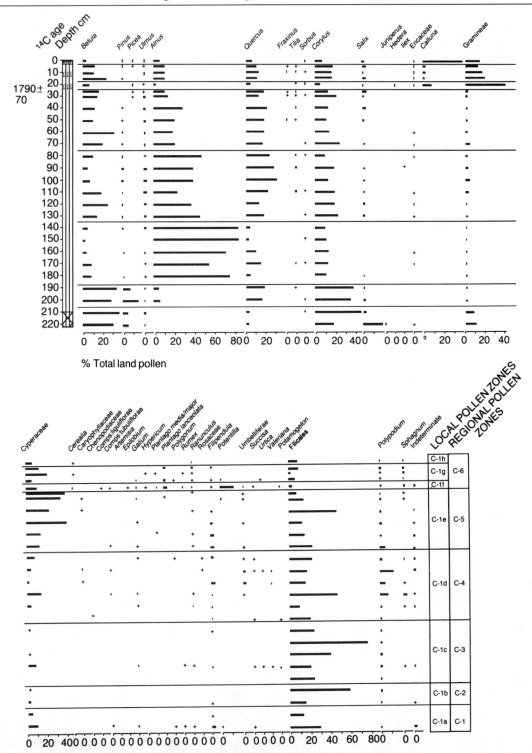

Fig. 15.2 Percentage pollen diagram from Carneddau 1 (NGR: SN 988 966).

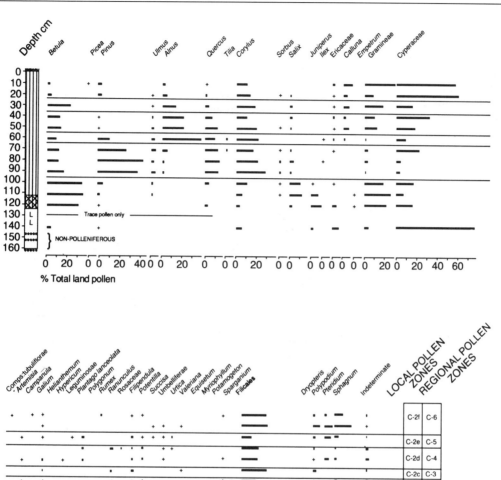

Fig. 15.3 Percentage pollen diagram from Carneddau 2 (NGR: SN 987 999).

C-6: Gramineae–Cyperaceae
C-5: *Betula–Alnus–Corylus*–Gramineae
C-4: *Alnus–Corylus–Quercus*
C-3: *Alnus*
C-2: *Corylus–Pinus–Quercus–Betula*
C-1: *Betula–Corylus*

These biozones are not recorded in every profile due to differences in timing of onset and cessation of peat accumulation and variations in rates of peat accumulation, and perhaps also to truncation of some of the profiles by peat cutting. They do, however, provide a

176 Vegetation change and human activity

basis for correlation between the sites and, in so far as they reflect regional vegetational changes, the boundaries of the zones may be regarded as broadly time-synchronous within the study area. On the RPAZ C-5/C-6 boundary, this has been confirmed by independent radiocarbon evidence. The percentage pollen diagrams for the RPAZs are shown in Figs 15.2 to 15.6.

15.6 VEGETATION AND LANDSCAPE CHANGE IN THE CARNEDDAU REGION

15.6.1 Early- and mid-Flandrian woodland development

The oldest pollen spectra are found in the Carneddau 1 and 2 profiles (RPAZ 1) and re-

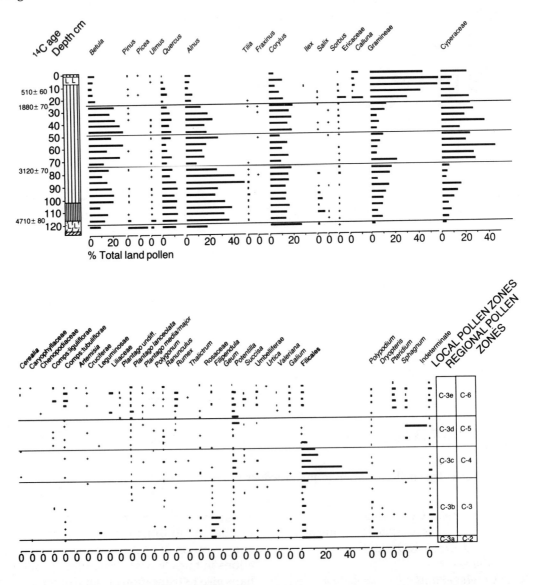

Fig. 15.4 Percentage pollen diagram from Carneddau 3 (NGR: SN 995 996).

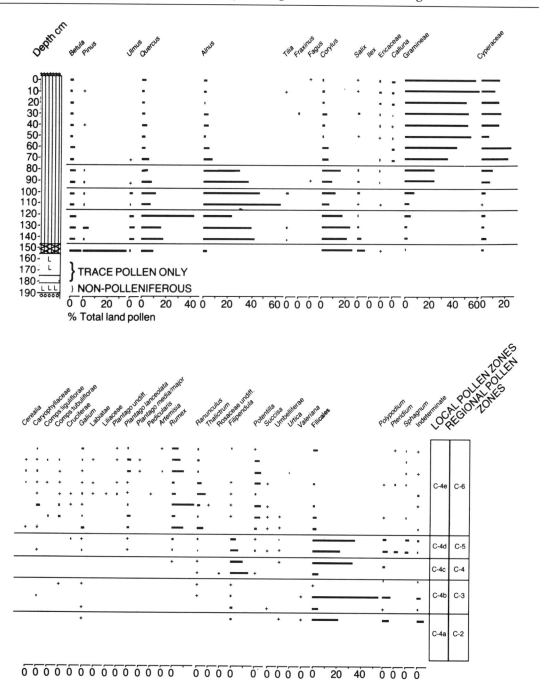

Fig. 15.5 Percentage pollen diagram from Carneddau 4 (NGR: SN 993 997).

flect a landscape of birch woodland into which hazel was rapidly expanding. Radiocarbon dates from sites elsewhere in Wales and in the Welsh borderland suggest that *Co-*

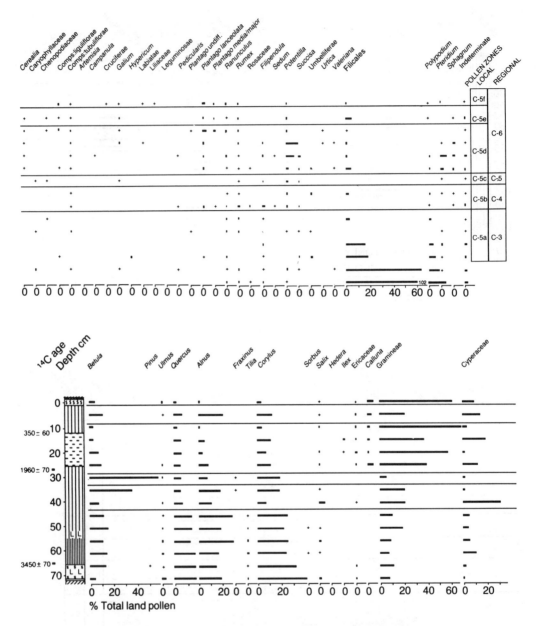

Fig. 15.6 Percentage pollen diagram from Carneddau 5 (NGR: SN 993 996).

rylus spread into west Wales around or even prior to 9500 BP, into the borderland region by *c.* 9000 BP and into the uplands of north and mid-Wales shortly thereafter (Hibbert & Switsur, 1976; Beales, 1980; Chambers, 1982a).

Isopollen mapping indicates that the main episode of hazel expansion into western Britain occurred between 9500 and 9000 BP (Huntley & Birks, 1983; Birks, 1989). Accordingly, the beginning of RPAZ C-1 in the Carneddau region may date from around 9500 BP while the age of the upper boundary is c. 9000–8800 BP.

Following the establishment of *Betula* and *Corylus*, mixed woodland taxa expanded into the Carneddau region (RPAZ 2). This episode is reflected in the Carneddau 1, 2 and 4 profiles, in which it is characterized by relatively high counts for woodland taxa including *Pinus*, *Betula*, *Quercus* and *Corylus*, lower frequencies of other trees including *Alnus* and *Ulmus*, and a significant reduction in pollen of open-habitat taxa, especially Gramineae. The evidence suggests a landscape of closed woodland in which pine was an important component. Indeed, at two of the sites, pine pollen values exceed 40% TLP during this zone. The distribution of pine in upland Wales during the Flandrian appears to have varied markedly (Walker, 1982), perhaps reflecting the influence of local controls (e.g. soil moisture levels) as opposed to regional controls (e.g. climate). At Crose Mere in Shropshire, the early-/mid-Flandrian pine episode has been dated to between 8500 and 7300 BP (Beales, 1980), while at Moel y Gerddi near Harlech on the Cardigan Bay coast, a similar pine–oak–hazel pollen assemblage is bracketed by radiocarbon dates of c. 8600 and 7500 BP (Chambers *et al.*, 1988c). If this chronology is applicable to the uplands of mid-Wales, RPAZ C-2 in the Carneddau profiles probably dates from around 8700 to shortly before 7000 BP.

A distinctive feature of all five pollen diagrams is the abundance of *Alnus* in sediments of mid-Flandrian age (RPAZ C-3). Other arboreal taxa are still represented in the pollen spectra of this biozone, but in reduced frequencies, the fall being most marked in the curve for *Pinus*. The evidence from the Carneddau 1, 2 and 4 profiles shows that alder was already present in the area at the beginning of the zone but, as in many areas of the British Isles, subsequently expanded from its previously more restricted habitats to colonize sites formerly occupied by pine, particularly on waterlogged sites such as valley bottoms (Bennett, 1984). Oak and hazel remained significant elements in the local woodland, although birch, like pine, appears to have become more localized in its distribution.

The expansion of alder during the Flandrian (Holocene) has attracted a considerable amount of attention from palaeoecologists in recent years (Chambers & Price, 1985; Chambers & Elliott, 1989; Bennett & Birks, 1990), for it seems that, unlike other trees in the post-glacial woodland, the spread of alder was patchy and erratic both in space and time. Its marked increase in abundance, which is such a distinctive feature of British post-glacial pollen diagrams, appears to reflect the gradual appearance of suitable habitats through hydroseral successions and floodplain development (Brown, 1988), along with rare weather events which produced the necessary conditions for its reproduction (Bennett & Birks, 1990). Radiocarbon dates from north and mid-Wales on the main phase of *Alnus* expansion range from c. 8500 BP in upland Ardudwy, 4 km north-east of Harlech (Chambers & Price, 1985), to c. 6800 BP in the uplands of Snowdonia (Hibbert & Switsur, 1976). In Shropshire, the *Alnus* expansion dates from around 7300 BP (Beales, 1980), while in Dyfed, dates of between 7000 and 6800 BP have been reported from both lowland and upland sites (Hibbert & Switsur, 1976; Chambers, 1982a). In the light of these age determinations, a date of at least 7000 BP for the expansion of alder in the Carneddau area might be anticipated.

In the context of these dates from other sites, the basal horizons of the Carneddau 3 profile merit further consideration. The minerogenic sediments beneath the blanket peat yielded a single pollen spectrum, dominated by *Pinus* and *Corylus* and, to a lesser extent, *Alnus*. This contrasts with the lowermost horizon in the overlying peat, which shows a significant increase in *Alnus* and an abrupt fall in *Pinus* pollen. At first sight, this appears to correspond with the pollen stratigraphic boundary between RPAZs 2 and 3. However, the date from the basal peats (4710 ± 80 BP (CAR-1240)) is considerably younger than the dates on the alder rise. Despite the erratic behaviour of *Alnus* referred to above, it is difficult to envisage expansion of alder in the Carneddau area having been delayed by some 2500 years relative to other localities in Wales and the Welsh borders. Hence, either a depositional hiatus is recorded in the basal sediments of the Carneddau 3 profile, or the lowermost pollen spectrum includes a considerable quantity of secondary pollen (especially *Pinus*) perhaps derived from soils that developed during the previous mixed woodland episode. What is clear from this profile, however, is that blanket peat began to form in the Carneddau region a considerable time after the regional expansion of *Alnus*, but somewhat earlier than the onset of peat accumulation in many other areas of upland Wales (Caseldine, 1990).

15.6.2 Mid- and late-Flandrian woodland decline: evidence of human impact

The first indications of woodland clearance occur during RPAZ C-4. In all five profiles a significant decline in *Alnus* frequencies (typically by 15–20% TLP) is recorded, which is often followed by a recovery in *Betula*. In Carneddau 3, 4 and 5, the abrupt decline in *Alnus* is accompanied by falls in *Corylus* and *Quercus*, while a gradual reduction in counts for these two arboreal genera is also apparent in Carneddau 2. In the Carneddau 1 profile, by contrast, both *Quercus* and *Corylus* increase at the RPAZ 3/4 boundary, although evidence from the other sites suggests that this may be, at least in part, a statistical artefact resulting from the unusually high counts for *Alnus* recorded in LPAZ C-1c at Carneddau 1. Overall, the evidence points to a contraction in areas of alder, oak and hazel, with birch perhaps expanding intermittently into open areas within the woodland. A radiocarbon date from Carneddau 3 gives an age of 3130 ± 70 BP (CAR-1239) for the RPAZ C-3/4 boundary.

The precise nature of this vegetational change remains to be firmly established, but there is considerable circumstantial evidence to suggest that human activity may have been a major causal factor. First, it is clear from the archaeological evidence that human groups were active in this part of the Welsh uplands during the Bronze Age. Indeed, the size of the excavated cairn and some of the other monuments, coupled with abundance of possible prehistoric structures recorded throughout the area, may be indicative of a relatively large Bronze Age population. Hence, a considerable impact on the vegetation cover, both in terms of the use of forest resources and clearance for pastoral agriculture, might reasonably be expected. The radiocarbon date from Carneddau 3 places this episode of woodland decline firmly in the middle Bronze Age, although it is apparent from the pollen record in this and, indeed, in other profiles that woodland clearance may have begun well before that time.

Secondly, although it is possible to invoke a climatic explanation for woodland decline (notably a shift to cooler and wetter conditions), the fact that the fall in arboreal pollen is most marked in the curve for *Alnus*, the one tree whose ecological affinities are most attuned to wetter substrates, would seem to falsify this hypothesis. Thirdly, there is abundant evidence for Bronze Age clearance throughout the uplands of Wales (Moore 1968; Moore & Chater, 1969; Chambers, 1982b, 1983; Wiltshire & Moore, 1983; Chambers & Price, 1988; Chambers et al., 1988; Smith & Cloutman, 1988), although not all of these pollen records are supported by radiocarbon dates. The scale of the clearances seems to have varied, however, with human impact being relatively minor in some localities (Mighall & Chambers, 1989) but more extensive in others (Chambers, 1982a).

One feature of a number of the pollen diagrams, however, is the marked decline in *Tilia* in those areas where Bronze Age human impact has been recorded (Price & Moore, 1984; Moore *et al.*, 1984), and this trend is also apparent at the RPAZ C-3/4 boundary, most notably in Carneddau 3 and 5. At Whixall Moss in Shropshire, where the fall in *Tilia* pollen has been linked to human activity, a date on the event of 3237 ± 115 BP (Turner, 1962, 1964) has been obtained, which is very similar to CAR-1239. Moreover, the clearance episode recorded in the Carneddau profiles is accompanied by a gradual increase in pollen frequently associated with pastoral activity, most notably *Plantago lanceolata*. Overall, therefore, human impact could explain the first woodland decline reflected in the Carneddau pollen diagrams, although the scale of the clearances is difficult to establish on present evidence.

Evidence for a second phase of woodland clearance in the Carneddau area may be reflected by the RPAZ C-4/5 boundary where there is a further decline in *Alnus* and *Quercus* pollen, often accompanied by an increase in Gramineae and/or Cyperaceae. *Betula* values susequently rise for a short time, probably reflecting a successional response to forest clearance. Interpolation from dated horizons in the Carneddau profiles suggests an age of around 2500 BP for this later clearance episode. Again, while it is possible to invoke a climatic explanation for this event, there is abundant evidence of Late Bronze Age/Early Iron Age human activity from a range of sites in upland Wales (e.g. Moore, 1968; Moore & Chater, 1969; Walker & Taylor, 1976; Chambers, 1982b; Moore *et al.*, 1984; Seymour, 1985; Smith & Cloutman, 1988; Mighall & Chambers, 1989), which suggests that it may be a reflection of continued clearance of the uplands that extended through into the Romano-British period (Caseldine, 1990).

Although the Bronze Age and Early Iron Age impact on the Carneddau woodland may have been considerable, it is clear from the pollen records that the most extensive episode of forest destruction occurred in the Romano-British period. This is reflected in the RPAZ C-5/6 boundary, which is marked by an abrupt decline in arboreal pollen (*Betula*, *Alnus*, *Quercus*, *Corylus* and *Ulmus*), and a significant increase in pollen of open-habitat plants (Gramineae, Cyperaceae, Ericaceae, *Calluna*, Compositae, *Potentilla*, *Galium*, *Ranunculus*, *Succisa*). Indeed, by the end of the clearance phase early in the historic period, the landscape of the Carneddau region must have closely resembled that of the present day. Radiocarbon dates on this horizon from three of the Carneddau profiles range from 1790 ± 70 BP to 1960 ± 70 BP, which provides a mean calibrated age for the clearance episode of between Cal. AD 30 and Cal. AD 230 (Table 15.1).

By comparison with earlier clearance phases, data from the Welsh uplands relating to the Romano-British period are less abundant (Caseldine, 1990). In south Wales there are indications of intensive activity in the area to the west of the Black Mountains prior to 1550 BP, i.e. during the late Romano-British period (Price & Moore, 1984). In Llangorse Lake to the north of the Brecon Beacons, changes in sedimentation rates, which are believed to reflect increased soil erosion associated with intensification of arable agriculture following Roman occupation, have been dated to c. 1790–1980 BP (Jones *et al.*, 1985). In the Carneddau profiles, however, only isolated grains of Cereal-type pollen have been recovered from deposits that accumulated during this time period, and the most likely interpretation of the pollen evidence is that clearance was associated with more extensive use of the uplands for pastoral farming, perhaps because of increasing population pressure in lowland areas.

The uppermost horizons in the Carneddau profiles contain a record through the Dark Age, Medieval and post-Medieval periods, although the stratigraphic resolution is such that only generalized comments are possible about patterns of landscape change. Nevertheless, glimpses are afforded of land use in this area, particularly over the course of the last millennium. In the Carneddau 5 section (Fig. 15.6), a distinct burnt horizon (which contains abundant charcoal) is bracketed by dates of 1960 ± 70 BP and 350 ± 60 BP. This site is located in the enclosed fields less than 100 m from the abandoned farm buildings, and the charcoal-rich horizon may therefore reflect burning of former moorland as part of a landscape management strategy similar to that employed in heathland areas at the present day (Gimingham, 1975). Some Cereal-type pollen grains were recovered from this profile, although their very low representation in all the pollen diagrams suggests only limited cultivation. The radiocarbon evidence indicates that this episode of agricultural activity occurred some time prior to the 15th century AD, in all probability during the medieval period.

In the upper levels of the Carneddau 3 section (Fig. 15.4), a distinct minerogenic horizon occurs within what are otherwise uniform blanket peats. The peats immediately beneath this horizon yielded a date of 510 ± 70 BP (Cal. AD 1410–1420), which indicates a post-Medieval age for the silt-clay deposits. The presence of minerogenic horizons in blanket peats has been noted in a number of upland areas (Wiltshire & Moore, 1983; Simmons & Innes, 1988), and has often been related to human activity (Edwards *et al.*, 1991). Hence, the silt-clay horizon in the upper levels of the Carneddau 3 profile could reflect further clearance of wood and scrub during the Medieval period that might have led to increasing instability and runoff upslope from the blanket peat, a process that would, perhaps, have been accelerated under the deteriorating climatic conditions of the Little Ice Age (Grove, 1988). However, evidence of peat cutting and the presence of ditches suggests that the minerogenic deposits might also be related to attempts to drain the peat surface, although the date of this activity cannot be established on present evidence.

At similar levels within the Carneddau 1 and 5 profiles, there are indications of short-lived expansions of wood and scrub in the post-Medieval period (C-1g and C-5e, respectively) that could reflect abandonment of the area during the Little Ice Age and the subsequent localized expansion of *Betula*, *Alnus*, *Corylus* and *Quercus* onto land that had previously been cleared for agricultural purposes. With the re-establishment of upland farming from the 18th century onwards, this brief woodland resurgence came to an end and the present landscape of open grass and heather moorland was finally established, as is clearly reflected in the uppermost levels of all the pollen diagrams.

15.7 CONCLUSIONS

Pollen-stratigraphic evidence from the Carneddau area suggests the following sequence of landscape/vegetational changes during the course of the last nine millennia. Colonization of the area by tree birch began some time shortly after the amelioration of climate at the beginning of the Flandrian, and this was followed by expansion of hazel into the birch forests between 9500 and 9000 BP. Immigration of mixed woodland species from 9000 BP onwards led to the establishment of an extensive forest cover of pine, oak, birch, elm, and hazel. This phase may have lasted for up to 1500 years. From around 7000 BP, alder spread from its previous habitats in the region to form the dominant element in the woodland cover, and this was followed by the gradual disappearance of other tree types, notably pine. Blanket peats began to form in the area during the Neolithic period, but the first indications of human impact on the vegetation cover date from the Bronze Age. The earliest clearance episode occurred around 3350 BP with a later clearance phase at *c.* 2500 BP. The most widespread clearance episode, however, dates from the Romano-British period between Cal. AD 30 and 230.

Subsequently, there is evidence for Medieval farming activity based largely on pastoralism, although there may have been some limited cultivation of cereals. Farm abandonment during the post-Medieval period led to a short-lived re-expansion of shrub and tree cover, but the contemporary cultural landscape of open grassland and heather moor dates from the 18th century AD.

In the wider context, there are implications in this environmental reconstruction for future research strategies relating to late-Flandrian vegetational change and associated studies of human impact. Previous investigations of this nature have frequently relied on a single pollen diagram to provide information on both local and regional vegetational developments. However, given the marked spatial and temporal variation in late-Flandrian vegetational change, particularly where human activity is involved, it is questionable whether a single profile can always provide a sufficiently detailed or reliable record. Not only are there problems relating to the sedimentary history of individual sites (episodes of non-deposition, poor stratigraphic resolution due to slow rates of sediment accumulation, human interference through peat cutting, etc.), but variations in pollen recruitment to sites of different character may skew the palynological record, while errors in radiocarbon dates may go undetected because of a lack of other local chronological evidence.

In the present study the history of landscape changes has been based on a synthesis of data from five separate pollen profiles located within a geographical area of less than 2 km^2. The sites, moreover, are of markedly different character, ranging from soligenous and topogenous mires to blanket peats, and hence errors arising from individual site histories should, in large measure, have been eliminated. 'Events' recorded in the five profiles have been correlated initially on biostratigraphic evidence, and subsequently underpinned by radiocarbon dating. Thus the radiocarbon chronology has been set against the relative timescale provided by the pollen record, while individual age determinations have been checked against others from common biostratigraphic horizons. In this way a coherent sequence of vegetational change and a history of human impact has been established within a relatively secure temporal framework.

It is suggested that this type of multi-site approach may well hold the key to deciphering the pattern of prehistoric and historic landscape change in the uplands, especially where the aim is to establish a detailed environmental context for archaeological investigations within relatively restricted geographical areas.

Acknowledgements

The research was funded by CADW (Welsh Historic Monuments) through the Clwyd-Powys Archaeological Trust, and the production of pollen diagrams was aided by a grant from the PantyFedwen Fund, St David's University College, Lampeter. I thank Bob Silvester and Alex Gibson for archaeological advice; Quentin Dresser for the radiocarbon dates; Astrid Caseldine, Ian Clewes, John Crowther, Malcolm Grant and Dave Weston for field and laboratory assistance; Astrid Caseldine, Alex Gibson and Bob Silvester for valuable comments on earlier drafts.

16

Early land use and vegetation history at Derryinver Hill, Renvyle Peninsula, Co. Galway, Ireland

Karen Molloy and Michael O'Connell

SUMMARY

1. Results are presented of palaeoecological investigations at Derryinver Hill, an important archaeological site in the Renvyle Peninsula, Connemara, western Ireland.
2. Four short radiocarbon-dated pollen diagrams have enabled land-use history on the Hill during late prehistory and the early historical period to be reconstructed in considerable detail.
3. It is shown that pre-bog walls on the Hill – the first to be dated in Connemara – were laid out at 2400 BP, but it was not until 1700 BP that peat accumulation began locally.
4. The evidence indicates that, though grassy heath conditions existed prior to the stone wall construction, peat initiation was retarded through land-use management. When management ceased, the grassy heathland gave way to blanket bog.
5. These findings are discussed in the light of results from a long radiocarbon-dated peat pollen profile from the base of Derryinver Hill and of other palaeoecological investigations in north-western Connemara.

Key words: pollen analysis, early human impact, blanket bog, western Ireland

16.1 INTRODUCTION

Palaeoecological investigations were undertaken to reconstruct the history of vegetation and human activity at Derryinver Hill (grid ref. L695 615), which lies at the eastern side of Tully Mountain in the Renvyle Peninsula (Fig. 16.1). Archaeologically, the Renvyle Peninsula is one of the most important early prehistoric areas of Connemara. It has megalithic tombs of the court, portal and wedge types (Fig. 16.1), the first two types datable to the Early Neolithic and the third to the later Neolithic or possibly the Early Bronze Age (ApSimon, 1986). Two hilltop cairns, probably Neolithic or Bronze Age, are recorded on Tully Mountain. The most conspicuous prehistoric monument is a stone alignment on the flat summit of Derryinver Hill (Fig. 16.2), known locally as The Tulach (Robinson, 1990). The Hill has several other less con-

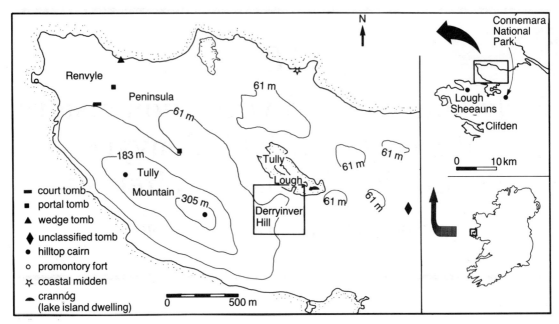

Fig. 16.1 Map of Renvyle Peninsula, north-western Connemara showing the more important archaeological field monuments, and the 61 m (200 ft), 183 m (400 ft) and 305 m (1000 ft) contours. For details of the Derryinver Hill area (within box), including location of sites where palaeoecological investigations have been carried out, see Fig. 16.2.

spicuous, but important, archaeological features that, combined with suitable material for palaeoecological investigations, namely peat and peat-covered mineral soils, made it an appropriate site for studies to complement those already carried out on lake sediments at Lough Sheeauns, 6.5 km to the south-west (Molloy, 1989; Molloy & O'Connell, 1991).

Derryinver Hill is considered to be a large drumlin deposited in the last (Midlandian) glaciation (W.P. Warren, personal communication). Peat, which in places is only a few centimetres thick presumably due largely to peat cutting, serves partially to conceal the archaeological monuments other than the six erect stones of the alignment and the three large recumbent stones that lie 35 m to the south-west of the alignment (Fig. 16.2). The latter may be the poorly preserved remains of a megalithic tomb (see Gibbons & Higgins, 1988) or, as suggested by Ó Nualláin (1988), a stone circle. Ó Nualláin (1984, 1988) favours, on archaeological grounds, a Late Neolithic/Early Bronze Age date for stone circles and alignments whilst a mid to Late Bronze Age date is put forward by Lynch (1981) for these monument types.

Associated with the alignment, in so far as it runs through it, is a pre-bog stone wall of substantial proportions that continues down either side of the Hill (W1 in Fig. 16.2). The relationship of wall and alignment at the point of intersection could not be determined without excavation. It is assumed that the stone wall post-dates the alignment. On the north-western side of the Hill another pre-bog wall delineates a wide arc (W2 in Fig. 16.2). There is no evidence that these walls meet nor that either formed an enclosure.

To the north-east of the alignment and towards the eastern edge of the flat summit, is a low-lying circular feature about 36 m in diameter and almost hidden by peat.

It consists of a substantial outer earthen bank with the occasional stone protruding through the thin peat cover and, immediately inside the bank, a relatively deep and uninterrupted fosse filled with peat that, at most points, attains a thickness of 50 to 70 cm (Figs 16.2 and 16.3). Stratigraphical investigations revealed a dark charcoal-rich basal peat overlain by a lighter coloured fibrous peat (Fig. 16.3). A wide gap in the bank appears to be a post-construction feature and is, most probably, the result of attempts at drainage in relatively recent times. The peat in the fosse at the gap has a fluid consistency and appears to lack the stratigraphical features seen elsewhere. A feature that may be a drainage channel passing through the gap can be seen in a 1973 Ordnance Survey aerial photograph. These observations support the idea that the gap is the result of recent disturbance. Since the monument does not appear to have an entrance, it is referred to as an embanked enclosure with fosse. Such monuments are usually regarded as dating to later prehistory

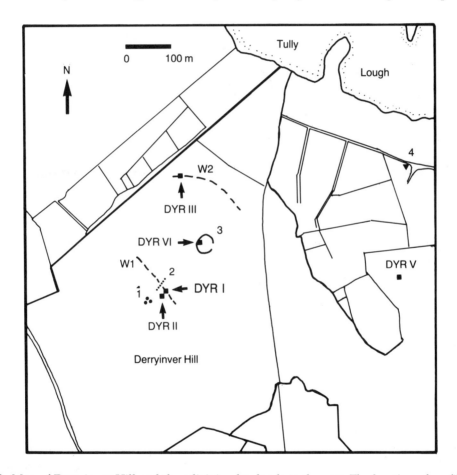

Fig. 16.2 Map of Derryinver Hill and the adjoining lowlands to the east. The location of profiles DYR I, DYR II, DYR III and DYR VI on the Hill, and profile DYR V from the bog at the base of the Hill are shown. Archaeological features are as follows: (1) megalith of three recumbent stones; (2) stone alignment; (3) embanked enclosure with gap; (4) cist burial; (W1): pre-bog stone wall at alignment; and (W2) curved pre-bog stone wall. Field boundaries are as in the Ordnance Survey 6-inch sheet 9, Co. Galway.

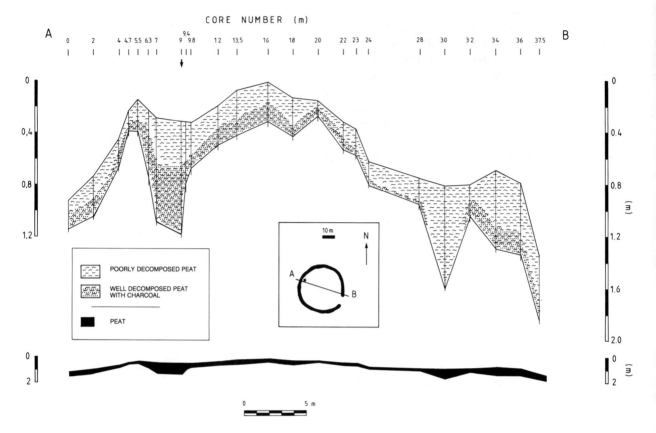

Fig. 16.3 Stratigraphy of the embanked enclosure ((3) in Fig. 16.2) along the transect A–B (see inset). The peat, which now largely conceals the circular mound and its inner fosse, is underlain by drift. In the upper sketch, the vertical axis is exaggerated by a factor of ten; the lower sketch is without vertical exaggeration. The core DYR VI was taken immediately to the north of the transect at a point corresponding to the position marked by an arrow on the transect. The coring position is indicated by a filled-in square in the inset.

(O'Kelly, 1989; cf. Smith *et al.*, 1981). The core DYR VI was removed from the fosse where 86 cm of peat had accumulated (Fig. 16.3).

Sandwiched between the peat-covered Hill and an extensive basin bog to the east are fields that, through careful management, support fertile pasture. Here a cist burial, dating probably to the Bronze Age, is recorded (Fig. 16.2). At the margin of the basin bog that adjoins the pasture, a small area of uncutover peat provided a 5.2 m core (DYR V) that spans the Holocene.

16.2 SITES INVESTIGATED

In addition to the long core (DYR V) taken from a basin bog, four profiles from the Hill have been investigated (Fig. 16.2). Details are as follows:

DYR I: Monolith of mineral soil from beneath the pre-bog stone wall that runs across the line of the alignment. The sampling point lay 18.6 m to the south-east of the alignment on the sloping hillside.

DYR II: Monolith of mineral soil and overlying peat removed from sloping ground, 7.5 m to the south of DYR I where a hag of peat 44 cm thick survived peat cutting

DYR III: Monolith of mineral soil from beneath the curved wall that runs more or less diagonally to the contours of the hillside at the sampling point. The sampling point lies c. 250 m to the north of the alignment and 30 m from the present-day wall that divides the bog-covered hillside from the fields at a lower elevation to the north-west.

DYR VI: A 90 cm long, 5 cm diameter core from the ditch in the embanked enclosure (Figs 16.2 and 16.3). This core included 4.5 cm of the underlying mineral soil and the overlying peat.

16.3 METHODS

Pollen sample preparation methods are as in Molloy & O'Connell (1991). Samples from DYR VI and the uppermost metre of DYR V were treated differently in that after KOH treatment, 0.1 mm as against 1.9 mm mesh sieves were used to remove debris. This should be borne in mind when interpreting the charcoal analyses. Conventions in naming taxa are as in Molloy & O'Connell (1991). *Corylus* and *Myrica* are differentiated, so that the *Corylus* curves are considered to represent *Corylus avellana* exclusively. The material retained in the sievings after KOH treatment was examined for macrofossils and charcoal.

In the short profiles, levels are designated with respect to the mineral soil surface, negative values being used to indicate distance above the mineral/peat interface. Local pollen assemblage zones (PAZs) are recognized on the basis of changes mainly in the percentage curves but also taking into account the concentration curves.

16.4 RESULTS AND INTERPRETATION: SHORT PROFILES ON DERRYINVER HILL

16.4.1 Description of soil profiles

The mineral soils consist of highly podsolized silty loams but showed no evidence of iron-pan formation. The mineral soils at DYR I and DYR II were similar. An organic and charcoal-rich A_0 horizon was followed by a poorly defined A_2 horizon and a well defined B_2 horizon (Fig. 16.4). Peat, 44 cm thick, and having a diffuse boundary with the mineral soil beneath, overlay the mineral soil at DYR II. The basal 11 cm was dark with charcoal; above this a greasy brown peat was recorded. The present-day rooting zone extended to -22 cm.

At DYR III, a 0.5 cm thick layer of charcoal (A_0) underlay the stone wall. An A_2 horizon, 6.5 cm thick and with a low organic content, followed. The B_2 horizon, with a higher organic content and freckled with charcoal especially in the upper 3 cm, extended to 30 cm where it was bounded by the C horizon. Mottling, indicative of gleying, was evident in the lowermost 11 cm of the B_2 and also in the C horizon. There appeared to be a positive lynchet present on the uphill side of the stone wall and a negative lynchet on the downhill side but, without excavation, it was not possible to establish this with certainty.

In DYR VI the basal mineral layer sampled (4.5 cm) consisted of a dark brown organic-rich clayey fine silt (Fig. 16.6). The basal peat was dark with charcoal and well decomposed. A small stone was recorded at -7 cm. Above -51.5 cm the peat was progressively more fibrous and charcoal appeared to be confined to occasional bands. A fibrous rooting zone extended from -90 cm (the surface) to -87.5 cm.

16.4.2 Pollen and macrofossil analyses

The evidence for vegetational and land-use change is first presented (Figs 16.4, 16.5 and

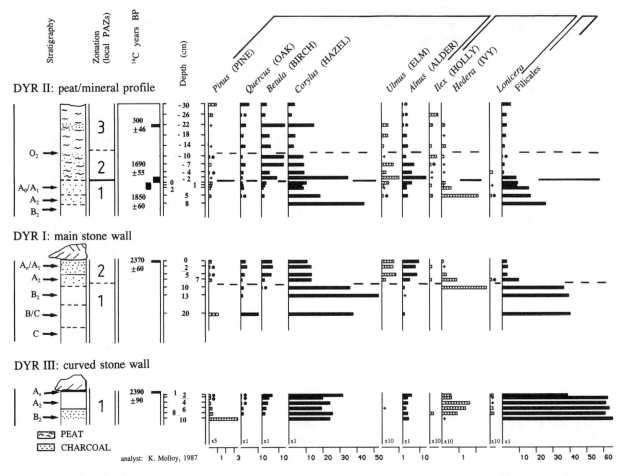

Fig. 16.4 Percentage pollen diagrams showing the main pollen curves for profiles DYR I, DYR II and DYR III from Derryinver Hill. The horizontal scale values are given at the base of each curve and the magnifications employed, e.g. ×10, are also indicated to highlight the different scales used. A closed circle is used to indicate low representation not readily legible because of the horizontal scale employed; '+' is used to indicate presence outside the count. Groups of taxa shown at the top of the diagram are as follows: (1) trees; (2) tall shrubs; (3) ferns; (4) herbs (pasture); (5) herbs (disturbed biotope). Curves for the bog/heath taxa are presented in Fig. 16.5.

16.6) and then the chronology attaching to the various profiles is considered.

(a) DYR I: profile from mineral soil beneath stone wall at alignment
Pollen was preserved at deeper levels in the mineral soil in this profile than elsewhere. In the lowermost spectrum (20 cm), where differential pollen decay has probably taken place, the high *Corylus* (38%) and substantial *Quercus* representation suggest a hazel-dominated scrub with the possible presence of oak. At 13 cm, hazel scrub with a substantial fern component is represented. At the top of DYR I-1, the rise of Gramineae and *P. lanceolata*, the increased representation of taxa such as *Ranunculus* and *Cerastium*-type, and the decline of *Corylus* point to opening of the hazel scrub as pasture expands.

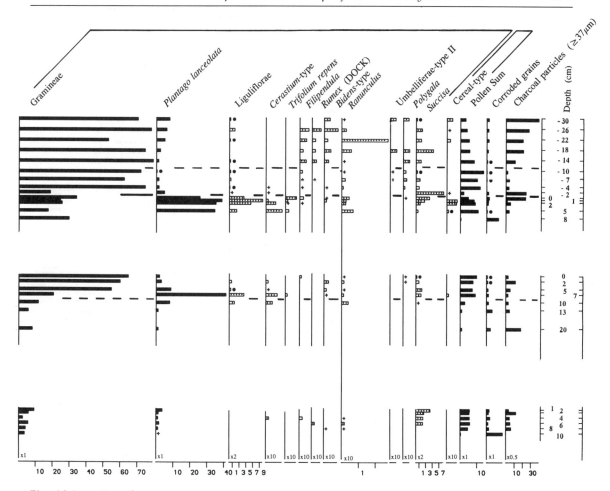

Fig. 16.4 continued

In DYR I-2, which includes the A soil horizons, the transition from scrub to heathland is recorded. At the base of the PAZ, *P. lanceolata* attains exceedingly high values (41%). Such high values are often a feature of old ground surfaces (cf. Pilcher & Smith, 1979). Liguliflorae and *Cerastium*-type also peak and *Corylus* declines to 13%. Scrub is replaced by pasture; Cereal-type pollen is recorded indicating that there may also have been arable farming. In the uppermost three spectra, the Gramineae curve expands to reach 65% and the herbaceous curves, and especially *P. lanceolata*, fall. On the other hand, bog/heath taxa expand. These changes suggest the development of heath. The high Gramineae values probably arise largely from locally present bog/heath grasses such as *Molinia caerulea* and *Nardus stricta*.

(b) *DYR III: profile from mineral soil beneath the curved wall*
This profile contrasts sharply with that from beneath the wall at the alignment, DYR I. At 10 cm, severe corrosion is evident. This has undoubtedly altered the pollen composition and may partly explain the high *Pinus* values in the basal sample. Throughout the profile, ferns (*Polypodium* and Filicales without perine) and *Corylus* dominate, suggesting that at the time of and prior to wall construction, hazel scrub with a rich fern layer was present locally. The low but continuous record for *P. lanceolata*, *Succisa* and *Calluna* indic-

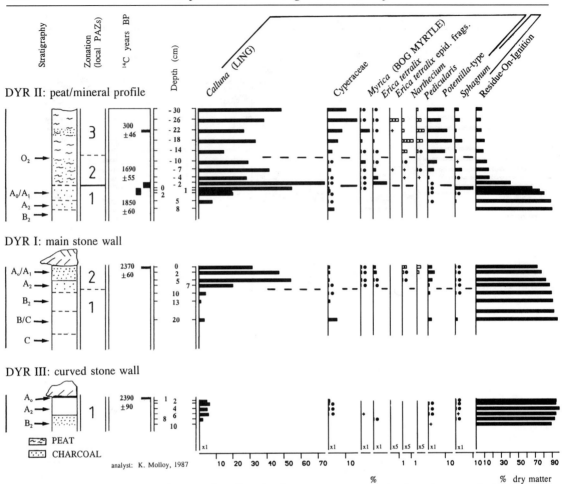

Fig. 16.5 Diagrams showing the main bog/heath percentage curves and selected concentration curves for profiles DYR I, DYR II and DYR III from Derryinver Hill. The conventions followed are as in Fig. 16.4. In the Gramineae and *Calluna* curves, the concentration values divided by 10 are shown by open histograms. Residue-on-ignition (ash content) is shown as a percentage of dry matter.

ates that grassland and heath were present, probably in the vicinity (see DYR I), prior to wall construction. In contrast with DYR I, hazel scrub persisted here until or shortly before wall construction. The layer of charcoal, 0.5 cm thick, present immediately beneath the wall, suggests a significant burning event immediately prior to wall construction.

(c) DYR II: soil/peat profile from near the alignment

Although the basal spectrum (8 cm) is most probably affected by corrosion (see concentration diagram), the high representation of *Corylus* (44%) suggests hazel-dominated scrub. The next four spectra (5 to 0 cm) record the clearance of hazel in the context of a farming phase that included both pastoral and arable elements. The latter may be responsible for possible lynchets noted at DYR III (see above). The exceptionally high levels of *P. lanceolata* and Liguliflorae (maximum values of 39% and 10%) and the substantial representation of *Ranunculus*, *Trifolium repens* and *Cerastium*-type indicate the presence of grasslands with a high species diversity at the

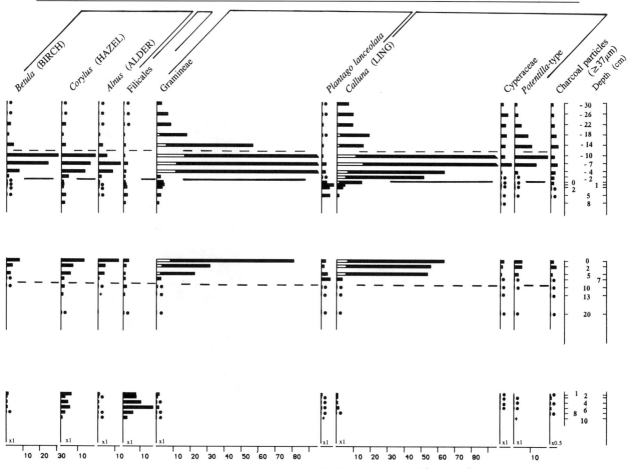

Fig. 16.5 continued

1 horizontal scale unit = 10⁴ grains cm⁻³

sampling site. The rise in *Calluna* and *Succisa* suggests progressive acidification, culminating in the local establishment of *Sphagnum*.

At the base of the peat (DYR II-2, −2 cm), *Corylus* and also *Alnus*, *Betula* and *Ulmus* expand. In the three spectra above this, the *Corylus* percentage curve has fallen substantially while Gramineae achieves values as high as 75%. The concentration diagram (Fig. 16.5) suggests that the depression in the AP percentage curves is largely an artefact of the percentage method of calculation.

The changes recorded in DYR II-2 suggest that hazel, and to a lesser extent birch and alder, expanded in response to a decline in farming activity. Simultaneously, peat accumulation began at the sampling site. Not only the tall shrubs and trees but also *Calluna* and *Myrica* appear to have responded positively to decreased grazing and fire frequency (see charcoal curve). The Gramineae curve probably consists largely of *Molinia* and *Nardus*. As edaphic conditions deteriorated locally, scrub continued to be important in the wider area, though birch now assumed greater importance than hazel.

In DYR II-3, NAP taxa increase in percentage representation and in diversity as woody taxa decline, providing evidence for renewed farming activity. In the uppermost spectra, cereal-type pollen and the secondary rise in *Pinus* are recorded; the latter suggests that recent centuries are represented (see below).

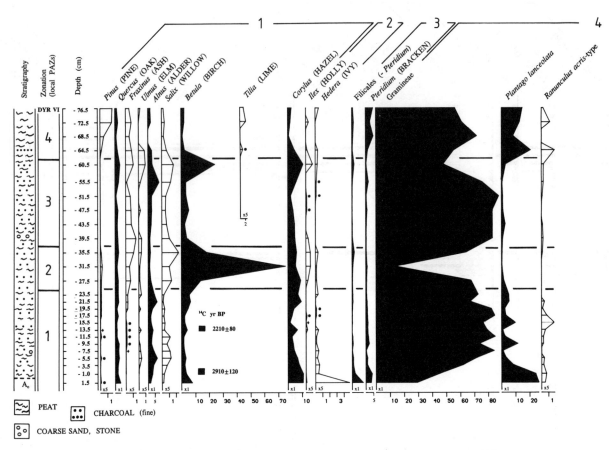

Fig. 16.6 Percentage pollen diagram for profile DYR VI from within the enclosure ((3) in Fig. 16.2; see also Fig. 16.3). Conventions followed are as in Fig. 16.4. Groups of taxa shown at the top of the diagram are as follows: (1) trees; (2) tall shrubs; (3) ferns; (4) herbs (pasture); (5) herbs (disturbed biotype); (6) bog/heath taxa.

(d) DYR VI: profile from embanked enclosure
The percentage pollen diagram only is presented (Fig. 16.6) as the percentage and concentration curves show a similar pattern. Overall pollen concentration is, however, greatly reduced in the uppermost PAZ.

A feature of the diagram is the exceptionally high Gramineae representation with values of between 60% and 80% in many spectra. The low representation of tall canopy trees, and especially of *Pinus*, *Quercus* and *Alnus*, is also noteworthy. Though the percentage AP values are probably depressed by the high Gramineae representation (but see DYR V, Fig. 16.7), the evidence strongly suggests that tall canopy trees were not present on the hilltop in the period spanned by the profile. The substantial representation of *Calluna*, Cyperaceae, *Potentilla*-type (presumably *P. erecta*) and also *Erica tetralix* and *Narthecium*, indicates that the peat is formed from blanket bog vegetation. This vegetation no doubt also included *Molinia*, which is probably mainly responsible for the high Gramineae values.

There are clear indications in the diagram of fluctuations in the level of human activity. In DYR VI-1, a suite of pollen taxa that are indicative of pasture, including *P. lanceolata*,

Results and interpretation: short profiles on Derryinver Hill

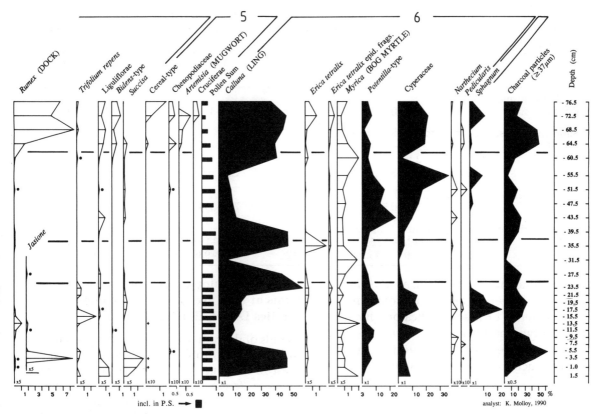

Fig. 16.6 continued

Ranunculus acris-type, Liguliflorae, *Succisa* and *Ophioglossum*, are consistently represented. The occasional Cereal-type pollen suggests some arable farming. In the upper half of the zone (see especially −11.5 to −17.5 cm), high values of *R. acris*-type, *Rumex* and *Trifolium repens* indicate substantial pastoral farming on the hilltop.

In DYR VI-2, the expansion of *Betula* and the decrease in Gramineae indicate a strong local expansion of birch. Holly (*Ilex*) also regenerated. Since there is no wood layer in the peat within the enclosure (see Fig. 16.3), this must represent scrub regeneration in the locality. The decline in Cyperaceae values suggests that the bog surface within the enclosure may have become drier.

The expansion of scrub is seen as a consequence of a decline in farming activity in the locality. While the representation of microscopic charcoal particles is low in samples −35.5 and −31.5 cm, substantial amounts of macroscopic charcoal (≥ 100 μm) were noted in the sievings from the pollen samples. This suggests that local fires (which presumably produced the macroscopic charcoal) persisted, but that firing in the region declined as a consequence of decreased human activity on the peninsula as a whole (see section 16.5).

The decline in *Betula* at the base of DYR VI-3 is presumably caused by human activity. A small piece of grit recorded during preparation of the sample from −39.5 cm (base of DYR VI-3), points to disturbance at the embanked enclosure, presumably in the form of grazing. Farming activity had once again extended onto the hilltop, which was now peat covered (see below).

The uppermost zone, DYR VI-4, is characterized by exceptionally high representation

Table 16.1 Radiocarbon dates from Derryinver, short profiles. Peat samples (as distinct from organic-rich mineral samples) are indicated by negative values in the depth column; these values show distance above the peat/mineral soil interface. The age given in the final column refers to the mid point of the 1σ range.

Profile	Lab. no. (KI–)	Depth (cm)	Age (BP)	Age (1σ range) (Cal. BC/AD)	Age (mid point) (Cal. BC/AD)
Wall at alignment					
DYR I	2748	0 to 0.5	2370 ± 60	730–390 Cal. BC	570 Cal. BC
Short mineral soil/peat profile DYR II from near DYR I					
DYR II	2749.01	–22 to –21	300 ± 46	Cal. AD 1514–1650	Cal. AD 1580
DYR II	2749.02	–2 to 0	1690 ± 55	Cal. AD 258–410	Cal. AD 330
DYR II	2749.03	0 to 3	1850 ± 60	Cal. AD 90–226	Cal. AD 160
Curved wall near base of Hill (W2)					
DYR III	2750	0 to 0.5	2390 ± 90	758–394 Cal. BC	580 Cal. BC
Fosse within embanked enclosure					
DYR VI	3012.02	–14 to –12.5	2210 ± 80	378–198 Cal. BC	288 Cal. BC
DYR VI	3012.01	–2.0 to –0.7	2910 ± 120	1302–938 Cal. BC	1120 Cal. BC

of pastoral indicators such as *P. lanceolata*, *Rumex*, *Bidens*-type and Liguliflorae and also the presence of arable indicators, including Cereal-type pollen and *Artemisia*. These features are interpreted as reflecting substantially increased farming activity on the peninsula generally (see below; also McCormack, 1991).

An indicator of local environmental change at, and in the vicinity of, the embanked enclosure is provided by the *Juncus* seed record from the sievings. In DYR VI-1, *Juncus* seed (all *J. effusus/conglomeratus*, except in –22.5 cm where *J. bufonius* was also recorded) is consistently present in considerable quantities. In samples above –25.5 cm (this pollen sample not counted), *Juncus* is not recorded. *Juncus* (rushes) may have grown in or at the edge of the bog within the embanked enclosure or in the pastures on the hilltop. In DYR VI-2 and DYR VI-3, pastoral indicators are poorly represented in the pollen record. This, and the absence of *Juncus* seed, would suggest that from the base of DYR VI-2 onwards, peat was widespread on the Hill (see below). The pastoral indicators in DYR VI-3 are interpreted as reflecting farming activity elsewhere on the peninsula rather than on the Hill itself.

16.4.3 Chronology, and significance in terms of land use, of events recorded in profiles DYR I, II, III and VI

Details of the radiocarbon dates are presented in Table 16.1. Profile DYR VI goes back furthest in time, i.e. to about 3000 BP. This provides a *terminus ante quem* for the construction of the embanked enclosure.

The radiocarbon dating evidence from both DYR I and DYR III indicates that the two walls sectioned were laid out at the same time, i.e. shortly after 2400 BP. On the hilltop near the alignment (DYR I), a heathy vegetation dominated by grasses and ling was present at the time of wall construction. At the north-west side of the Hill (DYR III), hazel scrub dominated until it was apparently cleared by fire immediately prior to wall construction. Hazel scrub was also present in earlier times at DYR I but had been cleared for pasture that eventually gave way to heathy grassland.

In DYR VI, the period of wall construction is represented by the spectra in the middle of DYR VI-1. As already pointed out, these spectra suggest considerable pastoral-based farming activity. This activity appears to have been sustained over several centuries (peat accumulation rate in this part of the profile is estimated to be as low as 50 yr/cm^{-1}).

Results and interpretation: short profiles on Derryinver Hill

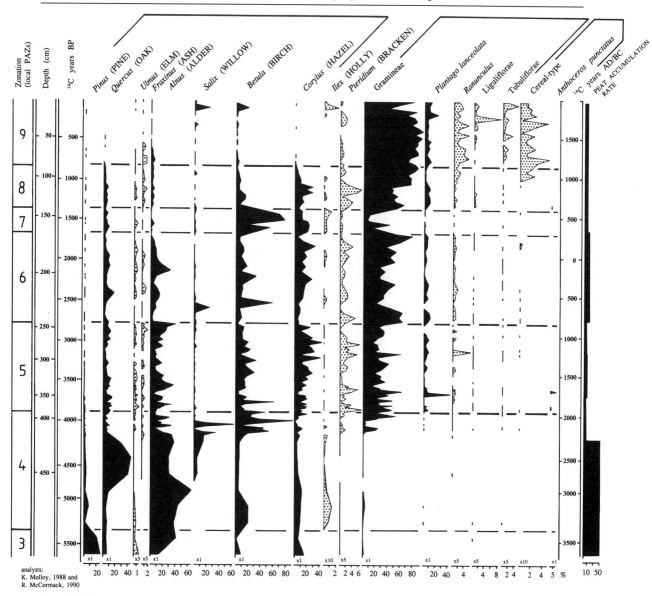

Fig. 16.7 Percentage pollen diagram showing the main curves in DYR V from the basin bog. The lowermost part of the profile has been omitted because of the poor resolution it provides as a result of a slow peat accumulation rate. The pollen sum is as in the other percentage diagrams, i.e. bog taxa are excluded. The timescale in radiocarbon years (BP), which is used instead of depth as the vertical axis, is based on eight radiocarbon dates. The estimated peat accumulation rates are also shown.

In DYR II, peat initiation is dated to *c.* 1700 BP, some 700 radiocarbon years after wall construction. The length of time intervening is unexpected, particularly since heathy conditions were already present here at 2400 BP (see DYR I-2). The management regime associated with the intensive farming that was pursued in the post-wall construction period (cf. *P. lanceolata* and Cereal-type curves in DYR II-1; also upper part of DYR VI-1) appears to have retarded peat development. The evidence from DYR VI also suggests that peat initiation in the vicinity of, as distinct from in, the fosse also began at about the same time (see above).

Attaching a chronology to the mid and upper part of profile DYR VI is difficult without radiocarbon dates. The expansion of *Betula* recorded in DYR VI-2 is almost certainly the same event as that recorded in DYR V-7 (see below and Fig. 16.7). On the basis of radiocarbon dates from near the upper and lower boundaries of the latter PAZ, this event spans the period 1650 to 1350 BP. Indeed, the same event appears also to be recorded in DYR II-2, the base of which is dated to 1690 ± 55 BP. The evidence therefore strongly suggests that there was a decline in farming in the period Cal. AD 420 to 680 in the peninsula as a whole (see section 16.5). At the local level, the cessation of land-use management appears to have led to initiation of peat accumulation on the hilltop.

The upper spectra of profiles DYR II and DYR VI (note: the uppermost 14 cm have not been investigated in these profiles) are similar in that both record the secondary rise of *Pinus* and increased representation of indicators of human activity. There is uncertainty, especially at DYR II, as to whether peat has been removed in the past. Indeed, it appears likely that there is a hiatus in DYR II, caused by either peat cutting or cessation of peat accumulation between DYR II-2 and DYR II-3 (cf. concentration diagram). A hiatus in profile DYR VI is less likely. The rise of *Pinus* in DYR VI-4 suggests that the last two centuries at least are represented here. In the uppermost spectrum of the preceding PAZ, there is a rise in *Betula* and other AP curves. This is similar to the event recorded at −22 cm in DYR II and dated to 300 ± 46 BP, i.e. Cal. AD 1500 to 1650. This fits reasonably well with the dating suggested by the rise in *Pinus*.

16.5 DISCUSSION

The vegetation and land-use history described above are now placed in a broader temporal and spatial context by reference to the long pollen profile from the basin bog (DYR V; Fig. 16.7) and to palaeoecological investigations in the Connemara National Park and at Lough Sheeauns.

Rather surprisingly, in view of the archaeological evidence for an early Neolithic presence on the Renvyle Peninsula (Fig. 16.1), an Early Neolithic landnam such as at Lough Sheeauns (Molloy & O'Connell, 1991) is not recorded. The profile, however, appears to reflect mainly local vegetational changes so that clearance episodes, which may be taking place near the head of the peninsula, are not necessarily recorded.

A major change in woodland composition is recorded in the lower part of DYR V-4 with *Quercus* at first expanding and then declining. Further investigations in the basin peat have shown that substantial soil erosion occurred at, and subsequent to, the decline in *Quercus* (McCormack, 1991). The NAP curves, however, do not respond, which suggests that farming was not important. These changes date to shortly before 4000 BP, when the alignment may have been erected on the hilltop. The prominent position of the alignment suggests that this may be a ritual site and not necessarily connected with local farming and settlement. The scarcity in the pollen record of indicators of human activity relating to this period would support this hypothesis.

A local expansion of pastoral farming, which included an arable component, is recorded at *c.* 3800 BP (base of DYR V-5). About 200 radiocarbon years later, AP representation drops and disturbance indicators, especially *P. lanceolata* and *Anthoceros punctatus*, attain exceptionally high representation. These changes suggest particularly intensive farming activity centred probably nearby on the sheltered eastern side of Derryinver Hill.

Above this level, substantial values of *P. lanceolata* continue to be recorded until near the top of DYR V-5. Here *P. lanceolata* values fall and the curve is interrupted in one spectrum, indicating a short lull in farming activity. The construction of the embanked enclosure took place prior to this (see above).

The evidence reviewed above points to considerable human activity on Derryinver Hill during most of the period 4000–3000 BP. During the same period, human activity at Lough Sheeauns appears to be relatively weak while, from the Connemara National Park, there is strong evidence for farming activity especially towards the end of the millennium when widespread peat initiation began in response to woodland clearance and sustained burning of the vegetation (O'Connell, 1990a).

At the base of DYR V-6, a major phase of clearance involving mainly *Alnus* and *Corylus* and also other tall canopy trees is recorded. A similar clearance episode signalled by a decline in *Alnus* and a rise in *P. lanceolata* is also recorded at Lough Sheeauns at this time. These changes suggest that, towards the end of the Bronze Age, the wetter areas were cleared for the first time to make way for pasture. This appears to have occurred over a considerable part of north-west Connemara.

While indicators of human activity are generally high throughout DYR V-6, there are a number of fluctuations with corresponding changes in the AP curves. The time of wall construction on the Hill, i.e. 2400 BP, may be represented in DYR V at 212 cm. In this spectrum, *Quercus* falls and *P. lanceolata* values are high.

It is of interest to compare the course of the AP curves in this part of the long profile (DYR V-6) with DYR VI-1, the corresponding PAZ in the profile from the enclosure. Though these PAZs certainly correspond in time, the AP representation is exceptionally low in DYR VI and there is no hint of the fluctuations seen in the long profile. It must be concluded that DYR VI is reflecting the local events on the hilltop and that vegetational developments taking place near the base of the Hill to the east, i.e. downward of the prevailing wind, are not recorded.

A notable feature of the long profile is the regeneration phase indicated by the expansion of the AP curves, and especially *Betula*, that is recorded in DYR V-7. As indicated above, this is also reflected in DYR VI-2 and in DYR II-2. The available dates indicate that this regeneration spans the interval 1650 to 1350 BP, i.e. it extends into the period normally associated with the onset of renewed agricultural activity thought to be connected with the spread of Christianity (cf. Mitchell, 1965, 1986). At Lough Sheeauns, this horizon is much earlier, but here the relevant radiocarbon date may be affected by the inwash of old organic material (Molloy & O'Connell, 1991). In the Connemara National Park (FRK II), on the other hand, increased farming activity following a distinct lull is dated to *c.* 1385 BP (O'Connell *et al.*, 1988), which is again late (*c.* Cal. AD 653) though perhaps not as delayed as at Derryinver. It remains to be established if this important palynological horizon is indeed synchronous within Connemara but late with respect to other parts of Ireland.

Acknowledgements

M. Gibbons brought the site to our attention and provided information on the local archaeology. Field assistance was given by P. O'Rafferty, N. Lockhart, L. Van Doorslaer, W.H. Zimmermann and the B.Sc. Botany class of 1988. P. Gosling made available the results of the OPW/UCG Archaeological Survey of Co. Galway, and he and J. Waddell gave helpful comments on the draft manuscript. Access to the site was provided by the Nee and McDonagh families.

Part Four

Climatic Change and Human Impact:
Relationship and Interaction

Introduction

'Was it (a)? Was it (b)? Or Both? Or Neither?' (wording on four successive transparencies shown by A.G. Smith, IBG Conference, Lancaster, 1981).

Discriminating between effects caused by climatic change and those caused by human impact has been a recurrent theme in palaeoecology. Travellers and landscape observers have long speculated on the balance between human impact and the effects of climate on landscape: the theme clearly pre-dates the work of Iversen (1949, 1973) by centuries! Human impact may modify the effects of climatic change, exacerbate or speed the effects, mimic the effects, or delay the effects. There are clear lessons here for those speculating on current climatic trends (CCIRG, 1991).

First (Chapter 17), Dr Brian Huntley reviews the apparently rapid early-Holocene migration and high abundance of hazel (*Corylus avellana*) in north-west Europe, and considers seven hypotheses to account for these phenomena. Dr Peter D. Moore (Chapter 18) then reviews the origin of the ombrotrophic mires that blanket large areas of the landscape in the British Isles and Norway, but are also found elsewhere in the world (cf. Fig. 21.3) where human impact has been less pronounced. Recent and current work on isolating human impact and on refining the relationship between peat stratigraphy and climatic change in the Anglo-Scottish Borders is presented in Chapter 19 by Dr Keith E. Barber, Lisa Dumayne and Rob Stoneman. Dr Barry G. Warner then considers (Chapter 20) the origin of floating or quaking mires in the Great Lakes region of North America, and assesses the impact of European settlers on the hydrology and vegetation of such mires.

Finally, in Chapter 21, the editor emphasizes the fluidity of views on climatic change and human impact on landscape, puts Holocene environmental change in its late-Quaternary context and reconsiders the themes raised in Parts One to Four. The significance of studies in palaeoecology and environmental archaeology is emphasized with reference to continuing nature conservation and future environmental management. Chapter 21 also acts as a guide to other chapters in the volume.

17

Rapid early-Holocene migration and high abundance of hazel (*Corylus avellana* L.): alternative hypotheses

Brian Huntley

SUMMARY

1. The rapidity of the early-Holocene range expansion of *Corylus* and the relative abundance of this taxon during the early Holocene are striking features of north-west European pollen diagrams. Several hypotheses have been put forward to account for one or both of these phenomena.
2. Seven general classes of hypothesis are discussed and the evidence relating to each briefly reviewed.
3. It is concluded that the unique character of the early-Holocene palaeoenvironment can probably account for the earlier expansion of *Corylus*, as well as for its relative abundance at this time.
4. The apparent rapidity of the expansion may, however, in part be an artefact of 'plateaux' in the radiocarbon timescale at *c.* 10 000 and 9500 BP.

Key words: succession, migrational lag, glacial refuges, palaeoclimate, *Corylus*

17.1 INTRODUCTION

Smith (1970b) discussed one of the most striking features of Holocene pollen diagrams from north-west Europe, namely the extremely rapid and early increase to high levels of pollen values for *Corylus* (hazel). He drew attention to the apparent speed with which *C. avellana* expanded its geographical range across Europe during the first millennia of the Holocene and to the unusually high pollen values reached at that time. Although several hypotheses were outlined, he suggested that there was 'no reason... for discarding the possibility that the expansion of hazel may be connected in some way with human activity' (Smith, 1970b, p. 83). He also highlighted the possible role of humans in bringing about the high abundance of hazel during the early Holocene through the deliberate management of the landscape using fire.

The last 22 years have seen a considerable increase in the volume of data relating to the early Holocene in Europe. There have been parallel advances in our knowledge of the Holocene vegetation of eastern North America and of many other regions of the world, as well as of the vegetation of earlier interglacials. We have also gained new insight into the mechanisms and underlying forcing factors of Quaternary environmental changes. These in turn have led to independent assessments of Quaternary palaeoclimates. It therefore seems timely to re-evaluate the various hypotheses, including those discussed by Smith (1970b). The most frequently suggested hypotheses can be placed into six general categories. An additional seventh hypothesis can also be proposed in the light of recent evidence relating to 'plateaux' in the radiocarbon timescale. These seven general hypotheses can be labelled as

1. succession and soil development;
2. migrational lag;
3. the position of 'glacial refugia';
4. late-glacial expansion;
5. human assistance;
6. climate;
7. plateaux in the radiocarbon timescale.

Each of these hypotheses is elaborated below. An assessment is then presented of their likely relative importance in the light of present knowledge and understanding of early Holocene palaeoenvironments and palaeoecology.

17.2 THE HYPOTHESES

17.2.1 Succession and soil development

It is tempting to draw an analogy between the progression of Holocene forest development in northern Europe and the process of succession. Such an analogy implies that taxa such as *Corylus*, that expanded rapidly and reached high abundance in the early Holocene, were behaving as early successional species. Implicit in such an analogy also is the role of progressive autogenic soil development that characterizes primary succession.

The successional status of *Corylus* relative to *Ulmus*, *Quercus*, etc. can be assessed using observations of the development of stands of secondary woodland (Tansley, 1939) and of the recovery of forest stands following clear-felling or other disruption (e.g. storm damage). In such circumstances, *Corylus* is frequently obvious and abundant during the earlier stages of woody succession. However, these are secondary successions and the later canopy-dominant trees are generally present as saplings alongside the *Corylus*. Furthermore, the duration of the phase of peak *Corylus* abundance during such secondary successions is apparently only a few decades. The behaviour of *Corylus* as an early successional species in primary successions is much more difficult to observe, although it can be seen, for example, in woodland development in abandoned quarries. Here again the overall duration of the succession to woodland is generally only a century or so, and canopy trees are generally present as saplings during any phase of vigorous scrub development in which *Corylus* is prominent.

Unidirectional soil development is an integral part of the 'interglacial cycle' concept (Iversen, 1958; Birks, 1986). This model proposes that soils are in an immature state at the beginning of interglacial stages. Progressive maturation leads to an optimally fertile state, with mull humus development, in the middle of interglacials. Thereafter, leaching leads to declining fertility and eventual transformation to acidified podsols with mor humus. In the extreme, deterioration culminates in ombrogenous peat development during the later parts of interglacials. If this pedogenic sequence is to account for the rapid expansion of *Corylus*, then evidence is required that trees such as *Ulmus*, *Quercus*, etc. are less tolerant of immature, base-rich soils. *Corylus* is found today on a wide range of soils. Woodlands on base-rich soils of pH > 7.0, developed over limestone, typically have *Corylus* growing along with *Ulmus* and *Fraxinus*, whereas in forest communities on relatively acid Brown Earth soils of pH < 4.5 developed over sandstones and other more porous rocks, *Corylus* is accompanied by *Ulmus* and *Quercus* (Rodwell, 1991). The pure *Corylus* woods found locally in western Ireland and Scotland reflect extreme wind exposure and not edaphic conditions.

Furthermore, the 'interglacial cycle' model is not universally applicable. This model parallels observed post-glacial soil development in glaciated regions of north-west Europe, especially in areas of oceanic climate. Glacial activity 'rejuvenates' soils in these regions because glacial deposits have high levels of available cations (Ca^{2+}, Mg^{2+}, K^+, etc.) and a high pH (Crocker & Major, 1955; Viereck, 1966; Birks, 1980). Pedogenesis is re-initiated on the deglaciated landscape at the beginning of each interglacial. However, much of Europe, including the region between the Alps and the North Sea, was not directly influenced by last glacial ice-sheets. Loess, being derived primarily from fluvioglacial outwash, is also base rich, and areas of loess deposition show renewed pedogenesis following deglaciation. Elsewhere, periglacial activity may have sufficiently mixed superficial materials to 'rejuvenate' soils, although acidified soils with peat development characterize permafrost development. South of the Alps, however, the *terra rossa* soils of the Mediterranean region are widely considered to have a development history extending beyond the last glacial and no significant renewal of cation supply during glacial periods is likely.

Thus, immature, base-rich soils were not ubiquitous throughout Europe at the time of deglaciation. Furthermore, pedogenesis began during the late-glacial interstadial in many areas (Pennington, 1986). Evidence of pedogenic rates from newly deglaciated areas (Crocker & Major, 1955; Viereck, 1966; Birks, 1980) indicates that acidified, raw humus soils can develop in less than 1000 years. Thus, areas deglaciated by 12 000 BP should have had 'mature' soils in equilibrium with the prevailing environment by 10 000 BP. The extent of renewed soil 'rejuvenation' during the late-glacial stadial is unclear. However, the corollary to the argument presented by Pennington (1986), to account for apparently delayed expansion of *Betula* into the British Isles during the late-glacial interstadial, is that the relatively rapid Holocene expansion of *Betula* indicates that soil development did not begin again from immature substrates following the late-glacial stadial. It is hence difficult to sustain a hypothesis that *Corylus* was favoured by immature soils; furthermore, such a hypothesis is inconsistent with the later expansion of *Corylus* in previous interglacials (West, 1980).

17.2.2 Migrational lag

It has frequently been suggested that the order of appearance of 'thermophilous' taxa during the Holocene was a consequence of differential migrational lag (see e.g. Davis, 1983; Birks, 1986). It is presumed that environmental conditions were suitable for the growth of all such taxa from at least the time of appearance of the first 'thermophile'. The capacity of different taxa to track their environmental limits, however, depends upon both their longevity and their dispersal mechanisms. Thus, long-lived organisms with limited dispersal capacity, and especially long-lived sessile organisms with large propagules (e.g. nut-bearing trees), should lag behind shorter-lived organisms with better dispersal capacity and/or smaller propagules.

Evidence that beetles characteristic of 'warm' conditions reached the British Isles much earlier than any trees during the late-glacial interstadial (see e.g. Coope & Brophy, 1972; Coope & Joachim, 1980) is paralleled by evidence for the rapid early-Holocene expansion of warmth-demanding aquatic plants

(Iversen, 1954; Huntley & Birks, 1983), often to localities beyond their recent range limits. It has been proposed that, because of their short life-cycles, and capacity for flight in the case of beetles (Coope, 1975, 1977), and because of their supposed ready dispersal by migratory waterfowl in the case of aquatic plants (Sauer, 1988; Ridley, 1930; Vlaming & Proctor, 1968), these organisms have a greater capacity to track environmental changes closely than have longer-lived organisms such as trees (but see Huntley, 1991).

Several lines of evidence suggest that migrational lag cannot account for the sequential arrival of woody taxa during the Holocene (see Bradshaw, this volume). The order of first appearance of taxa at a given locality differs between different interglacials (see e.g. West, 1968, 1980). This can only be consistent with a hypothesis of migrational lag if taxa had markedly different areas of distribution during different glacial periods. Although such differences of locality of 'glacial refugia' have been hypothesized (West, 1968, 1980), they have yet to be reliably demonstrated using independent evidence. Furthermore, the likelihood of *major* differences in relative geographical distribution of taxa during different glacial periods is small. The principal climatic gradients within Europe during different glacial periods are likely to have been broadly similar because they are determined by the geographical setting of Europe in relation both to the North Atlantic and to the principal regions of ice-sheet development (Kutzbach & Guetter, 1986; COHMAP, 1988) as well as by topography within Europe.

As a result, the extent of the *overall* area of glacial distribution of woody taxa is likely always to have been small, offering little scope for differences in relative proximity to north-west Europe for taxa with glacial distributions that include western and central parts of southern Europe. In addition, although it has been hypothesized that taxa may have shown progressive evolutionary changes in ecological preferences during the Pleistocene (West, 1968), evidence from pollen–climate response surfaces (Bartlein et al., 1986) indicates that taxa have not shown adaptive changes in their environmental tolerances during this period (Huntley et al., 1989). The response of species to Quaternary environmental changes has been to migrate so as to maintain the position in environmental space to which they are adapted (Good, 1931; Huntley & Webb, 1989; Huntley et al., 1989).

Prentice *et al.* (1991) have provided a critical test of a hypothesis of migrational lag to account for late expansion of *Carya* into regions west of the Appalachians during the Holocene. They have demonstrated that such a late migration can be simulated using palaeoclimate reconstructions made from a limited range of independent taxa. Although such tests have not yet been made for other late migrating taxa, it is nonetheless clear that the climate change c. 10 000 BP did not result in conditions equally suitable for immediate expansion of all so-called 'thermophilous' taxa. Furthermore, it is potentially misleading to group taxa together arbitrarily in this way, considering their climatic tolerances to be sufficiently similar that their response to a climate change will be almost identical. The construction of pollen–climate response surfaces (Bartlein et al., 1986) for all major European pollen taxa (B. Huntley, P.J. Bartlein & I.C. Prentice, unpublished data; Huntley, 1990a, in press) has shown that no two taxa have identical climate responses. Thus no two taxa can be expected to show the same response to any given climate change (Webb, 1986).

In the case of *Corylus*, it is difficult to explain its more rapid expansion than, for example, *Ulmus* or *Quercus* in terms either of its age at reproductive maturity or of its relative dispersal capacity. Although *Corylus* reaches reproductive maturity more quickly than *Quercus*, *Ulmus* also reaches reproductive maturity quickly. Furthermore, *Corylus* produces a nut of similar dimensions to the acorns produced by *Q. robur* and *Q. petraea*, the north European *Quercus* spp.; *Ulmus*, in contrast, produces a winged fruit, adapted for wind dispersal. The most probable agents of long-distance dispersal of hazel nuts (apart from humans, as discussed below) are birds, especially the jay (*Garrulus glandarius* L.), and flowing water (Huntley & Birks, 1983). It is unlikely that either of these would be selective between hazel nuts and acorns. Johnson & Webb (1989) showed that the average dispersal distance for fruits of various eastern North American Fagaceae carried by the blue jay (*Cyanocitta cristata* L.) did not differ between taxa and concluded that differential dispersal distances by jays cannot account for any supposed migrational lag. The northward flow of many major European rivers may account for the observation of higher maximum migration rates of trees in Europe than in eastern North America (Firbas, 1949; Huntley & Birks, 1983), but waterborne dispersal could only be selective if acorns and hazel nuts differ in either their buoyancy or their waterlogging tolerance.

17.2.3 The position of 'glacial refugia'

This hypothesis is related closely to the foregoing, but emphasizes the geographical distribution of *Corylus* and other 'thermophilous' taxa *c.* 10 000 BP rather than their relative migration rates. It proposes that the geographical whereabouts of 'glacial refugia' determine relative arrival times of taxa at a given locality during an interglacial (West, 1968). Thus it requires that *Corylus* had a more northern last-glacial distribution than *Quercus*, *Ulmus* etc. However, the whole of the present land area of northern Europe, as well as much of lowland southern Europe, was apparently devoid of tree cover during the last glacial maximum (Huntley & Birks, 1983; Huntley, 1988, 1990b). The various 'thermophilous' woody taxa all appear to have had restricted glacial distributions in southern Europe (cf. Deacon, 1974), allowing insufficient scope for the geographical separation necessary to enable *Corylus* to reach north-west Europe more quickly following deglaciation. Even if *Corylus* had a glacial distribution that included the margins of the Bay of Biscay (Huntley & Birks, 1983), this could not explain the early and rapid migration of *Corylus* because both *Quercus* and *Ulmus* are also listed by Huntley & Birks (1983) as present in this area during the last glacial.

17.2.4 Late-glacial expansion

Another related hypothesis is that late-glacial interstadial conditions allowed *Corylus*, but not other 'thermophilous' woody taxa, to migrate northwards. *Corylus* pollen recorded from late-glacial interstadial sediments at a number of localities in northern Europe (Huntley & Birks, 1983) provides support for this proposal. However, in at least some of these localities the *Corylus* pollen, along with that of other 'thermophilous' taxa also present, probably results from re-working and thus does not indicate contemporary presence. In addition, at some localities a combination of stratigraphic misinterpretation and/or dating problems may have resulted in mistaken assignment of early Holocene sediments, containing *Corylus* pollen, to the late-glacial interstadial. The conclusion reached by Godwin (1975, p. 271) that 'The suggestion of local presence [during the late Weichselian] in favourable situations, especially in the north and west [of the British Isles], must be strong' is open to considerable doubt (Birks, 1989). It is noteworthy that whereas macrofossil remains of hazel nuts are known from the early Holocene, there does not appear to be any record of *Corylus* macrofossils from the late-glacial interstadial in northern Europe.

Furthermore, even if *Corylus* were sparsely present in northern Europe during the late-glacial interstadial, this could only contribute to its rapid early-Holocene expansion if some of these northern populations survived the subsequent late-glacial stadial. Although it is probably impossible to *prove* the case either way, it is unlikely that such a relatively thermophilous and oceanic species would survive significantly further north, at this time of renewed glaciation and intense periglacial activity, than it had done during the preceding glacial period. The evidence of pingos and ice-wedge pseudomorphs formed during the late-glacial stadial in Ireland (Mitchell, 1976) and elsewhere in southern Britain indicates permafrost conditions and mean annual temperatures below 0°C and probably below -6°C (Lowe & Walker, 1984). *Corylus*, however, is not found today in areas with mean annual temperatures lower than *c*. 2.5°C. Furthermore, its present northern and northeastward range limits lie only a little north of those for *Quercus robur* and somewhat south of those for *Ulmus* (Jalas & Suominen, 1976). The northern limits of all three taxa extend furthest north along the coast of Norway where the moderating influence of the Norwegian Sea, and especially of the North Atlantic Drift, brings relief from extremes of winter temperature experienced at similar latitudes further east. Thus some northward spread of these taxa during the late-glacial interstadial and perhaps even their local survival during the late-glacial stadial along the Atlantic margins of north-western Europe are possible. They almost certainly, however, could not have been present extensively in northern Europe prior to 10 000 BP (Birks, 1989; cf. Deacon, 1974).

17.2.5 Human assistance

The widespread introduction of familiar, as well as valued or useful, species of both plants and animals has frequently accompanied human settlement of new geographical areas. It has been hypothesized that migrating humans introduced *Corylus* to northern Europe during the early Holocene (Danielsen, 1970; Smith, 1970). It is apparent from archaeological finds of hazel nuts that they were probably used as a food source by early-Holocene Mesolithic immigrants into northern Europe. Thus these migrating humans may have accidentally introduced *Corylus* as a result of carrying supplies of nuts that were then not fully utilized. Alternatively, they may have deliberately carried hazel nuts with them in their migrations so as to plant them. However, these people are viewed as hunter–gatherers who probably moved seasonally (Piggott, 1954) so as to exploit natural products including fruits and nuts produced by trees. Given evidence of the behaviour of contemporary hunter–gatherers, it seems unlikely that Mesolithic people would have deliberately planted nuts of species that would take many years to reach maturity and produce a crop. They are more likely to have collected hazel nuts opportunistically, along with other natural produce (Iversen, 1973).

The possibility that such hunter–gatherers may have accidentally aided the dispersal and northward migration of *Corylus* is also at variance with the likely behaviour of such people. Humans, in common with many other species, spread northwards through Europe following deglaciation. As they did so their numbers probably increased. This is true of many organisms that similarly had relatively limited south European distributions during the glacial period. Although as a result of their increasing numbers they may have exploited new resources as they expanded northwards, the increasing abundance at this time of many familiar food sources probably rendered this unnecessary (Iversen, 1973). It is more likely that both the expanding range of humans and their increasing numbers were conditioned primarily by the increasing extent of available habitat and resources. In this case, human migration is likely to have been limited by the spread of habitat and food resources, and is unlikely to have contributed significantly to the rapid early-Holocene migration of *Corylus*.

Corylus, however, not only spread more rapidly than other taxa (Huntley & Birks, 1983), but was also relatively more abundant during the early Holocene than thereafter. Smith (1970b) discussed the possibility that humans contributed to this period of abundance, recognizing that this was independent of any conclusion reached in relation to the role of humans in aiding the dispersal of *Corylus*. He suggested that *C. avellana*, like its North American relatives *C. cornuta* Marsh. and *C. americana* Walt., is relatively fire-tolerant because of its ready ability to re-sprout from the rootstock, although Rackham (1980) has questioned this premise. Early-Holocene sediments often contain abundant charcoal indicating frequent fires during this period. Smith (1970b) argued that more frequent fires during the early Holocene than subsequently could have led to the unusually high abundance of *Corylus*.

Hunter–gatherer people might use fire either to drive game as part of their hunting technique, or else to maintain a more open landscape offering more browse and/or grazing and allowing easier hunting (Simmons & Innes, 1985). Fires, however, are not necessarily started by humans. Natural fires can result from lightning strikes. Such fires are more frequent in regions of continental climate with a combination of summer drought and high incidence of electric storms. Thus more frequent fires might result from a more continental climate rather than reflecting human activities. Even the evidence of fire from human settlements with Mesolithic artefacts might indicate humans settling in clearings resulting from natural fires rather than deliberate use of fire by people at this time (F.M. Chambers, personal communication).

17.2.6 Climate

The preceding hypotheses are all, to a greater or lesser extent, predicated upon a model of a single step-like climate change *c.* 10 000 BP, with no substantial change in climate thereafter. Thus climate during the first centuries of the Holocene was already suitable for the growth of all 'thermophilous' taxa up to their Holocene range limits. It is also generally presumed that the climatic tolerances/requirements of these 'thermophilous' taxa are sufficiently similar that they would show the same response to a given climate change.

It is in relation to these assumptions, and to our understanding of mechanisms of Quaternary climate change, that probably the greatest advances have been achieved since Smith (1970b) discussed the '*Corylus* problem'. Variations in quantity, as well as in latitudinal and seasonal distribution, of solar radiation received at the Earth's surface are now generally accepted to underlie long-term climate changes (Hays *et al.*, 1976; Imbrie & Imbrie, 1979). These variations result from periodic changes in the Earth's orbit as first hypothesized by Milankovitch (Imbrie & Imbrie, 1979). The resulting climate changes include the characteristic Quaternary glacial–interglacial cycles (Hays *et al.*, 1976; Imbrie & Imbrie, 1979). General circulation models (GCMs) have been used to simulate palaeoclimate patterns for solar radiation levels corresponding to particular times in the past (Kutzbach & Guetter, 1986). These simulations have provided insight into patterns and mechanisms of Quaternary climate change (COHMAP, 1988).

In addition, the changing spatial extent (Denton & Hughes, 1981) and global volume (Shackleton & Opdyke, 1973a) of glacier ice is now better known, especially for the period since the last glacial maximum. Changes in global sea-surface temperatures (CLIMAP, 1976, 1981), and especially those of the North Atlantic (Ruddiman & McIntyre, 1976), have also been reconstructed. Most recently, evidence of atmospheric composition changes, especially levels of the major natural 'greenhouse gases' CO_2 (Barnola *et al.*, 1987; Lorius *et al.*, 1988) and CH_4 (Raynaud *et al.*, 1988; Chapellaz *et al.*, 1990), has been provided by examinations of ice cores from the polar ice caps. All of these contribute to a greatly increased knowledge of global palaeoenvironments and to an improved understanding of likely patterns of climate change in Europe since the last deglaciation.

Four conclusions emerge that are relevant to the early-Holocene history of *Corylus* and other 'thermophilous' taxa. First, although deglaciation was marked by extremely rapid climate change at some individual localities (Dansgaard et al., 1989), the general pattern is apparently of progressive change over several millennia. This change was interrupted by the late-glacial stadial ('Younger *Dryas*'). None the less, deglacial climate change occurred more rapidly and was of greater magnitude than any subsequent change during the Holocene (Jacobson et al., 1987; Huntley, 1990c, 1992). Second, seasonal solar radiation distribution was at an extreme c. 9000 BP with a minimum of winter radiation and a maximum of summer radiation in the northern hemisphere (Kutzbach & Guetter, 1986; COHMAP, 1988; Bartlein, 1988). Third, the enhanced seasonality of climate resulting from this extreme of seasonal solar radiation distribution was modified by the presence of a substantial residue of the Laurentide ice-sheet in north-eastern North America, as well as by the smaller remainder of the Scandinavian ice-sheet (Kutzbach & Guetter, 1986; COHMAP, 1988). Fourth, the resulting early-Holocene climate of many parts both of Europe and of eastern North America was unlike any contemporary climate (Jacobson et al., 1987; Huntley, 1990b, c).

It seems likely that the early-Holocene climate of Europe was characterized by generally greater seasonality than today (Kutzbach & Guetter, 1986). This was accompanied, however, by greater northward extension of the winter influence of the North Atlantic along the western seaboard (Kutzbach & Gallimore, 1988). Lake levels indicate a less positive balance between precipitation and evaporation in northern Europe (Harrison & Digerfeldt, forthcoming). The local impact of the remains of the Scandinavian ice-sheet probably had an overall cooling influence upon the regional climate of northern Europe, although this would be felt primarily in eastern Fennoscandia.

Along with this unique early-Holocene climate, the individualistic climate response of pollen taxa revealed by studies of their climatic response surfaces (Bartlein et al., 1986; Huntley, 1990a; B. Huntley, I.C. Prentice and P.J. Bartlein, unpublished results) must be taken into account. These studies show, as do the biogeographical studies of Hintikka (1963), that taxa with broadly similar ecology and/or geographical distribution at the present day will not necessarily show the same response to a given climate change (Webb, 1986). Thus the optimum of the pollen–climate response surface for *Corylus* lies relatively close to that for *Ulmus*, but at much cooler summer and warmer winter temperatures than that for *Quercus* (Huntley, 1990a). However, *Corylus* extends further toward the continental climate extreme in northern Europe. It also declines in relative pollen-abundance values more slowly than either *Quercus* or *Ulmus* with decreasing winter temperature at any given summer temperature less than c. 18°C (B. Huntley, I.C. Prentice and P.J. Bartlein, unpublished results).

Hence, although *Corylus* pollen values today reach their optimum in relatively oceanic climates, *Corylus* shows greater overall climatic tolerance than either *Ulmus* or the north European *Quercus* spp. It is more tolerant of seasonal drought, cold winters and relatively cool summers than either of these taxa. *Corylus* is probably also relatively fire-tolerant. The more continental and drier climate regime of the early Holocene probably led to a greater incidence of electrical storms and hence of natural fires. A combination is envisaged of frequent natural fires, very cold winters, relatively cool summers, a marked seasonal temperature range and frequent development of summer water deficit. This combination of climatic conditions is consistent with GCM simulations (Kutzbach & Guetter, 1986) and would be especially favourable for both rapid expansion and high relative abundance of *Corylus*.

17.2.7 Plateaux in the radiocarbon timescale

Determinations both of the rate of early-Holocene climate change and of the migration rate of *Corylus* depend almost entirely upon the use of radiocarbon dates to establish a timescale. The non-linear relationship between the radiocarbon and sidereal timescales has been well established for the later Holocene using tree rings (Pearson & Stuiver, 1986; Stuiver & Pearson, 1986). The recent work of Lotter (1991) and of Zolitschka (1988, 1989) using annually-laminated lake sediments has shown marked 'plateaux' in the radiocarbon timescale c. 10 000 and 9500 BP (see also Ammann & Lotter, 1989). Both 'plateaux' last for several centuries and result in indistinguishable radiocarbon ages for diachronous events that occur during these periods. Thus the synchrony and apparent rapidity of early-Holocene changes should be independently tested using alternative dating techniques wherever possible.

The second of these 'plateaux' might lead to calculation of a spuriously rapid migration rate for *Corylus* during the first millennium of the Holocene. In combination with an exaggerated synchrony of the transition to the Holocene, resulting from the earlier 'plateau', this could account for much of the anomalous rapidity of the calculated migration rates for *Ulmus* as well as for *Corylus* during the first millennium of the Holocene. Such non-linearities in the radiocarbon timescale, however, can account neither for the *relative* rapidity of expansion of *Corylus*, nor for its anomalous early-Holocene abundance.

17.3 DISCUSSION

In evaluating these alternative hypotheses it is helpful, as Smith (1970b) recognized, to consider separately the relative rapidity of the early-Holocene range expansion of *Corylus* and its anomalously high early-Holocene abundance.

17.3.1 Rapid expansion of geographical range

All seven general hypotheses are relevant to this problem. Those most worthy of attention, however, are the early-Holocene climate and the non-linear radiocarbon timescale during this period.

The contributions of both succession and pedogenesis seem likely to have been minor. Evidence from recently deglaciated terrain indicates that one millennium is sufficient for the completion of these processes (Crocker & Major, 1955; Viereck, 1966; Birks, 1980). Furthermore, they are not relevant throughout Europe. Migrational lag is difficult to invoke because *Corylus* has such a large, heavy fruit, relative to some taxa that expand more slowly. In addition, evidence from eastern North America shows no systematic difference in dispersal distance between earlier and later migrating Fagaceae that produce nut-like fruits (Johnson & Webb, 1989). The possible compounding influences of relative proximity of 'glacial refugia' and/or of expansion during the late-glacial interstadial, with subsequent survival in northern Europe, are also hard to sustain. Widespread pollen evidence from the late Weichselian (Devensian), plus sparser evidence from the full Weichselian, provides no indication of greater northward extent of glacial distribution of *Corylus* than of other relatively 'thermophilous' woody taxa. Although the range of *Corylus* may have extended to parts of northern Europe during the late-glacial interstadial, it is extremely unlikely that it survived the subsequent stadial in areas significantly further north than the margins of its full-glacial range.

The possible role of humans in assisting the dispersal, and hence the early-Holocene range expansion, of *Corylus* is difficult to evaluate. It has been suggested that these people depended upon hazel nuts as a major element of their diet (Godwin, 1975). If they were dependent upon *Corylus* in this way, however, then they are unlikely to have strayed far beyond the limits of its growth and so accidental dispersal is likely to have made only a limited contribution to its range expansion. Furthermore, ethnographic models for peoples believed to be at the same cultural level of development provide no support for an hypothesis of *deliberate* planting of hazel nuts beyond their limits.

The most probable explanation for the earlier *relative* range expansion of *Corylus* is the character of the early-Holocene climate. Summer conditions had warmed enough to allow its northward expansion, whereas the regular occurrence of summer water deficits and the continuing prevalence of cold winter conditions combined to exclude other 'thermophilous' woody taxa. Frequent natural fires also tended to exclude these other species. The apparently anomalous rapidity of the expansion of *Corylus* (Huntley & Birks, 1983) most probably results, however, from 'plateaux' in the radiocarbon timescale (Ammann & Lotter, 1989; Lotter, 1991).

17.3.2 Anomalous early-Holocene abundance

Only the first six general hypotheses are relevant in this case.

Corylus today is usually a short-statured understorey tree or shrub that reaches peak abundance earlier during succession than do species that are normally components of the canopy by which it is eventually overtopped (Tansley, 1939). However, the duration of this abundance peak is short, perhaps a century at most, whereas during the early Holocene, *Corylus* pollen values remained relatively high for several millennia. It is unlikely that pedogenic processes could have led to such an extended duration of this peak. Delayed immigration of canopy species, however, could have delayed the impact of shading upon *Corylus* and thus have allowed it to persist with greater relative abundance. Increased flowering by *Corylus* when in the open would exaggerate this abundance peak in the pollen record.

It is unlikely either that delayed immigration of canopy species resulted from their slower dispersal or that greater proximity of late-Weichselian range led to earlier expansion, and hence the abundance peak, of *Corylus*. The most probable explanation of the earlier expansion of *Corylus* than of other 'thermophilous' woody taxa lies in the character of the early-Holocene climate. The differing climate responses of *Corylus* and of the other 'thermophilous' woody taxa is sufficient alone to account for the peak abundance of *Corylus* during the early Holocene. Furthermore, although both *Betula* and *Pinus* were present in many areas into which *Corylus* expanded, and *Pinus* at least forms a canopy that can overtop *Corylus*, these species are likely to have declined in vigour in response to the same climate changes that favoured *Corylus* range expansion. The role of natural fires in maintaining a relatively open landscape, favourable to *Corylus*, may also have been important. Climatic conditions were favourable for a high incidence of natural fires, and abundant charcoal in lake sediments from this period indicates fire was a frequent phenomenon. Deliberate use of fire by humans may also have contributed to this high fire frequency.

It is noteworthy that a pollen taxon representative of ecologically similar understorey trees (*Ostrya virginiana/Carpinus caroliniana*) shows parallel behaviour in eastern North America (Huntley & Webb, 1989). Both taxa probably benefited from a relatively continental climate regime and a higher incidence of natural fires. These factors operated upon both continents during the early Holocene, whereas the human population of North America was sparse and apparently had little impact upon the natural vegetation (McAndrews, 1988).

17.4 CONCLUSIONS

Although he discussed at length evidence that human activities may have contributed both to rapid spread and to early-Holocene abundance of *Corylus*, Smith (1970b, p. 83) concluded that 'the prevailing climatic conditions can hardly be left out of account' and that 'the Boreal hazel maximum ... could be connected in some way with ... the dry conditions'. During the subsequent 22 years a greater understanding of the underlying mechanisms of climate change has been achieved and further evidence has accumulated indicating the character of early-Holocene climate. It is now possible to restate these conclusions less tentatively and to propose that the unique combination of climate conditions during the early Holocene, perhaps along with the high fire frequency that resulted from these conditions, favoured the early and rapid expansion of *Corylus* and its high relative abundance for several millennia.

Acknowledgements

I am grateful to Dr F.M. Chambers for the invitation to contribute to this volume. The stimulus to address the '*Corylus* problem' emerged by chance from a conversation with Drs R.H.W. Bradshaw, M.B. Davis and F.G.J. Mitchell as we walked through an Italian woodland; I thank them for their contributions, both conscious and unconscious.

18

The origin of blanket mire, revisited

Peter D. Moore

SUMMARY

1. Palynological and stratigraphic studies of blanket mires have provided an extensive source of evidence concerning their origin and development.
2. Suggestions (made in the 1970s) that prehistoric human activity in the form of forest clearance and modifications of upland hydrology were of importance in blanket mire origins are reconsidered here in the light of recent work.
3. Support for the hypothesis is strong and evidence from outside the British Isles demonstrates that human impact has been widespread.
4. It is not considered possible, however, to generalize concerning the precise date of origin since this varied with the relative human impact and with local topography.

Key words: blanket mire, human impact, palaeohydrology, forest clearance

18.1 INTRODUCTION

Blanket mires are an important feature of the vegetation of the British Isles and are more extensively developed on these islands than in any other part of Europe. In the extreme west, particularly in western Ireland and north-west Scotland, blanket mires are developed at low altitude (Tansley, 1939; Goode & Ratcliffe, 1977; Doyle, 1982, 1990), while in more easterly parts of the islands such mires are found only at higher altitudes, usually over 400 m. The main factor that appears to determine the extent of blanket mire in an area is the climatic humidity, perhaps best expressed in terms of precipitation/evaporation ratio. There is a reasonably good correlation between the distribution of blanket mire and the 1200–1250 mm isohyet (O'Connell, 1990a; Godwin, 1981). Atmospheric humidity, ensuring low evaporation rates, is also clearly important in determining the limits of blanket mire development, and O'Connell (1990a) regards 200 rain days per annum as the minimum number required.

Perhaps it was the very evident climatic restraints on the geographical distribution of blanket mires that led to the early assumption that climatic factors played the major role in determining the onset of blanket peat development. Godwin (1981), for example, records some early discussions of this subject dating from his field excursion to the blanket mires of western Ireland with Tansley, Jessen, Mitchell and Osvald in 1935. The discovery of layers of wood peat beneath the fibrous, monocotyledonous peats of the true blanket bogs immediately led Godwin to the conclusion that such a change in vegetation was climatically induced. He recalls his thoughts at the time with the following words: 'One is consequently led to the conclusion that originally the landscape was forest-clad, and with a climatic change towards oceanicity it became increasingly water-logged and peat-clad' (Godwin, 1981, p. 56). He continues to conjecture that such a change may have taken place at the time of the 'Sub-Boreal/Sub-Atlantic transition', that is, during the late Bronze Age, since this was regarded as a time of deteriorating climate.

Prior to the advent of radiocarbon dating, speculation about the timing of such 'climatic changes' that were assumed to be responsible for the induction of blanket peat formation was based upon the discovery of archaeological artefacts beneath the peat. Mitchell in Ireland and Woodhead in England were particularly active in this field. Palynological investigations of blanket mire stratigraphy in the southern Pennines of England by Conway (1954) led her to the conclusion that the origin of blanket bog in that region dated from 'Atlantic' times, and the discovery of Mesolithic artefacts at the peat base in an exposed section of Pennine blanket mire (Walker, 1956) in association with charcoal, radiocarbon dated to 6500 ± 310 BP, provided supporting evidence for this view. The evidence was compatible with the accepted belief that the 'Atlantic' period was one of warm but wet climatic conditions, so the concept of the 'spreading of a continuous mantle of peat mire poor in plant nutrients, waterlogged and moderately or highly acidic in reaction, across great stretches of country' that 'exterminated from vast areas most of the woodland plants' (Godwin, 1975, p. 463) accorded well with Godwin's developing concept of vegetation change in the British post-glacial.

Following the work of Conway, there was some delay in further research on the pollen stratigraphy of blanket mires, mainly because of the very natural concern to establish well-dated long profiles of the Holocene changes in vegetation that were more appropriately obtained from raised mires and lakes. Research effort on blanket mire palynology was resuscitated in the 1960s with the work of Simmons (1964) on Dartmoor, Tallis (1964b) in the southern Pennines, Chapman (1964) in the northern Pennines, Pennington (1965) in the Lake District, Moore (1968) in mid-Wales and Case *et al.* (1969) in western Ireland. From all of these studies there emerged evidence of strong human impact on the vegetation of the uplands, much of which had been anticipated by the soil pollen work of Dimbleby (1961c). There remained, however, the tacit assumption that the initiation and development of the blanket bogs themselves were a consequence of climatic 'deterioration' in the sense of lowered overall temperatures and increased precipitation. The time of origin of the blanket mires became of interest to palaeoclimatologists, such as Lamb (1977a), who used this information as evidence in climatic reconstruction.

Archaeological and palynological evidence of human activity at the time of the inception of blanket peat formation, which had once been of interest mainly as a source of information on the date of origin of the peat, then became a point of interest in its own right. Simmons (1969a), for example, drew attention to the environmental impact of prehistoric communities in the uplands at that time, but the possible mechanism by which human land use could have initiated peat development remained obscure. Case *et al.* (1969) then made the proposal that the activity of 'ploughing' in the western Irish lowlands during Neolithic times had led to changes in the pattern of land drainage that could have resulted in podsolization and subsequent waterlogging. The spread of blanket bog over areas formerly used for arable agriculture was thus established. But soil disturbance by ploughing could not be a general cause of blanket mire spread, for the development of peat occurred over areas that bore no evidence of such arable activity.

Close-sampling of adjacent layers of peat at the boundary between forest and blanket peat in the stratigraphy of blanket mire shed further light on possible mechanisms. In north Wales, a site examined in this way provided evidence of a sequence of interference phases in the history of the forest immediately prior to its demise and the establishment of blanket bog vegetation (Moore, 1973). It became apparent that a human hand was operative in the destruction of 'pre-peat' forest and the possibility that this destructive process was a causal agent of bog development was put forward (Moore, 1975). In many respects this suggestion was an extension of the work of Dimbleby (1962) and Iversen (1964) on the destruction of lowland forest and heathland development on poor soils in prehistoric times.

Three further problems were raised by such a suggestion, however. First, what was the precise mechanism that resulted in waterlogging and peat formation consequent upon tree removal? Second, was the impact of early man sufficiently extensive and intensive to effect widespread forest destruction? Third, could such forces have been operative even in remote areas where forest cover may have been incomplete or even absent, such as the Western Isles of Scotland?

18.2 MECHANISMS OF BLANKET MIRE INCEPTION

The first of these problems can be solved by reference to a hydrological model developed by Moore (1975) in which the removal of tree cover, or even the thinning of a canopy is considered to enhance the supply of ground

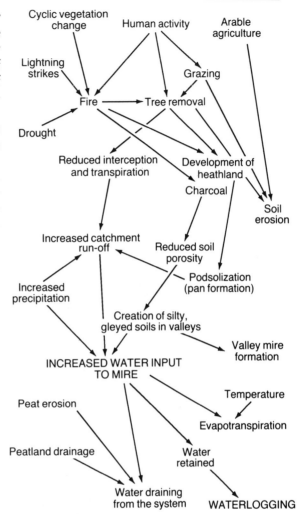

Fig. 18.1 Hydrological processes leading to waterlogging in a mire.

water by reducing transpiration losses and by reducing interception and re-evaporation of precipitation. The model has subsequently been elaborated and modified, and Figs 18.1 and 18.2 illustrate the complexity of the ecological and hydrological interactions potentially resulting from human intervention in the growth and development of upland forest.

This model also incorporates the concept of soil changes introduced by Taylor & Smith (1972) in which podsolization of the soil and

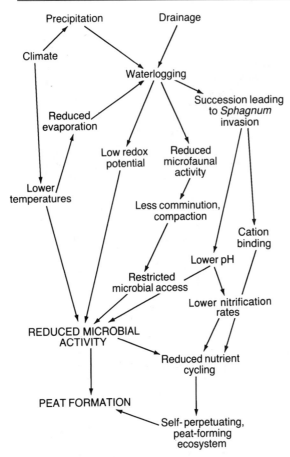

Fig. 18.2 Consequences of waterlogging on peat formation.

the development of an iron pan can further impede drainage and add to waterlogging within the soil. This is effectively an extension of the suggestion of Case et al. (1969) in their studies of Neolithic ploughing effects on soils and its influence on iron pan development. The schemes given in Figs 18.1 and 18.2 also incorporate the findings of Mallik et al. (1984), which have shown that the addition of charcoal to a soil following heathland fire alters its drainage properties. Effectively, the fine, inert particles of carbon reduce the porosity of a soil and make a sandy soil behave like a clay from the point of view of water retention. The process of conversion of forest to bog also has a self-perpetuating tendency in that the establishment of bog vegetation, especially *Sphagnum* mosses, renders the regeneration of most trees difficult, especially if grazing or burning management continues.

One important question that affects the application of the scheme shown in Figs 18.1 and 18.2 is the precise mechanism by which forest is removed from an area. Several possible methods could be employed by prehistoric cultures that would result in the demise of forest.

1. Trees could be cleared physically from a site using primitive tools for felling or for ring-barking the larger trees, as described by Iversen (1956).
2. The partial destruction of trees could be associated with the use of a woodland for pasture. Grazing animals would be herded or allowed to wander in woodland while pastoralists would lop the low branches from trees and shrubs to encourage browsing. This land use can still be observed in the foothills of the Himalayas. One result of this herding practice would be the modification of ground flora by grazing preferences and trampling, and a resultant reduction of tree and shrub regeneration, with the possible exception of grazing tolerant taxa, such as *Juniperus* and *Ilex* that can consequently become prominent elements of the flora (Pott, 1989).
3. Burning of woodland, or of partially damaged woodland and clearings would encourage young growth of some species and could benefit grazing animals. The outcome would be reduced regeneration for most trees and the overall effect on the nutrient capital of the ecosystem would be to diminish it. This type of practice could have been employed for the enhancement of game animal populations even prior to domestication (Mellars, 1976).
4. The cutting of wood for fuel or for construction purposes would modify the structure of the forest canopy. Species responding positively to coppicing and pol-

larding would be favoured, but pollen productivity and pollen circulation patterns would be strongly modified.
5. A combined impact involving several or all of the above techniques could have been employed in the management of upland woodlands, thus producing a significantly destructive force on forest already under climatic stress.

It is difficult to elucidate from the available pollen evidence which of these particular processes was most important in the loss of the pre-peat woodlands. Sometimes the evidence for a former woodland cover takes the form simply of pollen types persisting in the soils beneath blanket peat, as can be seen in the pollen diagram from Carneddau 3 (Fig. 15.4; Walker, this volume), and the Plynlimmon diagram of Moore (1968). Even when layers of woodland humus are preserved beneath blanket peats, they are often highly compressed and provide poor resolution for the detection of the processes at work in the conversion of forest to blanket bog.

The most appropriate sites for detailed study along these lines are ones in which some development of wood peat has occurred prior to the onset of blanket peat formation. One such is the site of Carneddau Hengwm in North Wales (not to be confused with Walker's site) where pollen analysis of the basal layer of wood peat has provided good resolution of the sequence of events leading up to the development of blanket bog over the site. These data have been published by Moore (1973). The conclusion from the pollen analysis at this site is that there is no evidence for wholesale local clearance of woodland by prehistoric people. Progressive exploitation of the forest for fuel resources or for grazing led to a stepwise retrogressive succession leading from a high biomass forest ecosystem to a low biomass blanket mire ecosystem. The latter evidently proved stable since it showed no sign in the subsequent stratigraphy of any reinvasion of trees. Presumably, the continued grazing of the area was adequate to ensure that successional development back to forest did not take place. Fire may have played a part in the vegetation changes described at Carneddau Hengwm, but in the absence of charcoal analyses from the site this cannot be certain. Many subsequent studies of blanket peat development, however, have shown that charcoal is abundant in the basal layers, implying that fire played an important role in the destruction of pre-peat forests (for example, Bostock, 1980; Kaland, 1986).

The pollen analyses that have been carried out on the basal peat layers of blanket mires have demonstrated that mechanisms can be postulated to account for the change in vegetation from forest to blanket bog and there is strong evidence that these mechanisms were of considerable significance in this type of development at many sites. It is conceivable that prehistoric human activity was generally responsible for blanket bog development and spread in many parts of the British Isles. The geographical extent of prehistoric influence now needs to be considered.

18.3 EXTENT AND INTENSITY OF PREHISTORIC FOREST MODIFICATION

The second question raised in the introduction to this paper concerns the ability of prehistoric populations to influence their environment to the degree demanded by the hydrological model. If the impact were entirely dependent on the hand felling of trees, then clearly the low population density in prehistoric times would make the model untenable. The use of fire and, subsequently, domesticated animals, however, opens up the possibility of a more extensive and a more intensive impact by human populations than might be expected on the basis of projected population levels. The question of which prehistoric culture or cultures were involved in blanket mire formation thus becomes an important one.

When the hydrological model was first formulated in 1975, there was a strong emphasis placed upon the correlation between blanket mire formation and the occurrence of the elm decline. Most profiles then available for blanket mires lacked radiocarbon dates and the elm decline provided a reasonably reliable datum horizon (Smith & Pilcher, 1973). It was proposed, therefore, that the origin of blanket mires was associated with Neolithic cultures around 5000 BP. This has subsequently proved to be only partially true. In Northern Ireland, A.G. Smith (1975) and his students provided a suite of radiocarbon dates for the basal layers of blanket peats and found that they were spread over a considerable time range, with a strong cluster during the Bronze Age (thus confirming the initial thoughts of Jessen and Godwin in 1935).

Work by Merryfield & Moore (1974) and Moore et al. (1984) showed that the deep plateau blanket bogs of Exmoor and the Black Mountains of South Wales, like those of Plynlimmon, contained evidence of an elm decline at their base and those radiocarbon dated were demonstrated to have originated around 5000 BP. However, other plateau sites and also peats from sloping sites have often proved to be of more recent origin. It is evident, therefore, that the actual date of origin of a blanket peat profile is a function of a number of factors, among which topographic position is important. This aspect of blanket peat development has been explored by several workers (Chambers, 1981, 1982b, 1982c, 1983, 1984; Moore, 1987; Smith & Taylor, 1989). The importance of topography in determining the date of onset of peat formation is well illustrated by the extensive peat profiles of Chapman (1964), who showed that upland basins may have accumulated peat throughout the Holocene, but that the development of peats over plateaux and slopes took place at a later period (in the case of the profiles Chapman studied at Coom Rigg in the Pennines, again at about the time of the elm decline). Since it is the hydrological balance leading to waterlogging that is the ultimate determinant of peat inception, the degree to which water is collected by or shed from a site is clearly a major factor in defining the time of local development of peat.

A systematic approach to the question of blanket bog spread was used by Smith & Cloutman (1988) in their very detailed study of the development of a blanket bog in south Wales at Waun-Fignen-Felen. Their work closely resembles that of Chapman in basic approach, but benefited from a greater number of stratigraphic profiles involving close interval corings, together with the valuable facility of radiocarbon dating. This placed on a firm basis the construction of a model of mire extension through time. Their site showed that peat formation began with hydroseral succession within a late-glacial (late-Devensian) lake basin and that the progression to ombrotrophic peat over this basin led to reduced drainage (secondary peat formation in the sense of Moore & Bellamy, 1973). The surrounding slopes bore a woodland in which *Corylus avellana* was an important component, but this was replaced by blanket bog between 6500 and 6000 BP. Virtually all of the basal layers of these blanket peat profiles contained charcoal, and this feature, combined with the abundant archaeological evidence for Mesolithic occupation of the site, suggests that early human land use involving the burning of vegetation was coincident with, and possibly a contributory cause of, blanket peat development and spread in the area.

The role of topographic factors is thus firmly established as a prime determinant of blanket peat spread within areas where climatic factors are appropriate for blanket peat formation. To speak of a date for blanket peat inception is thus meaningless without qualification as to the slope, aspect, location and morphology of the ground over which the peat has begun to form. The claim by Moore (1975) that blanket peats often began forming at the time of the Neolithic elm decline has proved true only for certain profiles, namely those peats in western and south-western Britain that developed on upland plateaux of low slope. But even profiles of this type may have begun to accumulate peat at different times (Chambers, 1981, 1984). Steeper slopes at lower altitude may not have undergone the change from woodland to bog until later times, while sites in receipt of greater quantities of drainage water in their early stages of peat development, such as those surrounding a lake basin as described by Smith & Cloutman (1988), may have begun blanket peat development at an earlier time. This is particularly true when such sites were subjected to pressures from the activity of Mesolithic peoples, as was the case in parts of upland south Wales. It is noteworthy that Mesolithic activity was also particularly intense in the Pennine area of Britain, where Conway had established an 'Atlantic' date for blanket peat formation at Ringinglow Bog. Jacobi et al. (1976, p. 317) regard the impact of these cultures on the local vegetation as considerable and feel that 'permanent suppression of a closed tree cover over large areas above c. 350 m altitude' took place in Mesolithic times. Such intensity of early human activity may well account for the early spread of blanket mires in this region.

18.4 REPLACEMENT OF FOREST BY BLANKET MIRE

The role of human activity in the establishment of blanket mires is now well established, especially in those parts of the British Isles where climatic conditions are marginal for blanket mire formation and where the pre-peat vegetation consisted of open canopy woodland already perhaps under climatic stress as far as regeneration was concerned. The hydrological model linking human activity to peat development that was propounded by Moore (1975) demands that tree or shrub-dominated vegetation existed before the onset of peat formation and that its loss at the hands of human activity resulted in hydrological changes and further reduced the regeneration potential of the woodland species. It presupposes, therefore, that woodland was somehow cleared by humans. One objection to the hypothesis has been that it could not apply in areas where such woodland may not have been developed because of adverse climatic conditions, such as high winds or very high rainfall. The far north of Scotland, for example, or parts of the Hebrides, or western Norway, may not have borne a woodland cover of sufficient density of trees and shrubs for their clearance to have produced significant hydrological impact.

Investigations in these regions, however, are producing increasing quantities of evidence to suggest that human modification of vegetation has been an important factor in the development of their blanket bog cover. In Norway, for example, Kaland (1986) has shown that the lowland, coastal blanket bogs of the Bergen region developed in association with palynological evidence for forest clearance and an increasing intensity of land use by prehistoric cultures. A similar correlation emerges in the coastal blanket bogs at Haramsoy, to the north of Bergen (Solem, 1989), where an intensification of human activity around 3000 BP led to extensive spread of blanket mire. At other sites in Norway, however, such as Nord-Trondelag (Hafsten & Solem, 1976) and Sor-Trondelag (Solem, 1986), the blanket bogs are believed to have a climatic origin. More detailed stratigraphic research is required to confirm this proposal.

In the Outer Hebrides of Scotland, an area once considered inappropriate for tree growth, there is now a body of evidence to indicate some woodland cover in the past. On the Isle of Lewis, for example, pollen analysis of blanket peats has shown the existence of *Betula/Corylus* woodland, which was cleared about 3600 BP at the time of blanket peat formation (Newell, 1988). Some clearance of forest from the Isle of Lewis may have preceded this Early Bronze Age clearance, perhaps dating back as far as 8400 BP (Bohncke, 1988).

The role of human activity in mire formation is being increasingly recognized in various types of peatland and in different parts of the world (see Moore, 1987). These range from the quaking mires of southwestern Ontario in Canada (Warner et al., 1989; Warner, this volume) to the upland mires of the Appenines in Italy (Cruise, 1990). Circumstantial evidence linking human prehistoric land use and mire formation is even more widespread, as in southern Africa, where there is an interesting correlation between the frequency of radiocarbon-dated archaeological materials and the frequency of dates for the initiation of peat formation in southern African mires (Meadows, 1988). Both show a marked peak in frequency between 5000 and 3000 BP.

The evidence for human modification of hydrological cycles and hence human involvement in peatland initiation and development is thus proving to be more widespread and varied than was originally expected.

18.5 CONCLUSIONS

The development of peat at a given location is ultimately dependent on the hydrological balance at that site. If the degree of waterlogging is such that microbial decomposition is suppressed to such an extent that it is unable to keep pace with the rate of litter formation, then peat accumulation becomes possible. Peat accumulation on water-shedding sites, which is characteristic of blanket mires, is clearly possible only under conditions of high precipitation/evaporation ratio. But even under climatic regimes where blanket peat development is feasible, the persistence of forest cover may so modify the local hydrology that no peat development occurs. Removal of the forest, or placing additional demands upon the forest that reduces the rate of seed-set, germination, establishment or survival of the individual trees can result in the loss of woodland cover and the establishment of peat-producing vegetation.

The evidence collated here suggests that the impact of early human cultures was effective in forcing vegetation development across a tolerance threshold, leading to the development of blanket mire over a wide area of the British Isles and beyond. It remains to be seen whether and where climate alone may have been responsible for such mire development within these Islands.

The work of Alan Smith in this area of palaeoecological research, as in many others, has been influential. His research team in Northern Ireland established the diachroneity of the date of blanket peat initiation and led to the development of topographic models in understanding such peat development. His work, together with that of his students Frank Chambers and Edward Cloutman in south Wales, established the role of pre-agricultural and subsequent cultures in peatland development in the uplands.

Acknowledgements

I thank referees for comments. Andrew Lawrence re-drafted the Figures.

19

Climatic change and human impact during the late Holocene in northern Britain

Keith E. Barber, Lisa Dumayne and Rob Stoneman

SUMMARY

1. Numerous ombrotrophic mires, 'raised bogs', in the Anglo-Scottish Border region provide an excellent opportunity for investigating late-Holocene environmental changes, natural and cultural.
2. The cultural record shows minimal human impact during the Bronze Age and most of the Iron Age, followed by massive deforestation during the Roman invasion and the construction of the Hadrianic and Antonine Walls, forts and roads.
3. The degree of clearance is related to distance from Roman structures; in most areas there is some post-Roman regeneration of woodland.
4. Later pollen-analytical events may be closely related to the history of settlement, war and agricultural innovation, as shown at Bolton Fell Moss.
5. At the same site a close relationship exists between the peat stratigraphy and climatic change. Detailed macrofossil analyses at a number of sites are refining and extending this relationship with the aim of reconstructing the climate of north Britain during the late Holocene.

Key words: peatlands, palynology, macrofossils, Romans, climatic change

19.1 INTRODUCTION

Alan Smith's earliest papers include two on the mires of south Yorkshire, north Lincolnshire and south-west Westmorland (now Cumbria) (Smith, 1958b, 1959). In both papers a good deal of attention is paid to peat stratigraphy, not only as an indicator of the local environment but also of regional climate, and to human impact as revealed through pollen analysis and consideration of the archaeology. Together with the research of Pennington (1947, 1970), Walker (1955, 1966), Oldfield (1960a, 1960c, 1963), Turner (1965) and Barber (1981), this work exploited the numerous lakes and mires of north-west England. What is reported here is a further development of this interest in late-Holocene environmental change, pre-eminently changes in climate, and human impact upon that palaeoenvironment. There is now a renewed impetus and urgency in such research as we realize both the scope for research and the threats to the peatlands themselves. In particular, environmental scientists are increasingly aware of the value of the record of change held in the peat of ombrotrophic raised mires – those growing above the ground-water table and therefore dependent upon precipitation for their nutrients and moisture. Besides the value of such mires for palaeoecological reconstruction, their ombrotrophic nature – the direct coupling of the bog surface vegetation with the atmosphere – means that they can also be used to reconstruct pollution history (Oldfield et al., 1978; Thompson & Oldfield, 1986) and that they record events such as past eruptions of Icelandic volcanoes (Dugmore, 1989).

This chapter is an account of past and continuing work on the natural record of local and regional vegetational and climatic change, and on the cultural record – the progressive clearing of the natural vegetation to create the present agricultural landscape.

19.2 THE LATE-HOLOCENE ENVIRONMENT OF NORTHERN CUMBRIA

It is not too gross a generalization to say that northern Cumbria was an agricultural and social backwater until the Roman occupation. In contrast to large parts of southern England, which were largely deforested in Neolithic and Bronze Age times (Waton, 1982), and compared to the massive clearances and soil erosion of the Iron Age evident in central England (Brown & Barber, 1985), both the archaeological evidence from northern Cumbria (Fowler, 1983) and the palynological evidence (Walker, 1966), point to a continuation of a basically Neolithic economy, with only small areas of cleared land on the more favourable soils.

In the rather damp and cool climate of the area, sandy soils derived from glacial deposits were likely to be the premium sites. For example, on such light soils, 83% arable cultivation was recorded in the 1841 Tithe Returns for the parish of Hayton (Bainbridge, 1943). One can also point to the large area covered by raised bog – Glasson Moss, Bowness Common, Wedholme Flow on the coast west of Carlisle, Solway Moss, Scaleby Moss, Walton Moss and Bolton Fell Moss to the north – land that was, of course, unavailable for settlement or agriculture, and to the heavy clay soils over large parts of the Carlisle plain (Walker, 1966) as disincentives to early farming communities. This is graphically borne out by the maps of known sites in Fowler (1983), where his figure 12, showing the distribution of known sites during the Neolithic, Beaker, Bronze and Iron Ages, has remarkable concentrations in the drier, hillier north-east, particularly in the Iron Age, and a distinct 'hole' in Cumbria north of the Vale of Eden (cf. Higham, 1986).

As Walker's (1966) diagrams hardly extend into the Iron Age, the main sources of palynological information are the work of Turner (1979) and her co-workers (Davies & Turner, 1979; Donaldson & Turner, 1977), and of Barber (1981). Davies & Turner (1979) document an almost complete lack of clearance activity during the Neolithic, followed by small temporary clearances during the Bronze Age at the four sites studied, and some Iron Age activity – less marked at their most westerly site, Fellend Moss (19 km east of Bolton Fell Moss), which is in accord with the archaeological evidence. The same lack of activity is evident in the diagram from Hallowell Moss, Co. Durham (Donaldson & Turner, 1977). The Roman invasion and occupation caused massive deforestation – for example, tree pollen fell to less than 5% of total pollen by late-Roman times at Hallowell Moss. The pollen diagram from Bolton Fell Moss (Barber, 1981) (Fig. 19.1) is calculated on a non-mire pollen sum (NMP), that is excluding Cyperaceae, Ericaceae and spores, which, it is felt, gives a truer picture of landscape change in that the

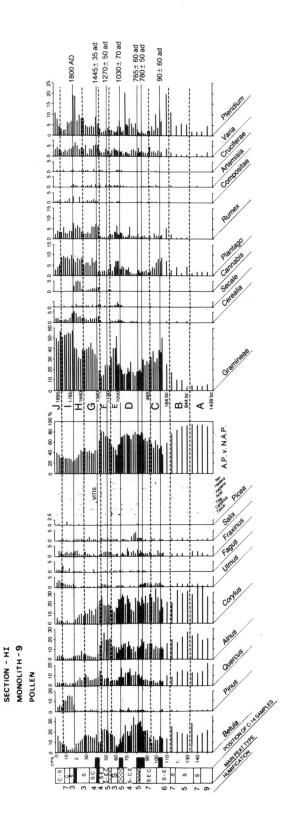

Fig. 19.1 Summary pollen diagram from Bolton Fell Moss section HI, monolith 9 (reprinted from Barber (1981), with radiocarbon dates uncalibrated as bc/ad). The AD 1800 level is based on correlation of pollen with records of land use.

curves of each tree taxon, and of the herbs, are not inflated as with a tree pollen sum.

It is clear that the grass pollen curves from Turner's sites and from Bolton Fell Moss are a response to forest clearance and are not the result of, for example, *Molinia* growth on the mires.

The correlation of the pollen stratigraphy of monolith HI9 with archaeological and historical events was given in full in Barber (1981, p. 112–123), and as calibration does not alter most of the radiocarbon ages significantly (see Appendix A), all in this chapter are again quoted uncalibrated. It is clear that zones A and B, covering the period *c.* 2450–2100 BP (1500–150 bc), agree with the other diagrams mentioned above by showing very little agricultural activity. There is some plantain and even cereal pollen, but arboreal pollen is generally over 80% of NMP. The Roman zone C, with a radiocarbon date of 1860 ± 60 BP (ad 90 ± 60), sees a fall in arboreal pollen to a low of 37% and a grass peak of 50% – a level not reached again for 1000 years – with higher levels of a number of open-ground indicators such as plantains, docks and bracken, though with almost no cereal pollen. This contrasts with later agricultural episodes when pollen of Cerealia – probably mainly *Triticum* (wheat) – and of *Secale* (rye) are prominent, and it seems therefore that during Roman times the land around Bolton Fell Moss was being used for pastoral purposes (cf. Behre, 1986).

There is a problem with making precise correlations with variations in Roman activity because of the temporal resolution of the pollen diagram at this depth. Between the presumed Roman levels and the major humification change at *c.* 1185–1170 BP (ad 765–780), the peat accummulation rate is slow, at 34 yr/cm, making it difficult to resolve events of less than a century duration; slicing the 'cheesy' peat into very fine samples might help here (cf. Garbett, 1981). The spatial registration of this first major human impact is discussed in the next section.

From the age/depth plot of the radiocarbon dates (Barber, 1981, figure 28), zone D covers *c.* 1485–900 BP (ad 465–1050) and is a long period of rather subdued agriculture in a generally worsening climate. The proxy climate curve from Bolton Fell Moss (Barber, 1981, figure 76) records four shifts from a fairly dry bog surface to wet lawns with more abundant and hygrophilous Sphagna between 1350–1150 BP (ad 600–800) and much wetter conditions with definite pools of open water forming from 1050–850 BP (ad 900–1100).

Scandinavian settlements built up during the latter part of zone D so that by 900 BP (ad 1050) and the beginning of zone E there was a distinct period of clearance with grass pollen again peaking at 50% and continuous curves of cereals and rye. This activity halted abruptly *c.* 760 BP (ad 1190) for almost 200 years, as a result of the Border Wars and later the Black Death, as is clearly shown in the pollen curves of zone F. Pollen types indicative of cultivation, such as cereals and hemp, are absent from this zone, and indicators of open ground and pastoral activity, such as grass pollen, are only present at very low levels. In contrast, the pollen of birch, alder and hazel increase markedly, indicating regeneration of scrubby woodland. This is, of course, in marked contrast to developments elsewhere in England in the High Middle Ages, generally a period of prosperity and flourishing agriculture up until AD 1350, with the Domesday population of about 2 million in AD 1086 doubling to 4 million (Taylor, 1984, p. 151). This dearth of agricultural activity in northern Cumbria is also in marked contrast to the evidence for the climate of the AD 1200s being at its warmest and driest since Roman times, as attested to by both documentary records and the peat stratigraphic and macrofossil data (Barber, 1981).

The two records, natural and cultural, are even more at variance during zone G, *c.* 570–310 BP (ad 1380–1640), when there is a remarkable revival of agricultural activity despite a deteriorating climate. Groundwater tables must have risen in response to the same climatic forcing that led to pools appearing in the peat stratigraphy just after 650 BP (after ad 1300), reaching their zenith c. 525 BP (ad 1425). According to Lamb (1977b) this was a period of high summer wetness and of declining temperatures at the beginning of the Little Ice Age. The higher agricultural activity also seems to be at variance with the century dominated by the Border raids (AD 1503–1603) until one recognizes what an organized, almost ritualistic, form of warfare the moss-troopers indulged in (Barber, 1981, p. 118). After the temporary setback of the Civil War, zones H, I and J document the making of the modern landscape of north Cumbria, with the pollen-analytical events such as the demise of *Cannabis* (hemp) and the rise in pine pollen due to planting on country estates, being in excellent accord with the documentary record (Barber, 1981, p. 118–123). The stratigraphic evidence of open pools (e.g. at 19 cm depth in monolith HI9) forming all across the bog c. 170 BP (ad 1780) is very much in accord with the climatic record of the cold wet decades either side of AD 1800 (Lamb, 1977).

Two main questions arise from the above work by Barber and that of Turner *et al.* To what extent did the major landscape change due to the Roman occupation affect the frontier zone as a whole, and to what extent can the climatic inferences from Bolton Fell Moss be tested and refined by examining other raised mires in the area?

19.3 THE ROMAN IMPACT ON THE LANDSCAPE OF THE FRONTIER ZONE

19.3.1 Historical accounts

In AD 43 southern Britain was invaded and conquered by the legions of the Roman Empire, which eventually moved north in AD 78 and, after a few years occupation of lowland Scotland, established a frontier along the Stanegate Road between the Solway and the Tyne. In AD 122 the Emperor Hadrian commanded that a permanent and obvious frontier, Hadrian's Wall, be built between Bowness-on-Solway and Wallsend-on-Tyne (Fig. 19.2).

Large amounts of timber and turf were needed for the construction of forts and other structures, and Keppie (1986, p. 67) has estimated that a fort 4 acres (1.6 ha) in area would have needed 22 000 cubic feet (622 m^3) of timber for the construction of its internal buildings, towers and gates. The major purpose of Hadrian's Wall was military. The Wall was a strategic attempt to defend and control a territory that had no obvious military coherence and marked the limit of Roman influence in Brigantian territory (Johnson, 1989). The Wall was also a political and administrative boundary. It barred north–south routes, except at forts and milecastles, which functioned as customs barriers where trade and movement of people were regulated rather than prevented. It also protected the south from raiding and attack by tribes and aided the peaceful economic exploitation of the province to the south (Breeze, 1982, p. 85). The Wall garrison has been variously estimated at between 12 000 and 17 000 men.

In AD 139 a full-scale Roman invasion of lowland Scotland was mounted (Keppie, 1986, p. 13). The natives were pushed across the Forth–Clyde isthmus and the war ended in AD 142 with the establishment of a new frontier, the Antonine Wall (Fig. 19.2). Small legionary detachments on Hadrian's Wall replaced the original garrisons, which were removed to the new line or to the forts between the two walls.

The Antonine Wall is a mainly turf wall 40 Roman miles (65 km) long and stretches from Carriden on the Firth of Forth (Bailey & Devereux, 1987) and Old Kilpatrick on the Firth of Clyde (Johnson, 1989), and is estimated to

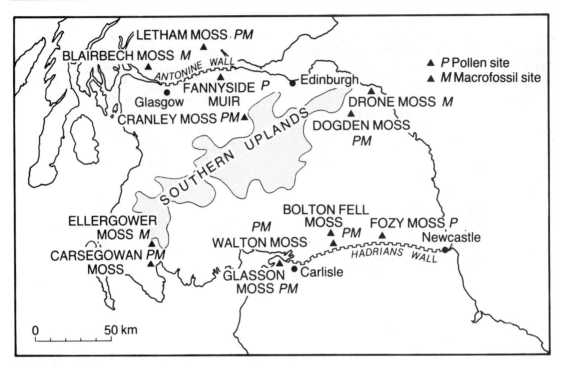

Fig. 19.2 Map showing sites under investigation.

have used up a corridor of turf 50 m wide from the front and rear of the frontier line (Keppie, 1986, p. 67). Approximately 20 000 men were needed to garrison the 19 forts on the Antonine Wall and the rest of Scotland; 7000 of these were probably located on the Antonine Wall itself. Between AD 142 and AD 208 the area was turbulent and unsettled with the Antonine Wall being abandoned in AD 150, repaired in AD 160, abandoned again in AD 180 and reoccupied in AD 195, but abandoned for the final time soon thereafter. The Hadrianic frontier was consolidated in AD 208 and throughout the 3rd century AD the frontier remained at peace. The villages, or *vici*, outside the forts grew and accommodated civilians – wives, families, veterans, traders and travellers (Branigan, 1980). In such settlements, agriculture and crafts were carried out to supply the Roman army, traders built shops and small-scale industrial activity developed.

At the beginning of the 4th century AD, part of the Wall garrison was withdrawn to fight against the central Roman government. The Picts took this opportunity to plunder the Wall, but by AD 305, the Wall garrison had reoccupied and rebuilt the Wall and its outposts. The 4th century AD was a time of instability in the Roman Empire. Troops were gradually withdrawn to defend Rome and the empire in Britain ended in AD 411, when the central government ceased paying salaries to the army and the food supply lines were cut. Behind the Wall, sites continued to be occupied by civilian farmers into the 5th century AD (Breeze & Dobson, 1987).

19.3.2 Palynological evidence

The effect of these activities on the environment of the frontier zone has not been intensively studied and few pollen diagrams concerned with human impact on the vegetation have been produced from the area between the Walls. The record from Steng Moss, Central Northumbria (NY 965 913), shows a substantial rise in Gramineae, herbs and cereals, from c. 1970–1490 BP (20 bc to ad 460), in an area well north of Hadrian's Wall (Turner, 1979). After the Roman withdrawal, arboreal pollen rose c. 1450 BP (ad 500) and tree regeneration continued to Anglo-Saxon times. In Ayrshire, a diagram from the edge of Bloak Moss shows that extensive clearance for pasture did not begin until c. 1500 BP (ad 450) and ended c. 1370 BP (ad 580), when arboreal pollen rose (Turner, 1965). Only local pollen analytical evidence is available for the Antonine Wall area, as the published literature is concerned with on-site vegetation reconstruction (Dickson, 1989; Dickson et al., 1985; Boyd, 1982, 1984, 1985). A regional reconstruction of vegetational history is therefore overdue (Dumayne, in preparation.)

According to the criteria of Jacobson & Bradshaw (1981), a large site greater than 1 km in diameter would be needed to obtain a regional, rather than a localized palynological record, and so eight large mires in the Hadrianic–Antonine frontier zone were chosen for study (Fig. 19.2). These were chosen in relation to known Roman structures and to obtain a wide spatial distribution of sites within the frontier zone. A 3 m core was obtained from the centre of each site using a 30×9 cm Jowsey-pattern corer (Barber, 1984) but at Letham Moss, samples were obtained from a cut peat face using $50 \times 10 \times 10$ cm monolith tins.

Pollen analysis is so far complete on the cores obtained from Walton Moss (NY 504 667), Glasson Moss (NY 238 603), Fozy Moss (NY 830 714), Carsegowan Moss (NX 454 555), Dogden Moss (NT 705 500) and Letham Moss (NS 880 865). Counting was at 4 cm intervals from 300 to 100 cm and at 8 or 16 cm intervals from 100 to 0 cm, using a pollen sum of 250 NMP to ensure that the results could be compared with those from Bolton Fell Moss (Barber, 1981).

The Roman period in each diagram has been identified using the following evidence:

1. The radiocarbon-dated pollen diagrams from the area (Donaldson & Turner, 1977; Davies & Turner, 1979; Barber, 1981) all show that the first major rise in forest clearance indicators, particularly grasses, date from the Roman period.
2. The evidence, both archaeological and palynological, for Bronze Age clearance is sparse, as it is for Iron Age activity except in northern Northumberland; whereas the presence of thousands of Roman troops and their military activities are well documented.
3. Recent unpublished data from around the Cheviot Hills show massive increases in clearance indicators at levels radiocarbon-dated to late Iron Age/Romano-British times (Tipping, personal communication, 1991).
4. The clearance 'signatures', as represented by the grass pollen curves, all show a sudden increase, implying a cultural invasion, rather than a steady increase, implying local population growth.
5. The dramatic response of the grass pollen curves, and the relatively low level of arable indicators (Behre, 1986), is in accord with archaeological evidence for the Roman impact being one of woodland clearance for military purposes – fort-building, clearing fields of view, etc. – rather than for settled agriculture.
6. There is a measure of correspondence between the four raised mires featured in Fig. 19.4 (Letham, Walton, Glasson and Carsegowan Mosses) at around 200 cm depth, representing peat accumulation rates of about 9.5 yr/cm. Fozy Moss is not a true raised mire, though the surface vegetation is oligotrophic, and it appears to have a lower accumulation rate of 13.5 yr/cm if the Roman level is placed at 140 cm. The radiocarbon-dated Roman level at Bolton Fell Moss (Fig. 19.1), at a

depth of 105 cm – accumulation rate therefore of 17.7 yr/cm – is from a peat section that has almost certainly dried and shrunk, rather than a freshly sampled core.

A programme of radiocarbon dating all the sites on Fig. 19.2 is in progress and is necessary to confirm these views, but for the reasons cited above, and based on the similarities between these pollen diagrams and those reviewed earlier in this chapter, there is a fair degree of confidence in the assumptions made.

Walton Moss is a large raised mire system, 184 ha in extent, lying 13 km north-east of Carlisle at an altitude of 95 m. It lies 3.5 km north of Hadrian's Wall and the same distance from Stanwix, Castlesteads and Netherby Roman forts and a Roman road to the west. The moss was chosen for coring because of its proximity to these Roman remains and because it lies 1 km south of Bolton Fell Moss (Barber, 1981), thus allowing comparison of the record from two sites in the same area.

The summary diagram for Walton Moss is shown in Fig. 19.3. We consider, based on the criteria outlined above, that the Roman period is probably recorded between 192 and 172 cm, and is marked by a sharp rise in non-arboreal pollen from 18% NMP at 192 cm to 60% at 180 cm, and followed by a decline to 42% at 172 cm. This is very similar to the record from Bolton Fell Moss with the degree of maximum clearance, and then the trend towards increased tree cover, being broadly the same at both sites. Together with the marked deforestation recorded at nearby Fellend Moss (Davies & Turner, 1979), these data demonstrate the intensity of the Roman impact in the area very close to this major frontier work. The question then arises as to whether this impact was widespread within the frontier zone or was related to the proximity of Roman walls and forts.

The Gramineae curves from five of the sites are shown in Fig. 19.4. The probable begin-

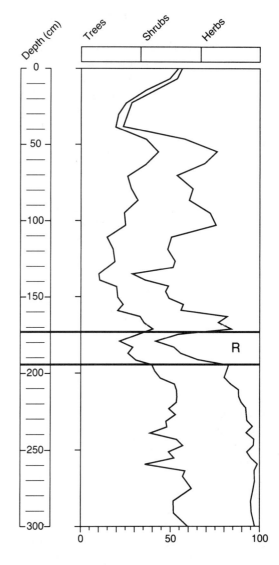

Fig. 19.3 Summary pollen diagram from Walton Moss, Cumbria. (R) assumed Roman period.

ning of the Roman period is represented in all the diagrams by the first major rise in Gramineae (marked R) and it is apparent that there is a degree of spatial variation in the intensity of clearance. The maximum Gramineae pollen (70%) occurs at Fozy Moss, which must be attributable to its situation adjacent to Hadrian's Wall itself and the nearby forts of Carrawburgh and Housesteads. Similarly,

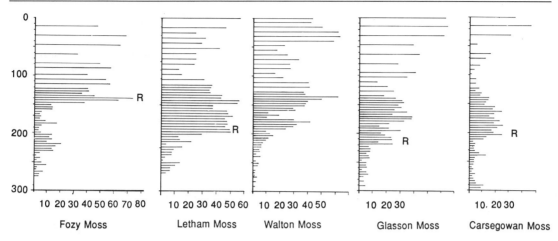

Fig. 19.4 Grass pollen curves (as percentage NMP) from five sites. (R) assumed Roman period.

relatively high Gramineae pollen, 50%, occurs at Letham Moss, situated close to the Antonine Wall and the forts at Camelon and Mumrills. Thus, as expected, the impact on the vegetation was greatest in the area immediately around the Roman walls and woodland was probably cleared for the construction of the walls themselves and their associated structures, as well as for agriculture.

At Walton Moss the Gramineae curve reaches a maximum of 40% in the Roman zone, indicating considerable areas of open ground, though less than at Fozy Moss and Letham Moss; and at Glasson Moss the grass level is lower still, at 28%. The Roman occupation of this area was not as intense as at the other sites, as there are no forts but only a series of mile-castles built along the coast. The lowest level of grass pollen, 25%, is found at Carsegowan Moss where there are no known Roman structures, though it is likely that the Roman presence in Dumfriesshire stimulated local native agriculture.

It is clear therefore that the simple hypothesis – that the proximity of an area to the Roman structures such as frontier walls and forts will be reflected in a greater degree of forest clearance – is upheld. This appears to be the first major human impact on the landscape of the region.

19.4 TESTING AND EXTENDING THE RECORD OF CLIMATIC CHANGE

The record of climatic change derived from peat stratigraphy and macrofossil analysis of raised bogs has been studied by Aaby (1976), Aaby & Tauber (1975), Barber (1981) and Haslam (1987). Smith (1985), working in the Humberhead Levels, and Wimble (1986), researching mires in south Cumbria, both recorded hydroclimatic changes synchronous across individual mires and which they interpreted as clear climatic signals, although sea-level change may have had some effect. Clearly there is a need for further testing and extension of the record (Stoneman, in preparation).

A pilot study of two cores a few metres apart on Walton Moss has shown good correlation of changes in the macrofossil assemblages, and the macrofossil curves at Bolton Fell Moss show a high degree of parallelism. The ten sites now being studied (Fig. 19.2) are in a relatively restricted area, minimizing temperature differences between sites as well as variation in vegetation composition, but there is a marked difference in precipitation totals, from c. 2250 mm at Ellergower Moss in Galloway, to c. 700 mm at Drone Moss in Berwickshire. Cores were taken from the geographical centre of each mire, where there was no evidence of surface flow or channels, and from present-day lawn environments, which are more sensitive to climatic change than hummocks or pools (Aaby, 1976).

The macrofossil results from Walton Moss, 1 km to the south of Bolton Fell Moss, are shown in Fig. 19.5. There are clear alternations in the *Sphagnum* species as well as interactive changes in the percentage of unidentifiable organic matter (UOM), monocotyledons and identifiable Sphagna. In such an ombrotrophic mire these changes must reflect changing surface wetness in the upper layers of the bog – the acrotelm – since the *Sphagnum* species show preferences in relation to water level (Ratcliffe & Walker, 1958; Ivanov, 1981; Boatman, 1983). *Sphagnum imbricatum* is considered to be indicative of relatively humid conditions (Casparie, 1972; van Geel & Middeldorp, 1988) and may reflect climatic fluctuations to more oceanic conditions (van Geel, 1978). However, this moss has a wide tolerance of water-level conditions (Green, 1968; Hill, 1988) and this may mask variations in local hydrology (Haslam, 1987). *Sphagnum* section *Acutifolia*, predominantly made up of hummock-top species may then indicate more continental conditions. The rise of *Sphagnum* section *Cuspidata* (44–20 cm) reflects a large increase in bog surface wetness and, as at a number of other British sites, at the top of the diagram *Sphagnum imbricatum* becomes locally extinct and is replaced by *Sphagnum magellanicum* and *Sphagnum papillosum*.

Clearly, there are strong signals in the data, both in composite indicators such as UOM and the total identifiable Sphagna, and in the taxon indicators such as the five *Sphagnum* species or sections. A common feature at all the sites is the general demise of *Sphagnum imbricatum*, a dominant peat-former, in post-Medieval times, though it is still recorded as present at Ellergower and Cranley Mosses (Nature Conservancy Council for Scotland, unpublished data). This date suggests that human interference through burning or draining the mires may be the cause, although this is contra-indicated at most sites by the fact that *Sphagnum* section *Cuspidata*, the wet-loving Sphagna, peaks as *S. imbricatum* is extinguished. Barber (1981, p. 199) and more recently van Geel & Middeldorp (1988) have reviewed this problem, the latter auth-

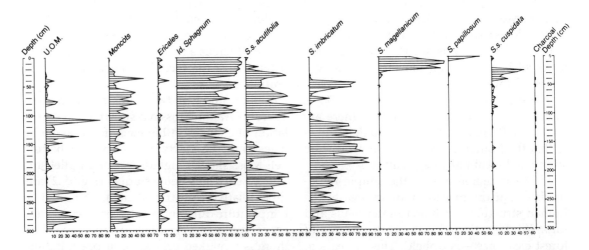

Fig. 19.5 Plant macrofossil diagram from Walton Moss, Cumbria. Scale is percentage of peat matrix. Presence/absence of charcoal shown. Cored 14 October, 1989. Analysis by Rob Stoneman.

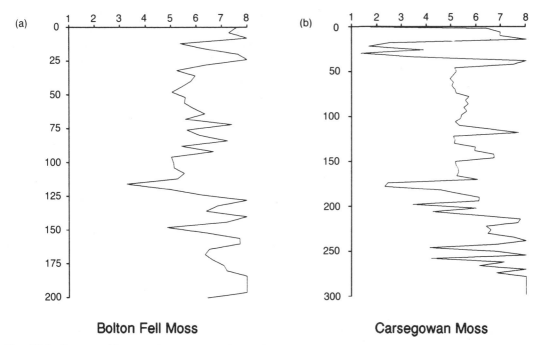

Fig. 19.6 Curves of bog surface wetness, based on weighted-average ordination (after Dupont, 1985) in (a) Bolton Fell Moss, and (b) Carsegowan Moss. Vertical scale is depth below bog surface in cm. Horizontal scale is index of wetness from 1 (wettest) to 8 (driest).

ors also favouring human influence. In the data from northern Britain there is also evidence to support a climatic cause (Barber, 1981).

It is possible to produce a relative hydroclimatic curve from these data by the application of weighted averages ordination (Dupont, 1985; Haslam, 1987), and Fig. 19.6 shows two such curves derived from macrofossil analysis of samples from Carsegowan Moss and from Bolton Fell Moss. These transformations of the raw macrofossil data must be regarded as only provisional attempts to express the data in a more accessible form. There are a number of approximations and assumptions built into this method, and there is a particular problem with assigning an index value to *Sphagnum imbricatum*. In these cases we have followed Haslam (1987) in giving the species a value of 5, but this may be too 'dry', and work is in progress to derive a transfer function by calibrating a contiguous dataset of *Sphagnum* leaf counts from the upper peat at Bolton Fell Moss with known climatic variation.

The two curves in Fig. 19.6 are undated except for the 200 cm level at Carsegowan being assigned to the Roman period by pollen-analytical correlation (Fig. 19.4). The Bolton Fell Moss surface-wetness curve is based on macrofossil data from a monolith taken from a section in old marginal peat-cuttings, and the peat may therefore have shrunk due to water loss. There are, however, sufficient similarities in the trends of these curves to justify further efforts in dating and correlating such data, with the ultimate aim of producing a series of high-quality sequential records, with a resolution approaching decadal, for a large part of northern Europe.

19.5 CONCLUSIONS

In the late Holocene of northern Cumbria there is clear evidence of periods of climate favourable to agriculture – for example AD 1200–1300 – but depressed farming activity due to warfare, and on the other hand, expansion of agriculture despite the evidence for a worsening climate – for example AD 1600–1800. Whilst this must caution us against simplistic and even deterministic views of prehistoric clearance and climatic change, one cannot ignore the evidence for a lack of human impact in north Britain during, say, the Iron Age, compared with the pollen-analytical and archaeological evidence for almost complete clearance in lowland and southern Britain (Turner, 1981; Brown & Barber, 1985) – but not without small-scale contrasts in the degree of clearance, as in Shropshire (Barber & Twigger, 1987).

This contribution has been framed in the light of Alan Smith's own early contributions to palaeoecology: a dual approach of research into human impact as revealed by pollen analysis, and into climatic change as revealed by peat stratigraphy. To the methods used by Alan Smith in the 1950s we can now add the powerful tool of macrofossil analysis of ombrotrophic peat to produce an integrated story of environmental change, natural and cultural.

Acknowledgements

We thank Professor Frank Oldfield, Dr Richard Tipping and Dr Frank Chambers for comments on an earlier draft, Nature Conservancy Council officers in northern England and southern Scotland for help in locating suitable sites, Dr Jim Milne for software for data manipulation and diagram production, the Cartographic Unit (Southampton) for the site map, Jane Barber for word-processing, and SERC and NERC for tenure of research studentships (for LD and RS).

20
Palaeoecology of floating bogs and landscape change in the Great Lakes drainage basin of North America

Barry G. Warner

SUMMARY

1. Floating bogs are a common feature of the landscape in the Great Lakes drainage basin of North America.
2. Palaeoecological studies confirm development and rapid horizontal growth of the quaking mat across kettle lakes during the historic period. These data refute the long-standing hypothesis that floating bog mats represent ancient relict habitats where northern disjunctions have persisted since early post-glacial time.
3. It is suggested instead that deforestation by European settlers upset the natural water balance of these systems either indirectly by altering the elevation of the regional ground-water table, or directly by accentuating seasonal extremes in the volume of atmospheric waters entering the basin, or both of these.
4. Human activity is therefore another cause, unrelated to climate, for bog formation in North America.

Key words: floating bogs, lake-fill bogs, kettle-hole mires, Great Lakes region, North America, deforestation

20.1 INTRODUCTION

We are beginning to realize that Europeans had a profound effect on landscape development in North America. In the Great Lakes drainage basin, the first Jesuit missions were established in the AD 1630s among villages of the Huron in south-central Ontario (Heidenreich, 1990). This event signalled a major change in the way humans manipulated the environment. Subsequently, the communal agricultural villages of the Huron Indians were gradually replaced by the European system of permanent colonial farms. The early Europeans had a minor impact on their environment in the 17th and 18th centuries AD compared to the great expansion of European agriculture and the establishment of a lumber industry in the 19th century (Gentilcore & Donkin, 1973).

Trees, initially only white pine (*Pinus strobus*), were cut for use both locally and for exportation to Europe. White pine was favoured for building furniture, houses, and other structures. The tall and straight trunks were shipped to Britain for construction of naval ships. With further population expansion and industrialization in later years, there was a greater dependence on forest resources. Other tree species were selected and the extent of forested land declined dramatically. Forested land was converted to fields. Forests probably became more dense in areas that remained unsettled because more summer sunlight was able to reach the once-shaded forest floor. The effect of massive deforestation on the environment reduced soil fertility, increased erosion, accentuated seasonal extremes in soil temperatures, and altered surface drainage patterns (Butzer, 1982).

The effect of Europeans on the hydrological balance of surface waters had consequences both locally in and around living areas, and on the regional water budget of the Great Lakes themselves. Historical documents from the middle to late 1800s record increases in volumes of runoff in the spring and in the frequency of floods along the major river and stream courses (Kapp, 1978; Janusas, 1987). Smaller streams and rivulets gradually became more ephemeral or disappeared. Water-powered mills ceased to function. These

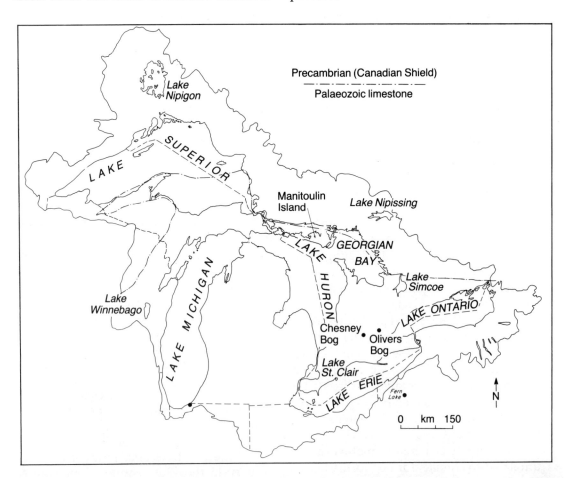

Fig. 20.1 Map of the Great Lakes drainage area, showing location of sites referred to in the text.

activities eventually led to the construction of dams to control frequent flooding and to conserve water (Janusas, 1987).

Landscape disturbance caused small wetlands to disappear and the vegetation on the larger ones changed in response to new supplies of water or to alterations in drainage (Snell, 1989; Warner, unpublished data). In this chapter, floating bogs (*sensu* Damman & French, 1987) in kettle-hole basins are examined as a special case because European alterations to the landscape seem to have had an effect by contributing to accelerated *Sphagnum* growth in these systems. I use the term floating bog in an informal sense to refer to any lake-fill bog (*sensu* Damman & French, 1987) or kettle-water wetland (Tarnocai *et al.*, 1988) with a quaking *Sphagnum* mat. Such bogs are referred to as 'schwingmoor' and 'kesselmoor' in Europe (Overbeck, 1975; Succow, 1988). General characteristics of all floating bog types throughout the Great Lakes drainage basin of central North America are explored, and the development of those in kettle-hole basins are considered in the context of landscape disturbance (Fig. 20.1).

20.2 EARLY INVESTIGATIONS

Floating bogs have long been of interest to naturalists and scientists in North America (e.g. Thoreau, 1837–1848; MacMillan, 1893; Bailey, 1896). Several ideas have been proposed that attempt to identify the ecological conditions for growth of these intriguing ecosystems; many of these ideas are still accepted today despite the lack of supporting scientific data.

In reviewing the early publications, three main explanations emerge. The first is phytogeographic. Distribution maps revealed disjunctions of plants and animals in the Great Lakes region, far south of their region of optimal growth in boreal North America. These disjunctions, many of which are found in bogs, were thought to represent northern relicts of intact boreal communities that have persisted since the northward wave of migration after deglaciation (Hooker, 1879; Transeau, 1903; Waterman, 1926; Deevey, 1949). The steep-sided and deep (kettle) basins were thought to have a local microclimate that maintained cool temperatures, moist air, and a shorter growing season, not unlike conditions in the boreal region.

The second theory considered physical and biological characteristics of the bog peat substrate. For example, it was felt that, if the causes for physiological xeromorphy in bog plants were understood, something might be learned about the nature and origin of the bog habitat as a whole (Dachnowski, 1912; Rigg, 1916; Waterman, 1926). Other features of the substrate that were explored included the temperature of the substratum and air immediately above it, soil pH, lack of aeration, presence of toxic substances, water-holding capacity of the peat, and lack of nutrients.

The last idea considered rates of change and direction of bog plant succession. Early workers observed, in general, a gradual continuum from open-water macrophytes, and loose and floating aquatic plants near the centre of bog basins, to terrestrial communities supported by a denser and drier substratum forming a ring around the periphery. Presumably, the former were the youngest communities and the latter were the oldest because of the time required for soil conditions to reach a stage suitable for upland plants to move onto the peat surface. As the floating mat continued to spread laterally, it thickened, and eventually filled in the entire basin (Transeau, 1903; Dachnowski, 1912, 1924). Concentric development of a well-zoned forest-to-floating-mat sequence originating at the edge of a lake basin remains the favoured explanation for bog formation in kettle basins today.

Although Waterman (1926, p. 269) agreed with the ideas of his contemporaries concerning the development of floating bogs, he observed that 'there has been a marked advance of the bog mat within the memory of man'. How then 'did a favourable habitat remain so long unoccupied, or why did an unfavourable habitat so suddenly change to a favourable one?'. Clearly there had to have been some sudden change in the water level in these bog basins. Waterman (1926, p. 270) continues: ' . . . there are indications that it may have been due to the activities of the first white settlers in the region, in clearing forests and in general drainage operations'.

20.3 ECOLOGICAL VARIATION

Floating bogs in kettles contain complexes of various wetland forms and types. Typical bog units exist where the water-table is at or slightly below the surface, and where acidic and nutrient-poor *Sphagnum* vegetation and peat is perched above minerotrophic ground waters. Fen and poor fen units develop where the peatland water is only slightly buffered from minerotrophic ground water. Open areas of deep water in the centre of the basin or in moats around the edge give to other parts of the floating bog complex the characteristics of shallow water wetlands (Tarnocai *et al.*, 1988). The bogs can vary in species composition depending upon water chemistry (i.e. alkaline or acidic; Crum, 1988). Yet if one follows a classification based on the source of waters feeding the system rather than plant composition, there are limnogenous, topogenous, and ombrogenous (*sensu* Damman & French, 1987) areas comprising the complete mire complex.

In an attempt to combine all of these features, Damman & French (1987) recognized four main types of floating bogs or peat bog–lake systems, all of which are represented in the Great Lakes drainage area (Figs 20.2, 20.3). Though poorly understood, there are common physical and ecological characteristics associated with floating bog mats, despite the variety of floating bog landforms, range of water types, and various vegetation

Fig. 20.2 Comparative photographs of Fern Lake (also known as Bradley Pond, Lake Kelso), Ohio. (a) From Dachnowski, 1912: a floating mat of shrubs, probably *Decodon verticillatus*, *Chamaedaphne calyculata*, and *Lysimachia ciliata*; little *Sphagnum*; thick cover of *Picea mariana* in background.

Fig. 20.2 (b) The same viewpoint, 1991: note absence of *Picea mariana*, believed cut for logs in the 1940s; *Larix laricina* is the only tree remaining in the grounded peat mat around the basin edge; lakeward is a floating *Sphagnum* mat with *Sarracenia purpurea* and *Vaccinium* spp. A ring of semi-submerged *Chamaedaphne calyculata* is anchored around the floating mat. Note little lateral growth of the mat, but dramatic change towards bog communities.

zonation patterns. One of the mysteries concerns the environmental factors differentiating the moat and pond border types (MacMillan, 1893; Bailey, 1896; Dachnowski, 1912; Waterman, 1926; Damman & French, 1987).

20.4 DISTRIBUTION

The distribution of floating bogs in the Great Lakes area is first a function of the topography and hydrological setting. It is important to differentiate between closed-basin and open-basin types. In the southern part of the Great Lakes drainage area, which is mainly covered by thick calcareous glacial drift deposits, closed basin systems are nearly always found in kettle holes. These bogs tend to occur on moraines, in glacial spillways, and on outwash and till plains. They are less common on glaciolacustrine deposits close to the existing Great Lakes, either because few ice blocks were left in these areas or possibly because later proglacial lake sediments filled up any kettle holes. The distribution of these types of floating bogs depends, then, upon the way late-Wisconsinan ice left disintegrating ice blocks during the final phases of deglaciation. Non-kettle type depressions in areas of thick glacial drift do not typically support floating bogs but have raised basin bogs or other wetland types.

Bedrock basins are the other major type containing floating bogs, most of which tend to

Fig. 20.3 Representative examples of floating bog or peat-bog lake systems in southwestern Ontario. (a) Little Turnbull Lake – an early stage of development of the moat bog type. Note dominance of shrubs and formation of remote island of vegetation. (b) Sudden Bog – a more advanced stage of the moat bog type. The centre is covered with *Picea marina* and *Larix laricina* with a mat of ericaceous shrubs and *Sphagnum*. Around the edge is an open water moat. (c) Oliver's Bog – a pond border bog, with a narrow open water moat, and wetland shrubs around the edge. A grounded mat is on the landward side, and a floating mat surrounds the central area of open water (d) Grass Lake – probably represents an early stage of lake–fill bog development. There is a narrow moat and an inner ring of low wetland shrubs. The basin centre is covered with sedges and herbs. A few stunted *Larix laricina* occur on the floating mat.

be part of an open drainage network. These floating bogs are most common on the crystalline bedrock of the Precambrian Canadian Shield in the northern part of the Great Lakes region. Such floating bogs are rare south of the Shield because there are few areas where the sedimentary Palaeozoic bedrock is exposed. However, floating bogs can occupy limestone bedrock basins, as on Manitoulin Island (Warner, 1988). Floating bogs may also occur in protected embayments along the shores of the Great Lakes and other large lakes such as Lake Simcoe and Lake Nipigon (Fig. 20.1).

Floating bogs are known to form in man-made reservoirs (Dachnowski, 1912; Swan & Gill, 1970) under circumstances not necessarily in harmony with the natural geological and hydrological conditions of the region.

Although all of the Great Lakes drainage area was covered by ice during the Wisconsinan, and floating bogs are a common feature of the landscape today, this does not mean that all floating bogs are confined to glaciated terrain. Deevey (1949) reminds us that bogs occur south of the glacial limit, particularly along the Atlantic coast and even as far south as Florida. Floating bogs in the southern part of the Great Lakes region contain not only northern boreal species, but also southern species. Examples of more southern species in floating bogs in the southern part of the Great Lakes drainage area are *Peltandra virginica* (arrow-arum), *Cephalanthus occidentalis* (buttonbush), and *Decodon verticillatus* (swamp-loosestrife) (Damman & French, 1987).

20.5 HYDROLOGICAL VARIATION

Given the appropriate topographic conditions for holding water, the hydrogeochemical regime will establish the course of succession for the developing bog. Unfortunately, the hydrology of floating bog systems is not well understood. Probably the most detailed study in North America is that of Hemond (1980) on a floating bog in a kettle-hole in Massachusetts. He found the basin was positioned in a ground-water recharge zone and precipitation was the sole source of water feeding the basin. Evapotranspiration was found to be the main mechanism for loss of water from the system, with runoff (surface and subsurface) of minor importance. He theorized that vertical movement downward through the mat occurs during precipitation and upward movement occurs by evapotranspiration. Runoff occurs during periods of high water when the bog is surrounded by a moat of water. Therefore, lateral runoff from the bog mat must be radially outward to the moat. Alternatively, some movement radially inward to the mat from the moat and rehydration of the mat must occur during periods of low water in the basin.

Hemond's (1980) study provides some important insights on the water budget of these systems, requiring further testing on other floating bog types. The seasonal, episodic (due to storms), or diurnal variations in lateral flow patterns could, in part, explain differences between the formation of moat and pond-border bog types in similar geological and hydrological settings. The complete annual water budget throughout the summer and winter must be included to evaluate the volume and geochemical importance of precipitation feeding these bogs. Seasonal variations in water volume could account for the floating mechanism of the bog mat. The existence of a non-stationary mat, which in itself has always been a source of puzzlement, suggests the ability to float up and down is an adaptation to dramatic changes in basin water volumes. The floating mechanism allows for maintenance of a constant chemical regime and protection from submergence of *Sphagnum* and other bog species growing on the surface (Crum, 1988).

20.6 STRATIGRAPHIC VARIATION

Most references to origin and development of floating bogs are casual observations based on dominant plant communities on the surface. The bogs that have been studied stratigraphically reveal underlying gyttja or other lake muds, indicating the existence of a lake during inception of the basin (Potzger, 1956; Kratz & DeWitt, 1986; Miller & Futyma, 1987; Winkler, 1988; Aravena et al., 1992). Dachnowski (1912, 1924) conducted the first detailed stratigraphic investigations of these bogs and published his classic diagram on the stages of basin in-filling. He identified the process of primary peat deposition and terrestrialization of closed basins.

The comparatively recent formation of peat in floating bog basins does not support the view that bogs are habitats where post-glacial relicts have persisted continuously since deglaciation (Deevey, 1949). Furthermore, fossil pollen profiles from underlying lake mud and peat show early post-glacial spruce assemblages interpreted to indicate 'boreal-type' vegetation, but the boreal pollen assemblages do not persist through the peat to the modern surface (Sears, 1935; Deevey, 1949; Potzger, 1956; Holloway & Bryant, 1985; Anderson et al., 1989).

Radiocarbon dating and detailed stratigraphic studies have identified the nature and timing of hydroseral succession leading to the formation of floating bogs (Kratz & DeWitt, 1986; Miller & Futyma, 1987; Wilcox & Simonin, 1988; Winkler, 1988). These authors essentially refine Dachnowski's original concept of hydroseral succession. They conclude that climate is an important determinant in producing hydrological conditions suitable for onset of bog formation during the latter part of the Holocene.

20.7 HUMAN ACTIVITIES

Observations over an additional 65 years or so since Waterman's (1926, p. 269) postulation of bog mats forming 'within the memory of man', indeed support the origin and rapidity of bog mat growth in kettle lakes and the influence of humans on formation of floating bogs (Buell et al., 1968; Swan & Gill, 1970; Hemond, 1980; Crum, 1988) (Fig. 20.3). If rates as rapid as 2.5–6.5 cm/yr are realistic for mat advancement (Swan & Gill, 1970; Hemond, 1980; Kratz, 1988), then an extrapolation backwards in time would suggest that the bog mats must be recent features, and clearly, are not post-glacial relicts.

Palaeoecological analyses on short cores through floating bog mats in kettle holes in southwestern Ontario confirm a recent origin (Warner et al., 1989; Bryan, 1991). Our studies have been conducted on both pond-border and moat bog types. From six sites studied to date, palynological analyses reveal a pre-*Ambrosia* pollen period (i.e. before European settlement) with abundant tree pollen of such forest species as *Pinus strobus*. There is a later *Ambrosia* pollen period with reduced proportions of tree pollen and abundant pollen of weedy species, plus cultivated taxa such as *Zea mays* (sweet corn) and *Nicotiana* (tobacco). The *Ambrosia* period (corresponding with the historic period) is accompanied by abundant spores and macrofossils of *Sphagnum*, and pollen of other typical bog species. We interpret these results as indicating that mat formation not only began within historical time as postulated by Waterman and later workers, but that rapid horizontal growth of the mat was a response to disturbance of the natural landscape by Europeans. There is no evidence for climatic control on bog growth.

Our current work on the hydrogeochemistry of Oliver's Bog, a pond-border bog in south-western Ontario, is attempting to identify processes operating within the mat that initiated and maintain it (Aravena et al., 1991; and unpublished data). The evidence using isotopic indicators agrees with Hemond's (1980) study and his hypothesis for the water balance in the bog mat. The water reaching the mat at Oliver's Bog is atmospheric in origin, being primarily melt waters in the spring, followed by summer rain water later in the growing season. There is little loss of water from the mat during periods of high water levels in the basin (i.e. spring). Greatest loss of water through evapotranspiration occurs during the late summer and early autumn dry spell.

Deforestation by Europeans could cause reduced soil aeration and water-holding capacity or infiltration rate, as well as accelerated surface runoff; all this would not permit as much precipitation to percolate into the subsoil and into the ground water. One result would be a drop in the ground-water table and a reduction in discharge to streams and surface water bodies; some would no longer be able to maintain a base flow during the dry part of the summer. A reduction in base flow certainly accounted for the demise of mills on smaller water courses in the 19th century. However, if Hemond's findings are representative of all floating bog systems in kettle-hole basins, then direct surface runoff into the basin is of minor importance. His speculations on the lateral transport of waters received as precipitation are probably more important, because an unnatural and persistent drop in the ground-water table in the basin would cause inward lateral flow due to evapotranspiration on the mat from basin edge to basin centre of waters received as precipitation. A reversal in lateral transport radially outwards would occur during the period of the spring melt when basin waters would be highest. This would be during the period of the spring melt as suggested by our preliminary results on the hydrogeochemistry. Deforestation by Europeans would have reduced the regulating effects of vegetation cover on snowmelt and possibly snow accumulation, raising basin water level much higher than normal high spring levels and putting the basin in a brief spring 'flood stage'. Waters would be raised either through direct internal surface runoff, or subsurface runoff into the basin through the ground-water flow system.

The duration and direction of lateral surface runoff in the basin would depend upon the size, shape, depth and nature of the enclosing geological deposits, the elevation of the basin with respect to the ground-water table, and the impact of landscape change on the local hydrological regime. The strength of each of these factors might determine the mode of floating mat development, either beginning at the edges (i.e. the pond-border type) or beginning as an island in the centre (i.e. the moat bog type) of the basin.

I have not mentioned the possible consequences of landscape disturbance on changes in basin hydrogeochemistry associated with the expected reduction in soil fertility, increased soil acidity, and leaching of nutrients that could produce favourable conditions for acid-loving bog plants. Changes in hydrogeochemistry are probably less important in the calcareous terrain of the southern part of the Great Lakes region than soils on the crystalline acid soils of the Canadian Shield in the north (Fig. 20.1).

Landscape disturbance identifies the time and cause for the origin of bogs in kettle basins. However, the mechanisms leading to rapid horizontal expansion of the floating mat, and the mechanisms for maintenance of a buoyant mat that changes elevation seasonally, need to be identified next. These are mechanisms created as a consequence of disturbances in the natural landscape. Further palaeoecological work is underway to characterize the hydrological system and prevailing ecological conditions in the lake basin before European time. Also, it is necessary to determine the origin of the peripheral forested bog zone, where it might be expected to have a history pre-dating Europeans.

Acknowledgements

I thank G. Bryan, K. Hanf, H. Kubiw and L. Lamb for field and laboratory assistance and for many useful discussions on floating bogs in southern Ontario. Versions of this manuscript benefited from critical readings by R. Aravena, L.R. Belyea, F.M. Chambers and D.J. Charman. N. Bahar prepared the figures. Funding support by the Natural Sciences and Engineering Research Council of Canada is acknowledged.

21

Late-Quaternary climatic change and human impact: commentary and conclusions

F.M. Chambers

SUMMARY

1. The subject matter of the volume is placed in late-Quaternary perspective; guidance is given to individual chapters and to recent literature.
2. Comments are made on some current themes, including precision and accuracy in dating, late-Pleistocene environments, the spread of Holocene temperate forest trees, volcanic records, peat stratigraphic studies, conflicting proxy climatic records, human impact on fauna in the late-Pleistocene, and human impact on vegetation in the Holocene.
3. Views on past climatic change, and on the magnitude and extent of human impact on past landscapes and present climate, have changed significantly over the years. Contemporary issues of climatic change, nature conservation and environmental management can be given added perspective through studies in late-Quaternary palaeoecology and environmental archaeology.

Key words: late-Pleistocene, Holocene, climatic change, human impact

21.1 INTRODUCTION

Nearly forty years have elapsed since A.G. Smith embarked on his scientific career. During the past four decades there have been improvements in palaeoecological techniques (Berglund, 1986b), increased collaboration between palaeoecologists and archaeologists, and application of Quaternary palaeoecology to address specific problems of the cultural landscape – past, present and future (Birks *et al.*, 1988; Berglund, 1991). Views concerning late-Quaternary climatic change and human impact have changed markedly since the 1950s.

This final chapter is intended (1) to draw the themes of the book together, (2) to act as a guide to individual chapters, and (3) for newcomers to the field, to place the contents of the volume in perspective by examining selected aspects of the record and mechanisms of late-Quaternary climate change and vegetational development. To achieve these ends, themes raised in Parts One to Four are first reconsidered. Comments emphasize developments in techniques, advances in understanding over the last 40 years, and remaining problems. Finally, the contemporary relevance of palaeoecology and environmental archaeology is considered in the context of nature conservation and environmental management.

21.2 PRECISION AND ACCURACY

21.2.1 Precision and accuracy in taxonomy and sampling

It is emphasized in Part One that a hallmark of good palaeoecological research is an insistence on good quality, precise (exact), but also accurate (correct) data. In Quaternary studies, many macrofossils or microfossils are now routinely precisely identified to species or genus – for example, pollen analysts in Europe and North America can use revised keys in Faegri & Iversen (1989) and Moore *et al.* (1991) – but careful comparison with detailed type collections is needed to ensure accuracy (cf. Chapter 1). In sampling, a profile can be precisely subsampled (Cloutman, 1987) – for example, for 'fine-resolution pollen analysis' (Chapter 10) – but to ensure that peat sections and sediment cores are labelled correctly, meticulous field and laboratory practices are essential.

Whilst precision may be improved, accuracy may still not be attained: '... measurements can be quite precise and at the same time not very accurate' (Bloom, 1989, p. 2). In palaeoecology, this paradox is best exemplified by a consideration of dating techniques.

21.2.2 Precision and accuracy in dating

In current studies of late-Quaternary climatic change and human impact, one of the most frequently used dating techniques is radiocarbon dating. Forty years ago, the technique was not available to Quaternary scientists. Over the past 20 to 30 years, it has been increasingly applied by researchers to date a range of sediments and organic materials. In the last two decades, in combination with dendrochronology, radiocarbon dating has revolutionized Holocene research. The lack of an effective dating method to separate and to correlate mid- to late-Pleistocene terrestrial records – and the problems that has created (Bowen, 1991) – illustrates how valuable radiocarbon dating has been for the Holocene.

There have recently been notable improvements: (1) accelerator mass spectrometry (AMS) dating now permits the use of very small samples (Hedges, 1991); (2) 'wiggle matching' may be used when several, closely spaced, samples are available (though resolution is better for some periods than others – see Chapter 3); (3) sample fractionation permits better estimates of the true radiocarbon age of sediments (Dresser, 1970; Walker & Harkness, 1990). Nevertheless, even with so-called high-precision calibration, individual radiocarbon dates are not sufficiently accurate to date material to within a few centuries, let alone decades. Beyond the tree-ring calibrated range, many radiocarbon dates from presumed late-Pleistocene interstadial deposits are not only imprecise but also probably highly inaccurate (cf. Evin, 1990), particularly when compared with estimates on the same or associated material by amino acid racemization (Bowen *et al.*, 1989). Despite such problems, a realistic approach to radiocarbon dating ensures it still has much to offer (see Chapters 3 and 8).

Radiometric techniques do not really provide dates; they provide estimated age ranges. They contrast with the precision and accuracy attainable in annually laminated sediments (varves), in ice-core research, and particularly in dendrochronology in which the precision of measurement is as close as the season of the year and accuracy can be as good as (and in some cases arguably better than) the sources available to the historian. However, even with such high precision and potential accuracy, dendrochronology and dendroclimatology are not without problems of interpretation (see Chapter 4).

New dating techniques, employing AMS for long-lived radioisotopes (e.g. ^{36}Cl), give imprecise but possibly reasonably accurate ages (see Elmore & Phillips, 1987; Phillips et al., 1990). New dating techniques in archaeology (cf. Aitken, 1990) give potentially accurate but somewhat imprecise ages (e.g. infra-red stimulated luminescence, used to date buried sediments are as yet both imprecise and inaccurate Romola Parrish, personal communication); improvements are awaited.

21.3 CLIMATIC CHANGE AND VEGETATIONAL RESPONSE

21.3.1 Climate changes of the last 500 000 years

Forty years ago, late-Quaternary research was still dominated by the notion of four ice ages. That view has been superseded. Climatic changes over the last 500 000 years are now frequently defined with reference to ocean-core oxygen-isotope stages (Fig. 21.1). Oxygen-isotope ratios are taken as a surrogate measure of ice volume and, by inference, of sea levels (Shackleton, 1987), and possibly of global temperature. The implications of the ocean-core record are far reaching: the data imply that: (1) post-glacial time, though elevated to the status of an epoch (the Holocene), can be regarded in mid to high latitudes as merely a temperate stage (or 'interglacial'); (2) that the Holocene is merely one of a large number of short-lived warm episodes in the Quaternary; (3) that inferred sea levels have seldom been higher over the last half million years than they are now; but (4) that a new ice build-up seems inescapable.

21.3.2 The last cold stage

Recent research has indicated that, although cold episodes – with inhospitable glacial or periglacial climates – have dominated the last 100 000 years in northern latitudes, there have been exceptions, both spatially and temporally. At the time of the last glacial maximum in North America (Wisconsinan), evidence suggests that Beringia, which incorporated a large land bridge across the Bering Straits, was largely ice free. Even the term periglacial is inappropriate there for times

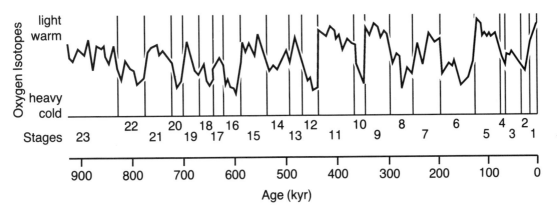

Fig. 21.1 Oxygen-isotope stages, with approximate timescale only (simplified from Mannion, 1991, based on Shackleton & Opdike, 1973b). The Holocene is stage 1.

Table 21.1 Marine oxygen-isotope stages, and suggested terrestrial equivalents in Britain, Denmark, France, Germany and The Netherlands (after Behre, 1989; Bowen, 1989, 1991; Bowen et al., 1989). North-Central Europe has c. 8 Weichselian interstadials compared with c. 4 or 5 (Devensian interstadials) recognized in Britain. Different techniques have been used for correlation and/or dating of these. Conflicting results highlight the difficulties in correlating relatively recent Quaternary terrestrial deposits. For guidance on Quaternary dating methods, see Smart & Frances (1991)

*Age (ka)	Oxygen-isotope stage	Britain	Denmark	France	Germany	Netherlands
80–90	5a	?†		St Germain II	Odderade	
90–100	5c	Chelford§	Brorup	St Germain I	Brorup	Amersfoort/Brorup
>120	5e	Ipswichian	——————————Eemian——————————			

* Approximate estimate of absolute age, not age span.
† The Upton Warren interstadial, with radiocarbon dates indicating an age of c. 40 000 BP, has been correlated on the basis of amino acid ratios with stage 5a (Bowen et al., 1989); but the flora of the Brimpton interstadial has been correlated with that of the Odderade and thus with stage 5a (see Ehlers et al., 1991 p. 494).
§ Radiocarbon dating of the Chelford interstadial produced an age of c. 60 000 BP, compared with a more recent estimate of an absolute age of 90–100 000 years inferred from thermoluminescence dating (Rendell et al., 1991).

that show evidence for high primary productivity and a range of large ice-age mammals (Hopkins et al., 1982). In Europe, the last cold stage (Weichselian, or Wurm; Devensian in Britain; Midlandian in Ireland) was interrupted by interstadials of reasonable warmth, some with forest development (Behre, 1989). Correlation of such interstadials with the oceanic record has proved problematical, largely owing to an over-reliance on radiocarbon dating that has resulted in underestimates of the age of terrestrial sediments. A revised scheme is suggested in Table 21.1.

In north-west Europe, interstadial forests seem to have been markedly different from Holocene temperate forest. For example, *Picea abies* ssp. *obovata* (*Picea obovata* Ledeb. – Siberian spruce) apparently has records in Britain (Chambers & Blackford, unpublished data), possibly of oxygen-isotope stage 5c (Table 21.1), yet its present range is 3000 km to the east, and there is now a distinct subspecies (or species) in between – the Norway spruce, *Picea abies* ssp. *abies* (*Picea abies* (L.) Karst.) – also not a Holocene native to Britain. Such significant changes in range in the late Quaternary have implications for future nature conservation (see below).

Major advances in understanding cold-stage environments have been made over the last 40 years, but a problem with reconstructing the vegetation of stadials is that there is apparently no modern analogue (cf. Delcourt & Delcourt, 1991) for either the steppe–tundra vegetation of continent-wide extent or for the large-mammal fauna that inhabited the land areas beyond the ice sheets. The landscape was apparently open and windswept, and has been described as the 'mammoth steppe' (Guthrie, 1982): mammoth in extent, from Beringia to Western Europe, and home of the mammoth (*Mammuthus*). Temperate trees were presumably sparse or absent. Their full-glacial refugia remain to be located precisely (cf. Chapter 17).

Recent research has shown that deglaciation was apparently relatively rapid. Once initiated, sea-level rise, floating and calving of ice margins, and downdraw of the ice are now believed to have been the principal deglaciation mechanisms. A complete record of vegetational change during deglaciation is thus subject to the survival of deposits from an environment characterized by periodic high energy. The terrestrial record, produced largely from low energy, lacustrine depositional environments, may be providing a biased picture of the landscape. Even with improved dating techniques (Walker & Harkness, 1990), it can be difficult to chart vegetation changes precisely. In Europe, particularly, there is the problem of the 'Younger Dryas' stadial – a sudden return to cold conditions – that immediately precedes the Holocene. Apart from the continuing debate over the cause of this cold snap (Berger, 1990; Zahn, 1992), it is not clear whether certain warmth-demanding taxa that may have become established during the preceding interstadial (e.g. *Alnus*; cf. Bush & Hall, 1987) then managed to survive through to the Holocene.

21.3.3 Holocene temperate forest development

(a) Isopoll mapping
Over the past two decades, the application of isopoll mapping has permitted tracing of the Holocene spread and establishment of major forest tree taxa into previous glacial and periglacial regions (cf. Huntley & Birks, 1983), though timing the first arrival of a taxon into any locality is presently impossible (cf. Bennett, 1986). Pollen values can be integrated to produce palaeovegetation maps (Huntley, 1990b). Difficulties are nevertheless encountered in space and time. Though radiocarbon dating has permitted the separation of changes previously believed to be synchronous over wide areas (cf. Smith & Pilcher, 1973), it has also produced spurious time-space correlations, revealed as artefacts of 'plateaux' in the calibration curve (see Chapters 7 and 17).

(b) Patterns of spread
It is clear that some tree taxa were well adapted for rapid Holocene spread (migration) and effective colonization (cf. Chapter 17). For eastern North America, a fascinating pattern of differing migration rates, directions of spread and apparent location of refugia has emerged, with individualistic behaviour by taxa (Davis, 1983; Davis *et al.*, 1986; Webb, 1988; Delcourt & Delcourt, 1981, 1991). Identifying the rates of spread has proved less easy at the scale of Britain and Ireland (Birks, 1989), and for some taxa there, e.g. *Alnus glutinosa* (Bennett & Birks, 1990), patterns of spread have proved very difficult to discern.

The main influences upon a taxon's arrival in any locality are (1) distance from refugia; (2) competitive ability; (3) prevailing climatic, ecological and edaphic conditions; (4) agencies of establishment – those that affect or alter local substrate and light conditions, notably disturbance agencies; and (5) vectors of spread (cf. Chambers & Elliott, 1989) – the means of effecting dispersal over short or long distance, such as wind, water, birds (Johnson & Webb, 1989) or mammals. Over recent years, reconsideration has been given to (4) and (5).

(c) Agencies of establishment and vectors of spread
Present-day vectors of *dispersal* are well-known, but these may not equate with past vectors of *spread*. There are several reasons for this. Vectors, such as the passenger pigeon of North America – now believed to be responsible for the rapid spread (migration) of nut trees on a sub-continental scale (Webb, S., 1986) – have become extinct. Present-day active agents of dispersal (such as some small mammals and birds) may have such localized territories that their significance in aiding rapid spread over long distances may be questioned. At least one mammal, *Homo sapiens sapiens*, implicated by some authors in the spread of certain taxa, has markedly changed both its subsistence base and its behaviour patterns in the Holocene!

Whilst it is well known that in late prehistory, tree taxa have been introduced to new regions of the same or adjacent continent by humans (e.g. transport of 'incense' trees to Egyptian Thebes *c.* 1500 BC; the assumed Roman introduction of the walnut (*Juglans regia*) into peripheral regions of the Roman Empire), evidence is lacking for active human dispersal of forest tree taxa in the early post-

Table 21.2. Some examples of volcanic eruptions, inferred from one or more of tephra layers, depressed growth of Irish oaks, depressed growth of North American oaks, ice-core acidity in Greenland, historical records (compiled from Hammer et al., 1980; Bradley, 1985; Baillie & Munro, 1988; Baillie, 1989; Burgess, 1989; Dugmore, 1989; Baillie, 1990b; Housley et al., 1991; Beget et al., 1992). Estimated ages are in calendar years.
The radiocarbon estimated ages of Santorini and Aniakchak overlap, whilst other notable eruptions in the 17th century BC are Mount St Helens and Vesuvius

Eruption	Estimated age	Evidence
Tambora	AD 1815	historical records
Laki	AD 1783	historical records
Hekla 1	AD 1104	historical records
?Rabaul	AD 536	tree rings; ice core
Mt Etna	44 BC	tree rings; ice core
Not known	207 BC	tree rings; ?ice core; Chinese historical
Hekla 3	1159 BC	tree rings; ice core; ?Chinese historical
?Santorini	1628 BC	?tree rings; ?ice core; ?Chinese historical
or: ?Aniakchak	1740 BC	radiocarbon dates around tephra layer
Not known	3195 BC	tree rings; ice core
Not known	4375 BC	tree rings; ice core
Mt Mazama (Crater Lake, Oregon)	5400 ± 100 BC	ice-core; ?tree rings

glacial. Humans could nevertheless then have been agents of disturbance (Moore, 1986b; Chambers & Elliott, 1989), if not vectors of spread. However, acknowledgement of the human vector in the very late arrival (i.e. in recent centuries) and subsequent success of introduced and naturalized taxa (e.g., *Rhododendron ponticum* and *Acer pseudoplatanus* (sycamore) in Britain and Ireland) surely raises again the notion of migrational lag – a concept that is tested in Chapters 6 and 17 for early-Holocene patterns of forest trees, but there finds little support.

(d) Climatic forcing of vegetation
Clearly, the prevailing climate will have been a major influence upon the post-glacial establishment of forest trees. Recent research now emphasizes the importance of climate *forcing* of vegetational change. This notion is hardly

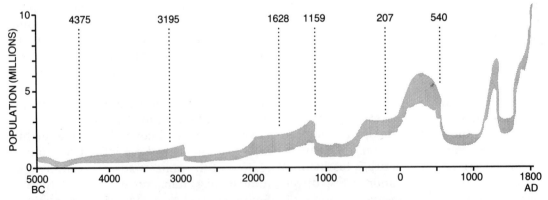

Fig. 21.2 A suggested population graph for Britain, with dates of postulated volcanic 'events' shown (modified from Burgess, 1989). Dates of volcanic 'events' are based on tree-ring dating, cf. Table 21.2. It has been suggested that short-term climatic deterioration (indicated by sequences of narrow tree-rings) might have been followed by population crashes after some, but by no means all, such 'events' (but see cautionary note in Baillie, 1991b).

new. It has underpinned most research into post-glacial vegetational history. Indeed, one could argue that only seminal papers, such as Smith (1965) on threshold and inertia, and Iversen (1949, 1956, 1964) on human influence, could have ever made researchers question the view that climate is the all-embracing forcing mechanism, mitigated by autogenic and allogenic successional processes. However, climate forcing has recently been re-emphasized, and presented in a rather different light (Huntley, 1990a, 1990c, 1992, in press); Prentice et al. 1991; see also Chapters 6, 7 and 17). Climate forcing may, for example, have influenced altitudinal tree limits during the Holocene (Kullman, 1992).

21.3.4 Volcanic activity and effects on climate

Hypotheses to account for past climatic change have recently been refined. Though, over long timescales, the major determinant of climate is believed to be variations in the earth's orbit (so-called 'Milankovitch' or 'astronomical' forcing; cf. Fig. 5.2), other mechanisms have also been suggested and may play a part (Bradley, 1985). There is now a growing view that, over shorter timescales, high-frequency variations in climate have been dictated, or at least strongly influenced, by two main factors: (1) orbital-related solar insolation and (2) volcanic aerosols (Nesje & Johannessen, 1992). The first can be calculated, and the second can be estimated from acidic gas concentrations found, for example, in Greenland ice cores (though not all eruptions leave a discernible signal, and signals are not necessarily proportional to the size of eruption nor to their climatic impact). The first now offers a predictive tool for future Holocene climatic change; the second – as yet relatively unpredictable – can be expected to perturb the calculated orbital-related climatic signal.

A major development in late-Pleistocene and Holocene palaeoecology has been the recognition of discrete volcanic ash (tephra) layers in sediments. Tephra layers (Table 21.3) have particular significance. They are time-stratigraphic markers; they may also imply short-term climatic changes, reflected in tree-ring widths (Table 21.2). The effect on human populations of past short-term climatic change is now a major, and controversial, area of Holocene archaeology (Fig. 21.2; see Burgess, 1989; Baillie, 1991b). Recently, microscopic tephra shards have been found in peats (Table 21.4), but have yet to be fully integrated with the climatic record inferred from peat stratigraphy (Chapters 5 and 19).

Table 21.3 Tephra layers in New Zealand, recorded in mire sediments (compiled from Hodder et al., 1991; Newnham & Lowe, 1991). Estimated ages are in uncalibrated radiocarbon years

Tephra Layer	Estimated age
Kaharoa	c. 770 BP
Taupo	c. 1850 BP
Mapara	c. 2160 BP
Whakaipo	c. 2685 BP
Egmont 2	c. 3700 BP
Egmont 4	c. 4100 BP
Hinemaiaia	c. 4510 BP
Whakatam	c. 4830 BP
Tuhua	6130 ± 30 BP
?Mamaku	7250 ± 20 BP
Rotoma	8530 ± 10 BP
?Opepe	9050 ± 40 BP
Maungarei	c. 9500 BP

Table 21.4 Icelandic tephra layers offering possible chronological markers in parts of north-west Europe (after Dugmore, 1989)

Tephra	Estimated age
Aska	AD 1875
Oraefajokull	AD 1362
Veidivotn	AD 900
Hekla 3	c. 2800 BP
Hekla 4	c. 4000 BP
Saksunarvatn	c. 9100 BP
Vedde	c. 10 600 BP

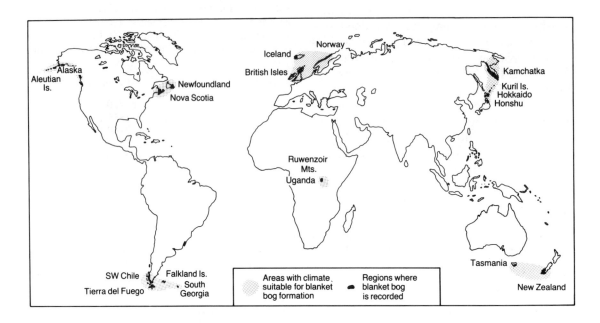

Fig. 21.3 Global distribution of blanket mires and related mire types (cf. Fig. 5.2). Diagram supplied by Jeff Blackford, based on Lindsey *et al.*, 1988).

21.3.5 Proxy records of climate from peat stratigraphy

Peat stratigraphy has been an influential source of proxy climate records for the Holocene – it gave rise to the Blytt–Sernander climatic periods (Chapter 5) – but, due to the persistence of the cyclic regeneration theory (Backeus, 1990), it remained one of the least-developed areas of climate research, and is today not readily listed by authors as a principal source of proxy palaeoclimatic data (e.g. Nesje & Johannessen, 1992). Yet it is claimed to yield continuous proxy climatic curves (Barber, 1981), and is potentially a global source of proxy climatic data (Fig. 21.3). Dupont (1985), Haslam (1987) and Stoneman (see Chapter 19) have applied techniques of plant macrofossil analysis to reconstruct past climates, though the 'Dupont index' (cf. Fig. 19.7) includes mainly Sphagna (bog mosses), not all mire taxa. An alternative technique for discerning proxy climate records is determination of peat humification (Aaby, 1976). Current research (by Barber *et al.*) indicates that though the direction, timing and frequencies of changes are not dissimilar from those indicated by the Dupont index, the magnitude of the peat humification changes may be markedly different (Chambers, unpublished data), depending on values assigned to individual taxa by the Dupont index (Maddy, personal communication). A 'multi-variable' approach is therefore advised in future peat-based climatic research (Chapter 5).

21.3.6 Conflicting evidence for past climatic change

Other conflicts of evidence exist. The Little Ice Age (Grove, 1988) of the 14th to the 18th centuries AD in north-west Europe has apparently been traced from southern Europe (Ponel *et al.*, 1992) to British Columbia (Gottesfeld & Gottesfeld, 1993). Yet, though the

Little Ice Age is apparently well established from research on glacier fluctuations in Norway (see Chapter 8), there is little confirmation of depressed temperatures from Scandinavian tree-rings (Briffa et al., 1990).

Such conflicts of evidence are not uncommon. It is a reminder that these are proxy climatic records, whose sensitivity may vary (as does climate itself) in time and space.

21.4 ASSESSMENT OF EVIDENCE FOR HUMAN IMPACT

21.4.1 Pleistocene impact

Today, as 40 years ago, the dominant technique for reconstructing human impact upon vegetation in the Quaternary is pollen analysis (Birks, 1988b). However, inferred human impact on vegetation is not now confined to the Holocene. In the southern hemisphere, human impact has been inferred from deposits tens of thousands of years old, whilst in the northern hemisphere the oldest palynologically recorded forest recession of some 250 000 BP has been attributed to *Homo erectus* (Chapter 9).

There are also lines of evidence that permit investigation of human impact upon fauna. It can be inferred from kill sites, and more graphically, from Palaeolithic art, that humans were having an impact on fauna at least by the late Pleistocene, not merely in Europe (Gamble, 1991) but in many other regions of the world (Bahn, 1991). How significant was that impact and how localized (e.g. of the neanderthals of southern Europe – *Homo sapiens neanderthalensis*) is difficult to determine. Towards the close of the Pleistocene, modern humans (*Homo sapiens sapiens*) may have had a significant and possibly crucial impact upon the ice-age fauna, notably in North America, where large-mammal extinctions have been partly attributed to human predation. The timing of the extinctions no longer appears to correlate with dated evidence for the arrival of the first humans (notably in North and South America), though this may not entirely invalidate models of Pleistocene overkill, both at the continental and global scale (cf. Martin, 1984).

21.4.2 Holocene human impact: a recurrent theme

For the Holocene, the story is rather different. A recurrent theme of Holocene environmental history in Europe has been the influence of human activity, not so much on fauna, but on flora – particularly on woodland (Thirgood, 1981). In north-west Europe, impacts have mainly been discerned through pollen analysis (see Chapters 12, 14, 15 and 16), though in chalkland areas, molluscan analyses have provided an alternative data source (Chapter 13). Human activity there has been implicated in the relative abundance of taxa, such as *Corylus avellana* (see Chapters 10, 11 and 17), in their establishment (e.g. *Alnus glutinosa*: Smith, 1984; Moore, 1986b), and in their decline (e.g. elm (*Ulmus*; cf. Smith, 1981b). Counter arguments include, *inter alia*, climate (for hazel), local site development and succession (for alder; Brown, 1988), and disease (for elm; Heybroek, 1963).

Studies in Holocene palaeoecology have tended to be site-specific and inductive. Recently, Birks (1988b, p. 463–466) wisely called for increasing use of hypothesis testing in palaeoecology and cultural landscape history (see Chapter 2), though he acknowledged that it can be difficult to test hypotheses in palaeoecology. As an example, consider the hypothesis of elm disease to account for the apparently near-synchronous decline of elm pollen in prehistory in north-west Europe. The hypothesis has some support: (1) a modern analogue in so-called Dutch elm disease that can be investigated palynologically (Perry & Moore, 1987); (2) a claimed parallel of disease-related decline in North America (the hemlock (*Tsuga*) decline; Davis, 1981); and (3) the vector of elm disease – the elm bark beetle *Scolytus scolytus* – has been found in deposits of Neolithic age (Girling, 1988). It is nevertheless difficult to **test** the disease hypothesis to account for the observed decline in the pollen curves of elm in prehistory (cf. Garbett, 1981). Without careful framing of hypotheses (Peglar & Birks, in prep.)

21.4.3 Mesolithic human impact in north-west Europe

Ideas of Mesolithic human impact on vegetation and landscape have changed over recent years. Forty years ago, it was believed that early-Holocene human populations in north-west Europe had minimal environmental impact: they were believed '... content to be confined by the forest to the margin of sea, lake and river' (Mitchell, 1953, p. 431). Human impact has since been invoked to account for a range of pollen stratigraphic changes in the early Holocene (Dimbleby, 1962, 1985; Simmons *et al.*, 1981; Bush, 1988), including the early-Holocene abundance of hazel (*Corylus avellana*). That *multiple working hypotheses* (cf. Birks, 1988b, p. 465) can be framed to account for palynological changes in this taxon's abundance, and then rigorously examined, is demonstrated in Chapter 17.

Researchers have increasingly acknowledged the level of technological development and potential for human environmental impact by Mesolithic communities (see Chapters 10 and 11). However, our understanding of early-Holocene environments has also changed recently. Particularly contentious is the association of charcoal evidence for fire with either artefact evidence for human presence and/or pollen evidence for forest recession. The association of human artefacts with stratified charcoal does not demand a causal connection, at least not in the sense of human cause and environmental effect. Indeed, the processes may be reversed: human groups may have moved into areas locally devastated by natural fire. This is a plausible, though not necessarily accurate, interpretation of charcoal evidence in environments dominated by readily combustible taxa such as *Pinus sylvestris* (Scots pine), but is less tenable as an interpretation in other forested environments in which the frequent occurrence of natural fire may be doubted (cf. Dimbleby, 1977), because the major tree taxa 'burn like damp asbestos' (Rackham, 1986, p. 72). This observation is not confined to ecologists: 'I don't know if you have ever thought what a rare thing flame must be in the absence of man [sic] and in a temperate climate.' H.G. Wells: *The Time Machine*.

The key question here concerns notions of temperate climate in the northern hemisphere. It is now believed that the early Holocene climate and vegetation were appreciably different from those of the late Holocene (Chapters 7 and 17). This may be due, in part, to increased summer solar radiation in the northern hemisphere, compared with the present (Kutzbach & Street-Perrott, 1985). Though understanding of Holocene climatic changes has advanced significantly over recent years (COHMAP, 1988; Wigley, 1988; Wigley & Kelly, 1990), there remains the difficulty of discriminating between natural and human-induced changes in vegetation. Though it may not be possible to test adequately the hypothesis of human impact (see Chapters 1, 9 and 10) in the conventional Popperian sense (see Chapter 2, Fig. 2.1), it is possible to examine critically the claimed supporting evidence. Nevertheless, for studies of early agriculture in forested environments – despite empirical studies (Chapter 14) and hypothesized models (Chapter 12) – we may still need, if not H.G. Wells's, then A.G. Smith's Time Machine (Smith, 1981b) to unravel the truth!

21.5 CONCLUDING REMARKS

21.5.1 Timescales of change

There is now considerable public concern over the rate and direction of climatic change, and of human impact upon climate. However, notions that human activity could influence weather and climate have permeated slowly. So-called 'acid rain' was first noticed last century, but research, particularly on lake acidification, only progressed over the last decade and a half (see Chapter 2). The 'Greenhouse Effect' was recognized last century, but only recently has human-induced 'global warming' gained scientific credence and demanded political attention (Fantechi & Ghazi, 1989; Leggett, 1990; IPCC, 1990; CCIRG, 1991).

Forty years ago, none of these ideas was being seriously considered. 'Accepted' wisdom on Quaternary climatic change has also changed: the 'Milankovitch' astronomical theory of the early decades of this century may have had its roots in James Croll's work of the 1860s, but was not widely accepted until the mid-1970s (Imbrie & Imbrie, 1979; Berger, 1988).

Though current ideas on climate change and human impact have taken a while to become established, it is salutary to note that theories of climate change and Quaternary earth history have been overturned within the quoted statistical precision that attaches to the average radiocarbon date. Consider an arbitrary radiocarbon sample of say, 75 ± 60 BP; 2σ (± 120 years) encompasses views from early 19th century notions of the Great Flood to as yet unexpressed views of the first part of the 21st century. In Quaternary timescales, how very fast have new ideas become accepted wisdom (cf. Chapter 2). It is a sobering thought that, 60 years from now, new, revolutionary ideas may replace current notions.

Contrast the ocean-core evidence and the predictions that can be derived from the Milankovitch-derived theory of astronomical forcing with present preoccupations concerning 'global warming'. Present-day temperatures are apparently towards (though not at) the warm extreme of records over the last half million years. It is claimed that human-induced 'global warming' might take the earth into a global climate regime that has not been recorded for the whole of the Quaternary, with a magnitude of change larger and at a rate faster than at any time since the end of the last cold stage (cf. IPCC, 1990; CCIRG, 1991; see Fig. 21.4). However, ideas on climate change are changing faster than climate itself: over a period of but five years, projected global mean temperature rises of 5.2°C for the mid-21st century have been scaled down to just 1.6°C (Table 21.5).

Clearly, Quaternary palynology (Huntley, 1990d), palaeoecology and palaeohydrology (Berglund, 1986b) have much to contribute to the study of global environmental change (Delcourt & Delcourt, 1987). Yet, it is easy to be seduced by the continuous climatic curves that can be derived or constructed from palaeoecological data; considerable smoothing of primary data is required to produce such curves and there are uncertainties attached to the dating of the deposits from which the proxy data have been obtained. Despite their power and attraction, global circulation models (GCMs) for palaeoenvironments have their limitations (Street-Perrott, 1991), as do those for future environments (Table 21.5).

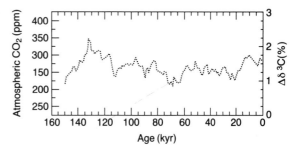

Fig. 21.4 Inferred changes in atmospheric CO_2 over the last 160 000 years, based on isotopic records in ocean sediments (redrawn from Delcourt, 1991; after Gammon et al., 1985). CO_2 doubling to 600 ppm has been predicted for the next century (Fantechi & Ghazi, 1989; IPCC, 1990).

Table 21.5 Projected rises of mean global temperature for next century, produced from successive modifications of the UK 'Met-Office Model', developed over the last five years, assuming a doubling of atmospheric CO_2 and with different cloud schemes (Mason, 1991). The original predicted rises (1) were claimed to be unprecedented in rate and magnitude; however, the revised figures show rates and magnitude of change that are not unprecedented – the last cold stage bears witness to several 'flip-flop' climatic shifts of considerable magnitude and rapidity. The incorporation of the effects of cloud type and reflectance (2), (3) reduced the projected rises, whilst the inclusion of deep ocean heat absorption (4) delays the rises further. Note that CO_2 concentrations next century would apparently be unprecedented in the late Quaternary (cf. Fig. 21.4). The role of peatlands in global carbon cycling, and release of CO_2 and CH_4 has yet to be incorporated in GCMs

Model	Cloud representation	Radiative properties of clouds	Temperature rise (°C)
UK MO (1)	Empirical-linked to relative humidity. All-water clouds.	Fixed	5.2
UK MO (2)	Computed liquid water and ice content.	Fixed	3.2
UK MO (3)	" "	Variable function of water and ice content	1.9
UK MO (4)	(as 3 but with deep-ocean circulation included)		1.6*

* Global atmosphere coupled to a deep global ocean with CO_2 increasing at 1% p.a. compound; data from Sir John Mason, personal communication, 1992.

21.5.2 Nature conservation and environmental management

Palaeoecologists and environmental archaeologists routinely describe and account for human impact over millennia. They can examine the rise and fall of great prehistoric civilizations and test whether mismanagement of the land or climatic change was responsible (e.g. Singh, 1971). They can test whether climatic change, pedogenesis or human influence gave rise to vegetational change and peat development (Chapter 18). The long timescales investigated in late-Quaternary palaeoecology and environmental archaeology permit estimates of the magnitude and frequency of major environmental disturbances, assessment of the degree of human impact, and evaluation of the rate and sustainability of such impacts, whether in prehistory (Chapter 16) or in recent centuries (Chapter 20). They give Quaternary scientists a greater and arguably clearer perspective (e.g. Roberts, 1989) on long-term climatic and landscape changes than contemporary climatologists and ecologists. The environmental histories and the record of human impact, inferred and chronicled in Parts Two to Four of this volume, can also be seen in terms of their implications for management of the resulting environments.

To manage effectively the landscapes produced as a result of millennia of human influence, it is necessary to be aware of the natural changes in climate that have apparently occurred, and of the predictable (orbitally derived) and unpredictable (e.g. volcanic) high-frequency changes that may be expected in the future, irrespective of human modification of the climate. In adding human modification of the global climate, present vegetation patterns may not survive, species may not be able to adapt or to 'migrate' fast enough, nature conservation areas may contain plants and animals that find themselves unable to adapt to changing climatic conditions (cf. Delcourt & Delcourt, 1991, pp. 212–218), and agricultural patterns may change significantly (cf. Parry, 1990). This has a familiar ring: the Quaternary palaeoecologist can, through various techniques, bear witness both to long-term climatic changes and to past environmental change of a sudden and possibly catastrophic nature. Studies demonstrate the dynamics of climate and landscape (particularly over Pleistocene timescales, cf. West, 1991). Indeed, the alternative future scenario – that of a new cold stage – emphasizes the point that in (presently) temperate regions, static nature reserves are not viable over long timescales.

Forty years ago, palaeoecology was a 'pure' historical science; its application – in archaeology – was then just beginning (at Star Carr, England), to be professionalized as 'environmental archaeology' in the 1970s. Today, the contribution that can be made by studies of long-term environmental change and past human impact to resolution of *current* environmental problems is still not widely appreciated. Despite the dangers of 'projective' research modes (see Chapter 2), palaeoecology and environmental archaeology have much to contribute to two major debates: (1) the direction, magnitude and rate of climatic change, and (2) the sustainability of human impact. Both are relevant for continuing nature conservation and future environmental management of all landscapes (cf. Mannion, 1991).

Acknowledgements

Thanks to Dave Alden, Jeff Blackford and Darrel Maddy for additional references, Andrew Lawrence for cartography, Jeff Blackford for Fig. 21.3 and for comments, Helena Chambers for advice, and finally Professor A.G. Smith for providing the stimulation for the whole volume.

Appendix

Calibration table of radiocarbon ages, abbreviated from the Belfast–Seattle calibration tables in Pearson & Stuiver (1986) and Stuiver & Pearson (1986) back to 3900 BP and from the University of Washington Quaternary Isotope Laboratory Radiocarbon Calibration Program (1987, Rev. 2.0) from 4000 BP to 8000 BP.

This table is intended as a guide only; see originals for full information. The bc/ad system, which has been used in several journals, is not recommended by the international radiocarbon community (see Preface), but is still used by some authors (cf. Chapter 19). Note that ages cited as lower-case bp in other publications, are equivalent to the ages in the first column in this table.

Uncalibrated (Years BP)	Uncalibrated (Years bc/ad)	Calibrated age (Cal. BC/AD)
200 BP	ad 1750	Cal. AD 1666, 1790, 1951, 1952
300 BP	ad 1650	Cal. AD 1636
400 BP	ad 1550	Cal. AD 1460
500 BP	ad 1450	Cal. AD 1422
600 BP	ad 1350	Cal. AD 1317, 1347, 1388
700 BP	ad 1250	Cal. AD 1279
800 BP	ad 1150	Cal. AD 1245
900 BP	ad 1050	Cal. AD 1159
1000 BP	ad 950	Cal. AD 1018
1100 BP	ad 850	Cal. AD 961
1200 BP	ad 750	Cal. AD 811, 847, 851
1300 BP	ad 650	Cal. AD 681
1400 BP	ad 550	Cal. AD 673
1500 BP	ad 450	Cal. AD 561
1600 BP	ad 350	Cal. AD 429
1700 BP	ad 250	Cal. AD 343
1800 BP	ad 150	Cal. AD 227
1900 BP	ad 50	Cal. AD 87
2000 BP	50 bc	1 Cal. BC
2100 BP	150 bc	151, 149, 117 Cal. BC
2200 BP	250 bc	353, 306, 236 Cal. BC
2300 BP	350 bc	392 Cal. BC
2400 BP	450 bc	408 Cal. BC
2500 BP	550 bc	765, 673, 667, 613, 608 Cal. BC
2600 BP	650 bc	801 Cal. BC
2700 BP	750 bc	838 Cal. BC

2800 BP	850 bc	976, 965, 933 Cal. BC
2900 BP	950 bc	1093 Cal. BC
3000 BP	1050 bc	1263 Cal. BC
3100 BP	1150 bc	1406 Cal. BC
3200 BP	1250 bc	1506, 1476, 1464 Cal. BC
3300 BP	1350 bc	1607, 1554, 1543 Cal. BC
3400 BP	1450 bc	1733, 1721, 1697 Cal. BC
3500 BP	1550 bc	1877, 1834, 1824, 1793, 1788 Cal. BC
3600 BP	1650 bc	1961 Cal. BC
3700 BP	1750 bc	2133, 2067, 2047 Cal. BC
3800 BP	1850 bc	2278, 2233, 2209 Cal. BC
3900 BP	1950 bc	2457 Cal. BC
4000 BP	2050 bc	2564, 2541, 2499 Cal. BC
4100 BP	2150 bc	2855, 2824, 2657, 2640, 2619 Cal. BC
4200 BP	2250 bc	2880, 2798, 2782 Cal. BC
4300 BP	2350 bc	2915 Cal. BC
4400 BP	2450 bc	3034 Cal. BC
4500 BP	2550 bc	3307, 3235, 3177, 3163, 3134, 3112, 3110 Cal. BC
4600 BP	2650 bc	3360 Cal. BC
4700 BP	2750 bc	3504, 3406, 3384 Cal. BC
4800 BP	2850 bc	3629, 3560, 3544 Cal. BC
4900 BP	2950 bc	3697 Cal. BC
5000 BP	3050 bc	3785 Cal. BC
5100 BP	3150 bc	3957, 3838, 3826 Cal. BC
5200 BP	3250 bc	3998 Cal. BC
5300 BP	3350 bc	4220, 4200, 4147, 4110, 4088, 4060, 4048 Cal. BC
5400 BP	3450 bc	4318, 4285, 4246 Cal. BC
5500 BP	3550 bc	4353 Cal. BC
5600 BP	3650 bc	4461 Cal. BC
5700 BP	3750 bc	4572, 4564, 4536 Cal. BC
5800 BP	3850 bc	4716 Cal. BC
5900 BP	3950 bc	4787 Cal. BC
6000 BP	4050 bc	4993, 4925, 4903 Cal. BC
6100 BP	4150 bc	5053, 5013, 5008 Cal. BC
6200 BP	4250 bc	5215 Cal. BC
6300 BP	4350 bc	5240 Cal. BC
6400 BP	4450 bc	5338 Cal. BC
6500 BP	4550 bc	5474, 5435, 5426 Cal. BC
6600 BP	4650 bc	5493 Cal. BC
6700 BP	4750 bc	5619, 5595, 5538 Cal. BC
6800 BP	4850 bc	5645 Cal. BC
6900 BP	4950 bc	5741 Cal. BC
7000 BP	5050 bc	5840 Cal. BC
7100 BP	5150 bc	5975 Cal. BC
7200 BP	5250 bc	6080, 6052, 6049, 6012, 6007 Cal. BC
7300 BP	5350 bc	6117 Cal. BC
7400 BP	5450 bc	6217, 6202, 6183 Cal. BC
7500 BP	5550 bc	6389 Cal. BC
7600 BP	5650 bc	6441 Cal. BC
7700 BP	5750 bc	6553, 6542, 6487 Cal. BC
7800 BP	5850 bc	6610 Cal. BC
7900 BP	5950 bc	6703 Cal. BC
8000 BP	6050 bc	7032, 6971, 6970 Cal. BC

References

Aaby, B. (1976) Cyclic climatic variations in climate over the last 5,500 years reflected in raised bogs. *Nature*, **263**, 281–4.

Aaby, B. (1978) Cyclic changes in climate during 5,500 yrs, reflected in Danish raised bogs. *Dansk. Met. Inst. Klimatologiske Meddelelser*, **4**, 18–26.

Aaby, B. (1986) Trees as anthropogenic indicators in regional pollen diagrams from eastern Denmark. In *Anthropogenic Indicators in Pollen Diagrams* (ed. K.-E. Behre), A.A. Balkema, Rotterdam, pp. 73–93.

Aaby, B. (1988) The cultural landscape as reflected in percentage and influx pollen diagrams from two Danish ombrotrophic mires. In *The Cultural Landscape: Past, Present and Future* (eds H.H. Birks, H.J.B. Birks, P.E. Kaland and D. Moe), Cambridge University Press, Cambridge, pp. 209–28.

Aaby, B. and Tauber, H. (1975) Rates of peat formation in relation to degree of humification and local environment, as shown by studies of a raised bog in Denmark. *Boreas*, **4**, 1–17.

Aaltonen, V.T. (1939) Zur stratigraphie des podsolprofiles II. *Communications Instituti Forestalis Fenniae*, **27**, 1–133.

Aitken, M.J. (1990) *Science-based dating in Archaeology*. Longman, London.

Ammann, B. and Lotter, A.F. (1989) Late-Glacial radiocarbon- and palynostratigraphy on the Swiss Plateau. *Boreas*, **18**, 109–126.

Ammerman, A.J. and Cavalli-Sforza, L.L. (1984) *The Neolithic Transition and the Genetics of Population in Europe*, Princeton University Press, Princeton.

Andersen, S. Th. (1979) Identification of wild grasses and cereal pollen. *Danmarks Undersogelse Arbog 1978*, 69–72.

Andersen, S. Th. (1988) Changes in agricultural practices in the Holocene indicated in a pollen diagram from a small hollow in Denmark. In *The Cultural Landscape: Past, Present and Future* (eds H.H.Birks, H.J.B. Birks, P.E. Kaland and D. Moe), Cambridge University Press, Cambridge, pp. 395–407.

Anderson, T.W., Mathewes, R.W. and Schweger, C.E. (1989) Holocene climatic trends in Canada with special reference to the Hypsithermal interval. In *Quaternary Geology of Canada and Greenland* (ed. R.J. Fulton), Geological Survey of Canada, Ottawa, pp. 520–8.

Andree, M., Oeschger, H., Siegenthaler, U., Reisen, T., Moell, M., Ammann, B. and Tobolski, K. (1986) ^{14}C dating of plant macrofossils in lake sediment. *Radiocarbon*, **28**, 411–16.

Andrew, R. (1984) *A Practical Pollen Guide to the British Flora*, Quaternary Research Association Technical Guide No. 1, QRA, Cambridge.

ApSimon, A. (1986) Chronological contexts of Irish megalithic tombs. *Journal of Irish Archaeology*, **3**, 5–15.

Aravena, R., Warner, B.G. and Gritz, P. (1991) Isotopic composition of peat and peatland plants and their implications in palaeoclimatic studies. *Abstracts, Chapman Conference on Continental Isotopic Indicators of Climate*. American Geophysical Union, Jackson Hole, Wyoming, p. 10.

Aravena, R., Warner, B.G., MacDonald, G.M. and Hanf, K.I. (1992) Carbon isotope composition of lake sediments in relation to lake productivity and radiocarbon dating. *Quaternary Research*, (**37**), 333–45).

Ashbee, P., Smith, I.F. and Evans, J.G. (1979) Excavation of three long barrows near Avebury, Wiltshire. *Proceedings of the Prehistoric Society*, **45**, 207–300.

Ashbee P., Bell, M. and Proudfoot, E. (1989) *Wilsford Shaft: Excavations 1960–62*, English Heritage, London.

Austad, I. (1988) Tree pollarding in western Norway. In *The Cultural Landscape: Past, Present and*

Future (eds H.H.Birks, H.J.B. Birks, P.E. Kaland and D. Moe) Cambridge University Press, Cambridge, pp. 11–29.

Averdiek, F.R. (1978) Palynologischer Beitrag zur Eutwicklungsgeschichte des Grosson Plöner Sees und der Vegetation seiner Umgebung. *Archive für Hydrobiology*, **83**, 1–46.

Backeus, I. (1990) The cyclic regeneration on bogs – a hypothesis that became an established truth. *Striae*, **31**, 33–5.

Bahn, P.G. (1991) Pleistocene images outside Europe. *Proceedings of the Prehistory Society*, **57**, 91–102.

Bailey G.B. and Devereux, D.F. (1987) The eastern terminus of the Antonine Wall: a review. *Proceedings of the Society of Antiquities of Scotland* **17**, 93–104.

Bailey, V. (1896) Tamarack swamps as boreal islands. *Science*, **3**, 250.

Baillie, M.G.L. (1979) Some observations on gaps in tree-ring chronologies. In *Proceedings of the Symposium on Archaeological Sciences*, (ed. A. Aspinall), University of Bradford, Jan 1978, pp. 19–32.

Baillie, M.G.L. (1982) *Tree-Ring Dating and Archaeology*, Croom Helm, London.

Baillie, M.G.L. (1983) Belfast dendrochronology – the current situation. In *Archaeology, Dendrochronology and the Radiocarbon Calibration Curve* (ed. B.S. Ottaway), Occasional Paper No. 9, University of Edinburgh, pp. 15–24.

Baillie, M.G.L. (1985) Irish dendrochronology and radiocarbon calibration. *Ulster Journal of Archaeology*, **48**, 11–23.

Baillie, M.G.L. (1988) The dating of the timbers from Navan Fort and The Dorsey, Co. Armagh. *Emania*, **4**, 37–40.

Baillie, M.G.L. (1989) Do Irish bog oaks date the Shang Dynasty? *Current Archaeology*, **117**, 310–13.

Baillie, M.G.L. (1990a) Checking back on an assemblage of published radiocarbon dates. *Radiocarbon*, **32**, 361–6.

Baillie, M.G.L. (1990b) Irish tree-rings and an event in 1628 BC. In *Thera and the Aegean World III*, Vol. 3 (ed D.A. Hardy), The Thera Foundation, London, pp. 160–6.

Baillie, M.G.L. (1991a) Marking in marker dates: towards an archaeology with historical precision. *World Archaeology*, **23**, 233–43.

Baillie, M.G.L. (1991b) Suck-in and smear: two related chronological problems for the 1990s. *Journal of Theoretical Archaeology*, **2**, 12–16.

Baillie, M.G.L. and Brown, D.M. (1988) An overview of oak chronologies. *British Archaeological Reports (British Series)*, **196**, 543–8.

Baillie, M.G.L. and Brown, D.M. (1991) Attempts to date Haughey's Fort by dendrochronology. *Emania*, **8**, 39–40.

Baillie, M.G.L. and Munro, M.A.R. (1988) Irish tree-rings, Santorini and volcanic dust veils. *Nature*, **332**, 344–6.

Baillie, M.G.L. and Pilcher, J.R. (1987) The Belfast 'long chronology' project. *British Archaeological Reports, International Series*, S333, 203–14.

Baillie, M.G.L., Pilcher, J.R. and Pearson, G.W. (1983) Dendrochronology at Belfast as a background to high-precision calibration. *Radiocarbon*, **25**, 171–8.

Bainbridge, T.H. (1943) Land utilisation in Cumbria in the mid-19th century as revealed by tithe returns. *Transactions of the Cumberland and Westmorland Antiquarian and Archaeological Society*, **NS XLIII**, 87–95.

Barber, K.E. (1981) *Peat Stratigraphy and Climatic Change – a Palaeoecological Test of the Theory of Cyclic Peat Bog Regeneration*, A.A. Balkema, Rotterdam.

Barber, K.E. (1982) Peat-bog stratigraphy as a proxy climate record. In *Climatic Change in Later Prehistory* (ed. A.F. Harding), Edinburgh University Press, Edinburgh, pp. 103–33.

Barber, K.E. (1984) A large-capacity Russian pattern sediment sampler. *Quaternary Newsletter*, **44**, 28–31.

Barber, K.E. (1985) Peat stratigraphy and climatic changes: some speculations. In *The Climatic Scene: Essays in Honour of Gordon Manley* (eds M.J. Tooley and G.M. Sheail), Allen and Unwin, London, pp. 175–85.

Barber, K.E. and Twigger, S.N. (1987) Late Quaternary palaeoecology of the Severn Basin. In *Palaeohydrology in Practice* (eds K.J. Gregory, J. Lewin and J.B. Thornes), John Wiley, Chichester, pp. 217–50.

Barker, G. (1985) *Prehistoric Farming in Europe*, Cambridge University Press, Cambridge.

Barnola, J.M., Raynaud, D., Korotkevich, Y.S. and Lorius, C. (1987) Vostok ice core provides 160 000-year record of atmospheric CO_2. *Nature*, **329**, 408–14.

Barnosky, C.W. (1981) A record of late Quaternary vegetation from Davis Lake, southern Puget lowland, Washington. *Quaternary Research*, **16**, 221–39.

Bartlein, P.J. (1988) Late-Tertiary and Quaternary paleoenvironments. In *Vegetation History* (eds B. Huntley and T. Webb III), Kluwer, Dordrecht, pp. 113–52.

Bartlein, P.J., Prentice, I.C. and Webb, T. III (1986) Climatic response surfaces from pollen data for some eastern North American taxa. *Journal of Biogeography*, **13**, 35–57.

Bartley, D.D., Jones, I.P. and Smith, R.T. (1990) Studies in the Flandrian vegetational history of the Craven district of Yorkshire: the lowlands. *Journal of Ecology*, **78**, 611–32.

Batterbee, R.W. (1978) Observations on the recent history of Lough Neagh and its drainage basin. *Philosophical Transactions of the Royal Society of London [B]*, **281**, 303–45.

Batterbee, R.W. (1986) The eutrophication of Lough Erne inferred from changes in the diatom assemblage of ^{210}Pb and ^{137}Cs dated sediment cores. *Proceedings of the Royal Irish Academy [B]*, **86**, 141–68.

Batterbee, R.W., Flower, R.J., Stevenson, A.C. and Rippey, B. (1990) Lake acidification in Galloway – a palaeoecological test of competing hypotheses. *Nature*, **314**, 350–2.

Baur, M. (1987) Richness of land snail species under isolated stones in a karst area on Öland, Sweden. *Basteria*, **51**, 129–33.

Beales, P.W. (1980) The Late Devensian and Flandrian vegetational history of Crose Mere, Shropshire. *New Phytologist*, **85**, 113–61.

Becker, B. and Schirmer, W. (1977) Palaeoecological study on the Holocene valley development of the River Main, southern Germany. *Boreas*, **6**, 303–21.

Becker, B. and Schmidt, B. (1982) Verlangerung der Mitteleuropaischen Eichenjahrringchronologie in das Zweite Vorchristliche Jahrtausand (bis 1462. Chr.). *Archaeologisches Korrespondenblatt*, **12**, 101–6.

Beckett, S.C. (1981) The Shaugh Moor project: environmental background. *Proceedings of the Prehistoric Society*, **47**, 245–73.

Beckett, S.C. and Hibbert, F.A. (1979) Vegetational change and the influence of prehistoric man in the Somerset Levels. *New Phytologist*, **83**, 577–600.

Begét, J., Mason, O. and Anderson, P. (1992) Age, extent and climatic significance of the *c.* 3400 BP Aniakchak tephra, western Alaska, USA. *The Holocene*, **2**, 51–56.

Behre, K.-E. (1978) Die Klimaschwankungen im europaischen Praboreal. *Petermanns Geographische Mitteilungen*, **2**, 97–102.

Behre, K.-E. (ed.) (1986) *Anthropogenic Indicators in Pollen Diagrams*, A.A. Balkema, Rotterdam.

Behre, K.-E. (1989) Biostratigraphy of the last glacial period in Europe. *Quaternary Science Reviews*, **8**, 25–44.

Bell, M. (1983) Valley sediments as evidence of prehistoric land-use on the South Downs. *Proceedings of the Prehistoric Society*, **49**, 119–50.

Bell, M. (1990) Sedimentation rates in the primary fills of chalk-cut features. In *Experiment and Reconstruction in Environmental Archaeology* (ed. D.E. Robinson), Oxbow Books, Oxford, pp. 237–48.

Bell, M. and Jones, J. (1990) Land Mollusca. In *The Stonehenge Environs Project* (ed. J. Richards), English Heritage, London, pp. 154–158.

Benn, D.I., Lowe, J.J. and Walker, M.J.C. (1992) Glacier response to climatic change during the Loch Lomond Stadial and early Flandrian: geomorphological and palynological evidence from the Isle of Skye, Scotland. *Journal of Quaternary Science*, **7** (in press).

Bennett, K.D. (1983) Devensian late-glacial and Flandrian vegetational history at Hockham Mere, Norfolk, England. I. Pollen percentages and concentrations. *New Phytologist*, **95**, 457–87.

Bennett, K.D. (1984) The post-glacial history of *Pinus sylvestris* in the British Isles. *Quaternary Science Reviews*, **3**, 133–55.

Bennett, K.D. (1986) The rate of spread and population increase of forest trees during the postglacial. *Philosophical Transactions of the Royal Society of London [B]*, **314**, 523–31.

Bennett, K.D. (1989) A provisional map of forest types for the British Isles 5,000 years ago. *Journal of Quaternary Science*, **4**, 141–4.

Bennett, K.D. and Birks, H.J.B. (1990) Postglacial history of alder (*Alnus glutinosa* (L.) Gaertn.) in the British Isles. *Journal of Quaternary Science*, **5**, 123–33.

Bennett, K.D., Fossitt, J.A., Sharp, M.J. and Switsur, V.R. (1990a) Holocene vegetational and environmental history at Loch Lang, South Uist, Western Isles, Scotland. *New Phytologist*, **114**, 281–98.

Bennett, K.D., Simonson, W.D. and Peglar, S.M. (1990b) Fire and man in post-glacial woodlands of eastern England. *Journal of Archaeological Science*, **17**, 635–42.

Bennett, K.D., Boreham, S., Sharp, M.J. and Switsur, V.R. (1992) Holocene history of environment, vegetation and human settlement on Catta Ness, Lunnasting, Shetland. *Journal of Ecology*, **80** (in press).

Berger, A. (1988) Milankovitch theory and climate. *Reviews of Geophysics and Space Physics*, **26**, 624–57.

Berger, W.H. (1990) The Younger *Dryas* cold spell – a quest for causes. *Global and Planetary Change*, **3**, 219–37.

Berglund, B.E. (1966) Late-Quaternary vegetation in eastern Blekinge, southeastern Sweden. II. Post-glacial time. *Opera Botanica*, **12** (2), 1–190.

Berglund, B.E. (1969) Vegetation and human influence in South Scandinavia during prehistoric time. *Oikos Supplement*, **12**, 9–28.

Berglund, B.E. (1985) Early agriculture in Scandinavia: research problems related to pollen-analytical studies. *Norwegian Archaeological Review*, **18**, 77–105.

Berglund, B.E. (1986a) The cultural landscape in a long-term perspective. Methods and theories behind the research on land-use and landscape dynamics. *Striae*, **24**, 79–87.

Berglund, B.E. (ed.) (1986b) *Handbook of Holocene Palaeoecology and Palaeohydrology*, John Wiley, Chichester.

Berglund, B.E. (ed.) (1991) *The Cultural Landscape During 6000 Years in Southern Sweden*, Ecological Bulletins 41, Munksgaard, Copenhagen.

Betancourt, J.L. and Davis, O.K. (1984) Packrat middens from Canyon de Chelly, northeastern Arizona: palaeoecological and archaeological implications. *Quaternary Research*, **21**, 56–64.

Beug, H.-J. (1961) *Leitfaden der Pollenbestimmung für Mitteleuropa und angrenzende Gebiete*, Gustav Fischer Verlag, Stuttgart.

Beug, H.-J. (1967) On the forest history of the Dalmatian coast. *Review of Palaeobotany and Palynology*, **2**, 271–9.

Beug, H.-J. (1977) Vegetations geschichtliche Untersuchungen im Küstenbereich von Istrien (Jugoslawien). *Flora*, **166**, 357–81.

Birks, H.H. (1972) Studies in the vegetational history of Scotland III. A radiocarbon-dated pollen diagram from Loch Maree, Ross and Cromarty. *New Phytologist*, **71**, 731–54.

Birks, H.H. (1975) Studies in the vegetational history of Scotland. IV. Pine stumps in Scottish blanket peats. *Philosophical Transactions of the Royal Society of London [B]*, **270**, 181–226.

Birks, H.H. and Mathewes, R.W. (1978) Studies in the vegetational history of Scotland. V. Late Devensian and early Flandrian pollen and macrofossil stratigraphy at Abernethy Forest, Inverness-shire. *New Phytologist*, **80**, 455–84.

Birks, H.H., Birks, H.J.B., Kaland, P.E. and Moe, D. (eds) (1988) *The Cultural Landscape: Past, Present and Future*, Cambridge University Press, Cambridge.

Birks, H.J.B. (1973) *Past and Present Vegetation on the Isle of Skye – a Palaeoecological Study*, Cambridge University Press, Cambridge.

Birks, H.J.B. (1977) The Flandrian forest history of Scotland: a preliminary synthesis. In *British Quaternary Studies – Recent Advances* (ed. F.W. Shotton), Clarendon Press, Oxford, pp. 119–35.

Birks, H.J.B. (1980) The present flora and vegetation of the moraines of the Klutlan Glacier, Yukon Territory, Canada: a study in plant succession. *Quaternary Research*, **14**, 60–8.

Birks, H.J.B. (1981) The use of pollen analysis in the reconstruction of past climates: a review. In *Climate and History* (eds T.M.L. Wigley, M.J. Ingram and G. Farmer), Cambridge University Press, Cambridge, pp. 111–138.

Birks, H.J.B. (1982) Holocene (Flandrian) chronostratigraphy of the British Isles: a review. *Striae*, **16**, 99–105.

Birks, H.J.B. (1985) Recent and possible future mathematical developments in quantitative palaeoecology. *Palaeogeography, Palaeoclimatology, Palaeoecology*, **50**, 107–47.

Birks, H.J.B. (1986) Late-Quaternary biotic changes in terrestrial and lacustrine environments, with particular reference to north-west Europe. In *Handbook of Holocene Palaeoecology and Palaeohydrology* (ed. B.E. Berglund), John Wiley, Chichester, pp. 3–65.

Birks, H.J.B. (1988a) Long-term ecological change in the British uplands. In *Ecological Change in the Uplands* (eds B.M. Usher and D.B.A. Thompson), Blackwell, Oxford, pp. 37–56.

Birks, H.J.B. (1988b) Conclusions. In *The Cultural Landscape: Past, Present and Future* (eds H.H. Birks, H.J.B. Birks, P.E. Kaland and D. Moe), Cambridge University Press, Cambridge, pp. 463–6.

Birks, H.J.B. (1989) Holocene isochrone maps and patterns of tree-spreading in the British Isles. *Journal of Biogeography*, **16**, 503–40.

Birks, H.J.B. and Berglund, B.E. (1979) Holocene pollen stratigraphy of southern Sweden: a reappraisal using numerical methods. *Boreas*, **8**, 257–79.

Birks, H.J.B. and Birks, H.H. (1980) *Quaternary Palaeoecology*, Edward Arnold, London.

Birks, H.J.B. and Gordon A.D. (1985) *Numerical Methods in Quaternary Pollen Analysis*, Academic Press, London.

Birks, H.J.B. and Williams, W. (1983) Late Quaternary vegetational history of the Inner Hebrides. *Proceedings of the Royal Society of Edinburgh*, **83**, 269–92.

Björklund, J. (1984) From the Gulf of Bothnia to the White Sea. *Scandinavian Economic History Review*, **1**, 17–41.

Blackford, J.J. (1990) *Blanket Mires and Climatic Change: a Palaeoecological Study Based on Peat Humification and Microfossil Analyses*, Unpublished Ph.D. Thesis, Keele University.

Blackford, J.J. and Chambers, F.M. (1991) Proxy records of climate from blanket mires: evidence for a Dark Age (1400 BP) climatic deterioration in the British Isles. *The Holocene*, **1**, 63–7.

Bloom, A.J. (1989) Principles of instrumentation for physiological ecology. In *Plant Physiological Ecology* (eds R.W. Pearcy, J. Ehleringer, H.A. Mooney and P.W. Rundel) Chapman and Hall, London, pp. 1–13.

Blytt, A. (1876) *Essay on the Immigration of the Norwegian Flora During Alternating Rainy and Dry Periods*, Cammermeyer, Kristiana.

Boag, D.A. and Wishart, W.D. (1982) Distribution and abundance of terrestrial gastropods on a winter range of bighorn sheep in southwestern Alberta. *Canadian Journal of Zoology*, **60**, 2633–40.

Boatman, D.J. (1983) The Silver Flowe National Nature Reserve, Galloway, Scotland. *Journal of Biogeography*, **10**, 163–274.

Bogen, J., Wold, B. and Østrem, G. (1989) Historical glacier variations in Scandinavia. In *Glacier Fluctuations and Climatic Change* (ed. J. Oerlemans), Kluwer, Dordrecht, pp. 109–28.

Bohncke, S.J.P. (1988) Vegetation and habitation history of the Callanish area, Isle of Lewis, Scotland. In *The Cultural Landscape: Past, Present and Future* (eds H.H. Birks, H.J.B. Birks, P.E. Kaland and D. Moe), Cambridge University Press, Cambridge, pp. 445–61.

Bortenschlager, S. (1982) Chronostratigraphic subdivisions of the Holocene in the Alps. *Striae*, **16**, 75–9.

Bostock, J.L. (1980) *The History of the Vegetation of the Berwyn Mountains, North Wales with Emphasis on the Development of Blanket Mire*, Unpublished Ph.D. Thesis, University of Manchester.

Bostwick Bjerck, L.G. (1987) *In-context Pollen Diagrams from 3 Archaeological Sites in Western Norway: Towards a Unified Model of Land Use in the Late Mesolithic and Neolithic I Periods*, Unpublished Magister Grad Dissertation, University of Bergen.

Bowen, D.Q. (1989) The last interglacial–glacial cycle in the British Isles. *Quaternary International* **3/4**, 41–7.

Bowen, D.Q. (1991) Time and space in the glacial sediment systems of the British Isles. In *Glacial Deposits in Great Britain and Ireland* (eds J. Ehlers, P.L. Gibbard and J. Rose), A.A. Balkema, Rotterdam.

Bowen, D.Q., Hughes, S., Sykes, G.A. and Miller, G.H. (1989) Land–sea correlations based on isoleucine epimerization in non-marine molluscs. *Nature*, **340**, 49–51.

Bowman, S. (1990) *Interpreting the Past: Radiocarbon Dating*, British Museum, London.

Boyd, W.E. (1982) Botanical remains of edible plants from an Iron Age broch, Fairy Knowe, Buchlyvie, Stirling. *Forth Naturalist and Historian*, **7**, 77–83.

Boyd, W.E. (1984) Environmental change and Iron Age land management in the area of the Antonine Wall, central Scotland: a summary. *Glasgow Archaeological Journal*, **12**, 75–81.

Boyd, W.E. (1985) Palaeobotanical evidence from Mollins. *Britannia*, **19**, 37–48.

Boyd, W.E. (1988) Methodological problems in the analysis of fossil non-artifactual wood assemblages from archaeological sites. *Journal of Archaeological Science*, **15**, 603–19.

Boyd, W.E. and Dickson, J.H. (1986) Patterns in the geographical distribution of the early Flandrian *Corylus* rise in southwest Scotland. *New Phytologist*, **102**, 615–23.

Boyd, W.E. and Dickson, J.H. (1987) A post-glacial pollen sequence from Loch a'Mhuilinn, north Arran: a record of vegetation history with special reference to the history of endemic *Sorbus* species. *New Phytologist*, **107**, 221–44.

Bradley, R.S. (1985) *Quaternary Paleoclimatology*, Allen & Unwin, Boston.

Bradshaw, R.H.W. and Browne, P. (1987) Changing patterns in the post-glacial distribution of *Pinus sylvestris* in Ireland. *Journal of Biogeography*, **14**, 237–48.

Bradshaw, R.H.W. and Webb, T. III (1985) Relationships between contemporary pollen and vegetation data from Wisconsin and Michigan, U.S.A. *Ecology*, **66**, 721–37.

Branigan, K. (ed.) (1980) *Rome and the Brigantes: The Impact of Rome on Northern England*, Department of Prehistory and Archaeology, University of Sheffield.

Breeze, D.J. (1982) *The Northern Frontiers of Britain*, Batsford, London.

Breeze, D.J. and Dobson, B. (1987) *Hadrian's Wall*, Pelican, London.

Brenninkmeijer, C.A.M., van Geel, B. and Mook, W.G. (1982) Variations in the D/H and $^{18}O/^{16}O$ ratio in cellulose extracted from a peat bog core. *Earth and Planetary Science Letters*, **61**, 283–90.

Bridge, M.C., Haggart, B.A. and Lowe, J.J. (1990) The history and palaeoclimatic significance of subfossil remains of *Pinus sylvestris* in blanket peats from Scotland. *Journal of Ecology*, **78**, 77–99.

Briffa, K.R., Jones, P.D., Wigley, T.M.L., Pilcher, J.R. and Baillie, M.G.L. (1986) Climate reconstruction from tree-rings: Part 1, basic methodology and preliminary results for England. *Journal of Climatology*, **3**, 233–42.

Briffa, K.R., Jones, P.D., Wigley, T.M.L., Pilcher, J.R. and Baillie, M.G.L. (1983) Climate reconstruction from tree-rings: Part 2, spatial reconstruction of summer mean sea level pressure patterns over Great Britain. *Journal of Climatology*, **6**, 1–15.

Briffa, K.R., Wigley, T.M.L., Jones, P.D., Pilcher, J.R. and Hughes, M.K. (1987) Patterns of tree-growth and related pressure variability in Europe. *Dendrochronologia*, **5**, 35–57.

Briffa, K.R., Bartholin, T.S., Eckstein, D., Jones, P.D., Karlen, W., Schweingruber, F.H. and Zetterberg, P. (1990) A 1400-year tree-ring record of summer temperatures in Fennoscandia. *Nature*, **346**, 434–9.

Brown, A.G. (1982) Human impact on the former flood plain of the Severn. In *Archaeological Aspects of Woodland Ecology* (eds M. Bell and S. Limbrey), British Archaeological Reports, International Series 146, Oxford, pp. 93–104.

Brown, A.G. (1988) The palaeoecology of *Alnus* (alder) and the postglacial history of floodplain vegetation. Pollen percentage and influx data from the West Midlands, United Kingdom. *New Phytologist*, **110**, 425–36.

Brown, A.G. and Barber, K.E. (1985) Late Holocene palaeoecology and sedimentary history of a small lowland catchment in central England. *Quaternary Research*, **24**, 87–102.

Brown, D.M., Munro, M.A.R., Baillie, M.G.L. and Pilcher, J.R. (1986) Dendrochronology – The absolute Irish standard. *Radiocarbon*, **28**, 279–83.

Browne, P. (1986) *Late-glacial and Post-glacial Pollen Stratigraphy and Vegetational History of the Nephin Begs, Co, Mayo*, Unpublished Ph.D. Thesis, Trinity College, Dublin.

Brugam, R.B. (1978) Pollen indicators of land-use change in southern Connecticut. *Quaternary Research*, **9**, 349–62.

Bryan, G. (1991) *The Age and Formation of the Sudden Tract Bog Complex*, Bachelor of Environmental Studies Thesis, University of Waterloo.

Buckland, P.C. and Edwards. K.J. (1984) The longevity of pastoral episodes in pollen diagrams – the rôle of post-occupation grazing. *Journal of Biogeography*, **11**, 243–9.

Buell, M.F., Buell, H.F. and Reiners, W.A. (1968) Radial mat growth on Cedar Creek bog, Minnesota. *Ecology*, **49**, 1198–9.

Burga, C.A. (1988) Swiss vegetation history during the last 18,000 years. *New Phytologist*, **110**, 581–602.

Burgess, C. (1989) Volcanoes, catastrophe and global crisis of the late 2nd millennium BC. *Current Archaeology*, 117, Vol. X, No. 10, 325–9.

Burleigh, R. and Kerney, M.P. (1982) Some chronological implications of a fossil molluscan assemblage from a Neolithic site at Brook, Kent, England. *Journal of Archaeological Science*, **9**, 29–38.

Burney, D.A. (1987) Late Holocene vegetational change in Central Madagascar. *Quaternary Research*, **28**, 130–43.

Bush, M. and Colinvaux, P.A. (1988) A 7000-year pollen record from the Amazon lowlands, Ecuador. *Vegetatio*, **76**, 141–54.

Bush, M.B. (1988) Early Mesolithic disturbance: a force on the landscape? *Journal of Archaeological Science*, **15**, 453–62.

Bush, M.B. (1989) On the antiquity of British chalk grasslands: a response to Thomas. *Journal of Archaeological Science*, **16**, 555–60.

Bush, M.B. and Flenley, J.R. (1987) The age of the British chalk grassland. *Nature*, **329**, 434–6.

Bush, M.B. and Hall, A.R. (1987) Flandrian *Alnus*: expansion or immigration? *Journal of Biogeography*, **14**, 479–81.

Butzer, K.W. (1982) *Archaeology as Human Ecology*, Cambridge University Press, New York.

Campbell, C.A., Paul, E.A., Rennie, D.A. and McCallum, K.J. (1967) Factors affecting the accuracy of the carbon-dating method in soil humus studies. *Soil Science*, **104**, 81–5.

Canti, M.G. (1987) Soil report on Lismore Fields, Buxton, Derbyshire. *Ancient Monuments Laboratory Report*, 216/87.

Care, V. (1982) The collection and distribution of lithic materials during the Mesolithic and Neolithic periods in southern England. *Oxford Journal of Archaeology*, **1**, 269–85.

Carter, B.A. (1986) *Pollen and Microscopic Charcoal Studies in South West Scotland*, Unpublished M.Phil. Thesis, University of Birmingham.

Carter, S.P. (1990) The stratification and taphonomy of shells in calcareous soils: implications for land snail analysis in archaeology. *Journal of Archaeological Science*, **17**, 495–507.

Case, H. (1969) Neolithic explanations. *Antiquity*, **43**, 176–86.

Case, H.J., Dimbleby, G.W., Mitchell, G.F., Morrison, M.E.S. and Proudfoot, V.B. (1969) Land use in Goodland Townland, Co. Antrim, from Neolithic times until today. *Journal of the Royal Society of Antiquaries of Ireland*, **99**, 39–53.

Caseldine, A. (1990) *Environmental Archaeology in Wales*, St. David's University College, Lampeter.

Caseldine, C.J. (1983) Pollen analysis and rates of pollen incorporation into a radiocarbon-dated palaeopodzolic soil at Haugabreen, southern Norway. *Boreas*, **12**, 233–46.

Caseldine, C.J. (1984) Pollen analysis of a buried arctic-alpine Brown Soil from Vestre Memurubreen, Jotunheimen, Norway: evidence for postglacial high-altitude vegetation change. *Arctic and Alpine Research*, **16**, 423–30.

Caseldine, C.J. and Maguire, D.J. (1983) A review of the prehistoric and historic environment on Dartmoor. *Devon Archaeological Society, Proceedings*, **39**, 1–16.

Caseldine, C.J. and Maguire, D.J. (1986) Lateglacial/early Flandrian vegetation change on northern Dartmoor, South West England. *Journal of Biogeography*, **13**, 255–64.

Caseldine, C.J. and Matthews, J.A. (1985) ^{14}C dating of palaeosols, pollen analysis and landscape change: studies from the low- and mid-alpine belts of southern Norway. In *Soils and Quaternary Landscape Evolution* (ed. J. Boardman), John Wiley, Chichester, pp. 87–116.

Caseldine, C.J. and Matthews, J.A. (1987) Podzol development, vegetation change and glacier variations at Haugabreen, southern Norway. *Boreas*, **16**, 215–30.

Casparie, W.A. (1972) Bog development in south-eastern Drenthe. *Vegetatio*, **25**, 1–271.

CCIRG (1991) *The Potential Effects of Climate Change in the UK: UK Climate Change Impacts Review Group – 1st Report for DoE*, HMSO, London.

Chambers, F.M. (1981) Date of blanket peat initiation in upland South Wales. *Quaternary Newsletter*, **35**, 24–9.

Chambers, F.M. (1982a) Environmental history at Cefn Gwernffrwd, near Rhandirmwyn, mid-Wales. *New Phytologist*, **92**, 607–15.

Chambers, F.M. (1982b) Two radiocarbon-dated pollen diagrams from high-altitude blanket peats in South Wales. *Journal of Ecology*, **70**, 445–59.

Chambers, F.M. (1982c) Date of blanket peat initiation – a comment. *Quaternary Newsletter*, **36**, 37–9.

Chambers, F.M. (1983) Three radiocarbon-dated pollen diagrams from upland peats north west of Merthyr Tydfil, South Wales. *Journal of Ecology*, **71**, 475–87.

Chambers, F.M. (1984) *Studies on the Initiation, Growth-Rate and Humification of 'Blanket Peats' in South Wales*, Department of Geography Occasional Paper No. 9, Keele University.

Chambers, F.M. and Elliott, L. (1989) Spread and expansion of *Alnus* Mill. in the British Isles: timing, agencies and possible vectors. *Journal of Biogeography*, **16**, 541–50.

Chambers, F.M. and Price. S.-M. (1985) Palaeoecology of *Alnus* (alder): early post-glacial rise in a valley mire, northwest Wales. *New Phytologist*, **101**, 333–44.

Chambers, F.M. and Price, S.-M. (1988) The environmental setting of Erw-wen and Moel y Gerddi: prehistoric enclosures in upland Ardudwy, North Wales. *Proceedings of the Prehistoric Society*, **54**, 93–100.

Chambers, F.M., Kelly, R.S. and Price, S.-M. (1988) Development of the late-prehistoric cultural landscape in upland Ardudwy, north-west Wales. In *The Cultural Landscape: Past, Present and Future* (eds H.H. Birks, H.J.B. Birks, P.E. Kaland and D. Moe), Cambridge University Press, Cambridge, pp. 333–48.

Chandler, C., Cheney, P., Thomas, P., Traband, L. and Williams, D. (eds) (1983) *Fire in Forestry. Vol.1. Forest Fire Behaviour and Effects*, John Wiley, New York.

Chapellaz, J., Barnola, J.M., Raynaud, D., Korotkevich, Y.S. and Lorius, C. (1990) Ice-core record of atmospheric methane over the past 160,000 years. *Nature*, **345**, 127–131.

Chapman, S.B. (1964) The ecology of Coom Rigg Moss, Northumberland. I. Stratigraphy and present vegetation. *Journal of Ecology*, **52**, 299–313.

Charles, D.F., Batterbee, R.W., Renberg, I., Van Damm, H. and Smol, J.P. (1990) Palaeoecological analysis of lake acidification trends in North America and Europe using diatoms and chrysophytes. In *Acid Precipitation Vol. 4: Soils, Aquatic Processes and Lake Acidification* (eds S.E. Norton, S.E. Lindberg and A.L. Page), Springer-Verlag, New York.

Chen, S.H. (1988) Neue Untersuchungen über die spät – und postglaziale Vegetsionsgeschichte im Gebiet zwischen Harz und Leine (BRD). *Flora*, **181**, 147–77.

Clark, J.S. (1988a) Particle motion and the theory of charcoal analysis: source area, transport, deposition and sampling. *Quaternary Research*, **30**, 67–80.

Clark, J.S. (1988b) Effect of climate change on fire regimes in northwestern Minnesota. *Nature*, **334**, 233–5.

Clark, J.S., Merkt, J. and Muller, H. (1989) Postglacial fire, vegetation, and human history on the northern Alpine forelands, south-western Germany. *Journal of Ecology*, **77**, 897–925.

Clark, R.L. (1982) Point-count estimation of charcoal in pollen preparations and thin sections of sediments. *Pollen et Spores*, **24**, 523–35.

CLIMAP Project Members (1976) The surface of the Ice-Age Earth. *Science*, **191**, 1131–7.

CLIMAP Project Members (1981) Seasonal reconstructions of the Earth's surface at the last glacial maximum. *Geological Society of America, Map and Chart Series*, **MC-36**, 1–18.

Cloutman, E.W. (1987) A mini-monolith cutter for absolute pollen analysis and fine sectioning of peats and sediments. *New Phytologist*, **107**, 245–8.

Cloutman, E.W. and Smith, A.G. (1988) Palaeoenvironments in the Vale of Pickering Part 3: Environmental History at Star Carr. *Proceedings of the Prehistoric Society*, **54**, 37–58.

Clymo, R.S., Oldfield, F., Appleby, P.G., Pearson, G.W., Ratnesar, P. and Richardson, N. (1990) The record of atmospheric deposition on a rainwater-dependent peatland. *Philosophical Transactions of the Royal Society, London [B]*, **327**, 331–8.

COHMAP Project Members (1988) Climatic changes of the last 18,000 years: Observations and model simulations. *Science*, **141**, 1043–51.

Coles, B. and Coles, J. (1986) *Sweet Track to Glastonbury. The Somerset Levels in Prehistory*, Thames and Hudson, London.

Coles, J.M. (1976) Forest farmers: some archaeological, historical and experimental evidence. In *Acculturation and Continuity in Atlantic Europe* (ed. S.J. De Laet), IV Atlantic Colloquium, Brugge, pp. 59–66.

Conway, V.M. (1947) Ringinglow Bog, near Sheffield. Part I. Historical. *Journal of Ecology*, **34**, 149–81.

Conway, V.M. (1954) Stratigraphy and pollen analysis of southern Pennine blanket peats. *Journal of Ecology*, **42**, 117–47.

Coope, G.R. (1975) Climatic fluctuations in northwest Europe since the last interglacial indicated by fossil assemblages of Coleoptera. *Geological Journal*, Special Issue 6, 153–68.

Coope, G.R. (1977) Fossil coleopteran assemblages as sensitive indicators of climatic changes during the Devensian (last) cold stage. *Philosophical Transactions of the Royal Society of London [B]*, **280**, 313–37.

Coope, G.R. (1986) Coleoptera analysis. In *Handbook of Holocene Palaeoecology and Palaeohydrology* (ed. B.E. Berglund), John Wiley, Chichester, pp. 703–13.

Coope, G.R. and Brophy, J.A. (1972) Late-glacial environmental changes indicated by a coleopteran succession from North Wales. *Boreas*, **1**, 97–142.

Coope, G.R. and Joachim, M.J. (1980) Lateglacial environmental changes interpreted from fossil Coleoptera from St. Bees, Cumbria, NW England. In *Studies in the Lateglacial of North-West Europe* (eds J.J. Lowe, J.M. Gray and J.E. Robinson), Pergamon, Oxford, pp. 55–68.

Crocker, R.L. and Major, J. (1955) Soil development in relation to vegetation and surface age of Glacier Bay, Alaska. *Journal of Ecology*, **43**, 427–48.

Cruise, G.M. (1990) Holocene peat initiation in the Ligurian Appenines, northern Italy. *Review of Palaeobotany and Palynology*, **63**, 173–82.

Crum, H.A. (1988) *A Focus on Peatlands and Peat Mosses*, University of Michigan Press, Ann Arbor, Michigan.

Cundill, P.R. and Whittington, G. (1983) Anomalous arboreal pollen assemblages in Late Devensian and early Flandrian deposits at Creich Castle, Fife, Scotland. *Boreas*, **12**, 297–311.

Cunliffe, B.W. (1984) *Danebury: An Iron Age Hillfort in Hampshire*, Council for British Archaeology, London.

Cwynar, Les C. and Watts, W.A. (1989) Accelerator-mass spectrometer ages for Late-glacial events at Ballybetagh, Ireland. *Quaternary Research*, **31**, 377–80.

Dachnowski, A. (1912) Peat deposits of Ohio: their origin, formation and uses. *Geological Survey of Ohio, 4th Series, Bulletin*, **16**, 1–424.

Dachnowski, A.P. (1924) The stratigraphic study of peat deposits. *Soil Science*, **17**, 107–23.

Damman, A.W.H. and French, T.W. (1987) The ecology of peat bogs of the glaciated northeastern United States: a community profile. *U.S. Fish and Wildlife Service Biological Report*, **85** (7.16), 1–100.

Danielsen, A. (1970) Pollen-analytical late Quaternary studies in the Ra district of Østfold, southeast Norway. *Årbog Universiteit Bergen*, **14**, 1–146.

Dansgaard, W., White, J.W.C. and Johnsen, J. (1989) The abrupt termination of the Younger Dryas climate event. *Nature*, **339**, 532–4.

D'Antoni, H.L. (1983) Pollen analysis of Gruta del Indio. In *Quaternary of South America and Antarctic Peninsula* (ed. J. Rabassa), A.A. Balkema, Rotterdam, pp. 83–104.

Davies, C. and Turner, J. (1979) Pollen diagrams from Northumberland. *New Phytologist*, **82**, 783–804.

Davis, M.B. (1964) Pollen accumulation rates: estimates from Late-Glacial sediments of Rogers Lake. *Science*, **145**, 1293–5.

Davis, M.B. (1981) Outbreaks of forest pathogens in Quaternary history. *IV International Palynological Conference, Lucknow*, **3**, 216–27.

Davis, M.B. (1983) Holocene vegetational history of the eastern United States. In *Late-Quaternary Environments of the United States*, Vol. 2, *The Holocene* (ed. H.E. Wright Jr.), University of Minnesota Press, Minneapolis, pp. 166–81.

Davis, M.B. (1984) Climatic instability, time lags, and community disequilibrium. In *Community Ecology* (eds J. Diamond and T.J. Case), Harper & Row, New York, pp. 269–84.

Davis, M.B., Woods, K.D., Webb, S.L. and Futyma, R.P. (1986) Dispersal versus climate: expansion of *Fagus* and *Tsuga* into the Upper Great lakes region. *Vegetatio* **67**, 93–103.

Dawson, A.G., Lowe, J.J. and Walker, M.J.C. (1987) The nature and age of the debris accumulation at Gribun, western Mull, Inner Hebrides. *Scottish Journal of Geology*, **23**, 149–62.

Deacon, J. (1974) The location of refugia of *Corylus avellana* L. during the Weichselian glaciation. *New Phytologist*, **73**, 1055–63.

Deevey, E.S. (1949) Biogeography of the Pleistocene. Part I: Europe and North America. *Bulletin of the Geological Society of America*, **60**, 1315–416.

Deevey, E.S. (1969) Coaxing history to conduct experiments. *Bioscience*, **19**, 40–3.

Deevey, E.S. (1978) Holocene forests and Maya disturbance near Quexil Lake, Peten, Guatemala. *Poskie Archiwum Hydrobiologii*, **25**, 117–29.

Delcourt, H.R. and Delcourt, P.A. (1985) Quaternary palynology and vegetation history of the southeastern United States. In *Pollen Records of Late-Quaternary North American Sediments* (eds V.M. Bryant and R.G. Holloway), American Association of Stratigraphic Palynologists Foundation, pp. 1–37.

Delcourt, H.R. and Delcourt, P.A. (1991) *Quaternary Ecology – a Palaeoecological Perspective*, Chapman and Hall, London.

Delcourt, P.A. and Delcourt, H.R. (1981) Vegetation maps for eastern North America: 40 000 yr BP to the present. In *Geobotany II* (ed. R. Romans), Plenum, New York, pp. 123–66.

Delcourt, P.A. and Delcourt, H.R. (1987) Late-Quaternary dynamics of temperate forests: applications of palaeoecology to issues of global environmental change. *Quaternary Science Reviews*, **6**, 129–46.

Delcourt, P.A., Delcourt, H.R., Cridlebaugh, P.A. and Chapman, J. (1986) Holocene ethnobotanical and palaeoecological record of human impact on vegetation in the Little Tennessee River Valley, Tennessee. *Quaternary Research*, **25**, 330–49.

Denton, G.H. and Hughes, T.J. (eds) (1981) *The Last Great Ice Sheets*, Wiley, New York.

Dickinson, W. (1975) Recurrence surfaces in Rusland Moss, Cumbria (formerly north Lancashire). *Journal of Ecology*, **63**, 913–35.

Dickson, C. (1989) The Roman army diet in Britain and Germany. *Archaobotanik: Dissertationes Botanicae*, **133**, 135–54.

Dickson, J.H., Dickson, C., Boyd, W.E., Newall, P.J. and Robinson, D.E. (1985) The vegetation of central Scotland in Roman times. *Ecologica Mediterranea*, **11**, 81.

Digerfeldt, G. (1988) Reconstruction and regional correlation of Holocene lake-level fluctuations in Lake Bysjön, South Sweden. *Boreas*, **17**, 165–182.

Dimbleby, G.W. (1957) Pollen analysis of terrestrial soils. *New Phytologist*, **56**, 12–28.

Dimbleby, G.W. (1961a) Soil pollen analysis. *Journal of Soil Science*, **12**, 1–11.

Dimbleby, G.W. (1961b) Transported material in the soil profile. *Journal of Soil Science*, **12**, 12–22.

Dimbleby, G.W. (1961c) The ancient forest of Blackamore. *Antiquity*, **35**, 123–8.

Dimbleby, G.W. (1962) *The Development of the British Heathlands and their Soils*, Oxford Forestry Memoirs, Clarendon Press, Oxford.

Dimbleby, G.W. (1965) Post-glacial changes in soil profiles. *Proceedings of the Royal Society of London [B]*, **161**, 355–62.

Dimbleby, G.W. (1977) *Ecology and Archaeology*, Arnold, London.

Dimbleby, G.W. (1985) *The Palynology of Archaeological Sites*, Academic Press, London.

Dodson, J.R. (1978) The influence of man at a quarry site, Nairn River Valley, Chatham Island, New Zealand. *Journal of the Royal Society of New Zealand*, **8**, 377–84.

Dodson, J.R. and Bradshaw, R.H.W. (1987) A history of vegetation and fire, 6,600 BP to present, County Sligo, Western Ireland. *Boreas*, **16**, 113–23.

Donaldson, A.M. and Turner, J. (1977) A pollen diagram from Hallowell Moss, near Durham City, UK. *Journal of Biogeography*, **4**, 25–33.

Donner, J.J. (1962) On the Post-glacial history of the Grampian Highlands. *Societas Scientiarum Fennica Commentationes Biologicae*, **24**, 1–29.

Doyle, G.J. (1982) The vegetation, ecology and productivity of Atlantic blanket bog in Mayo and Galway, western Ireland. *Journal of Life Sciences, Royal Dublin Society*, **3**, 147–64.

Doyle, G.J. (1990) Phytosociology of Atlantic bog complexes in north-west Mayo, Ireland. In *Ecology and Conservation of Irish Peatlands* (ed G.J. Doyle), Royal Irish Academy, Dublin, pp. 75–90.

Dresser, P.Q. (1970) *A Study of Sampling and Pretreatments of Materials for Radiocarbon Dating*, Unpublished Ph.D. Thesis, Queen's University Belfast, Belfast.

Dresser, P.Q., Smith, A.G. and Pearson, G.W. (1973) Radiocarbon dating of the raised beach at Woodgrange, Co. Down. *Proceedings of the Royal Irish Academy*, **73** (B 2), 53–6.

Drew, C.D. and Piggott, S. (1936) The excavation of long barrow 163a on Thickthorn Down, Dorset. *Proceedings of the Prehistoric Society*, **2**, 77–96.

Dubois, A.D. and Ferguson, D.K. (1985) The climatic history of pine in the Cairngorms based on radiocarbon dates and stable isotope analysis, with an account of the events leading up to its colonization. *Review of Palaeobotany and Palynology*, **46**, 55–80.

Dugmore, A.J. (1989) Icelandic volcanic ash in Scotland. *Scottish Geographical Magazine*, **105**, 168–72.

Dupont, L.M. (1985) *Temperature and Rainfall Variation in a Raised Bog Ecosystem: a Palaeoecological and Isotope Geological Study*, Thesis, University of Amsterdam.

Dupont, L.M. and Brenninkmeijer, C.A.M. (1984) Palaeobotanic and isotopic analysis of the late Sub-Boreal and early Sub-Atlantic peat from Engbertsdijksveen VII, the Netherlands, *Review of Palaeobotany and Palynology*, **41**, 241–71.

Durno, S.E. (1958) Pollen analysis of peat deposits in eastern Sutherland and Caithness. *Scottish Geographical Magazine*, **74**, 127–35.

Durno, S.E. (1959) Pollen analysis of peat deposits in the eastern Grampians. *Scottish Geographical Magazine*, **75**, 102–11.

Ednew, J.R., Kershaw, A.P. and De Deckker, P. (1990) A Late Pleistocene and Holocene vegetation and environmental record from Lake Wangoom, western plains of Victoria, Australia. *Palaeogeography, Palaeoclimatology, Palaeoecology*, **80**, 325–43.

Edwards, K.J. (1979) Palynological and temporal inference in the context of prehistory, with special reference to the evidence from lake and peat deposits. *Journal of Archaeological Science*, **6**, 255–70.

Edwards, K.J. (1982) Man, space and the woodland edge – speculations on the detection and interpretation of human impact in pollen profiles. In *Archaeological Aspects of Woodland Ecology* (eds S. Limbrey and M. Bell), British Archaeological Reports, International Series, S146, Oxford, pp. 5–22.

Edwards, K.J. (1983) Quaternary palynology: multiple profile studies and pollen variability. *Progress in Physical Geography*, **7**, 687–709.

Edwards, K.J. (1985) Chronology. In *Quaternary History of Ireland* (eds K.J. Edwards and W.P. Warren), Academic Press, London.

Edwards, K.J. (1988) The hunter–gatherer/agricultural transition and the pollen record in the British Isles. In *The Cultural Landscape: Past, Present and Future* (eds H.H. Birks, H.J.B. Birks, P.E. Kaland and D. Moe), Cambridge University Press, Cambridge, 255–66.

Edwards, K.J. (1989) The cereal pollen record and early agriculture. In *The Beginnings of Agriculture* (eds A. Milles, D. Williams and N. Gardner), British Archaeological Reports, International Series S496, Oxford, 113–35.

Edwards, K.J. (1990) Fire and the Scottish Mesolithic. In *Contributions to the Mesolithic in Europe* (eds P.M. Vermeersch and P. Van Peer), Leuven University Press, Leuven, 71–9.

Edwards, K.J. (1991) Using space in cultural palynology: the value of the off-site pollen record. In *Modelling Ecological Change* (eds D.R. Harris and K.D. Thomas), International Academic Projects, London, 61–73.

Edwards, K.J. and Hirons, K.R. (1984) Cereal pollen grains in pre-elm decline deposits: implications for the earliest agriculture in Britain and Ireland. *Journal of Archaeological Science*, **11**, 71–80.

Edwards, K.J. and MacDonald, G.M. (1991) Holocene palynology: II. Human influence and vegetation change. *Progress in Physical Geography*, **15**, 364–91.

Edwards, K.J. and McIntosh, C.J. (1988) Improving the detection rate of cereal-type pollen grains in *Ulmus* decline and earlier deposits from Scotland. *Pollen et Spores*, **30**, 179–88.

Edwards, K.J. and Ralston, I. (1984) Postglacial hunter–gatherers and vegetational history in Scotland. *Proceedings of the Society of Antiquaries of Scotland*, **114**, 15–34.

Edwards, K.J., Hirons, K.R. and Newell, P.J. (1991) The palaeoecological and prehistoric context of minerogenic layers in blanket peat: a study from Loch Dee, southwest Scotland. *The Holocene*, **1**, 29–39.

Ehlers, J., Gibbard, P. and Rose, J. (eds) (1991) *The Glacial Deposits of Britain and Ireland*, A.A. Balkema, Rotterdam.

Ellis, S. (1979) The identification of some Norwegian mountain soil types. *Norsk Geografisk Tidsskrift*, **33**, 205–12.

Ellis, S. (1980) Soil–environmental relationships in the Okstindan Mountains, north Norway, *Norsk Geografisk Tidsskrift*, **34**, 167–76.

Ellis, S. and Matthews, J.A. (1984) Pedogenic implications of a ^{14}C-dated palaeopodzolic soil at Haugabreen, southern Norway. *Arctic and Alpine Research*, **16**, 77–91.

Elmore, D. and Phillips, F.M. (1987) Accelerator mass spectrometry for measurement of long-lived radioisotopes. *Science*, **236**, 543–50.

Engelmark, O. and Hofgaard, A. (1985) Sveriges äldsta tall. *Svensk Botanisk Tidskrift*, **79**, 415–16.

Entwistle, R. and Grant, A. (1989) The evidence for cereal cultivation and animal husbandry in the southern British Neolithic and Bronze Age. In *The Beginnings of Agriculture* (eds A. Milles, D.

Williams and N. Gardner), British Archaeological Reports, International Series S496, Oxford, pp. 203–15.

Evans, J.G. (1971) Habitat change on the calcareous soils of Britain: the impact of Neolithic man. In *Economy and Settlement in Neolithic and Early Bronze Age Britain and Europe* (ed. D.D.A. Simpson), Leicester University Press, Leicester, pp. 27–73.

Evans, J.G. (1972) *Land Snails in Archaeology*, Seminar Press, London.

Evans, J.G. (1990) Notes on some Late Neolithic and Bronze Age events in long barrow ditches in southern and eastern England. *Proceedings of the Prehistoric Society*, **56**, 111–16.

Evans, J.G. and Jones, H. (1973) Subfossil and modern land-snail faunas from rock-rubble habitats. *Journal of Conchology*, **27**, 103–29.

Evans, J.G. and Rouse, A.J. (1992) Small vertebrate and molluscan analysis from the same site. *Circaea*, **8**, 75–84.

Evans, J.G. and Simpson, D.D.A. (1991) Giants' Hills 2 long barrow, Skendleby, Lincolnshire. *Archaeologia*, **109**, 1–45.

Evans, J.G. and Smith, I.F. (1983) Excavations at Cherhill, north Wiltshire, 1967. *Proceedings of the Prehistoric Society*, **49**, 43–117.

Evans, J.G., Pitts, M.W. and Williams, D. (1985) An excavation at Avebury, Wiltshire, 1982. *Proceedings of the Prehistoric Society*, **51**, 305–10.

Evans, J.G., Limbrey, S., Máté, I. and Mount, R. (1988a) Environmental change and land-use history in a Wiltshire river valley in the last 14 000 years. In *The Archaeology of Context in the Neolithic and Bronze Age: Recent Trends* (eds J.C. Barrett and I.A. Kinnes), University of Sheffield, Sheffield, pp. 97–103.

Evans, J.G., Rouse, A.J. and Sharples, N.M. (1988b) The landscape setting of causewayed camps: some recent work on the Maiden Castle enclosure. In *The Archaeology of Context in the Neolithic and Bronze Age: Recent Trends* (eds J.C. Barrett and I.A. Kinnes), University of Sheffield, Sheffield, pp. 73–84.

Evin, J. (1990) Validity of the radiocarbon dates beyond 35 000 years BP. *Palaeogeography, Palaeoclimatology, Palaeoecology*, **80**, 71–8.

Ewan, L. (1981) *A Palynological Investigation of a Peat Deposit near Banchory: Some Local and Regional Environmental Implications*, O'Dell Memorial Monograph 11, Department of Geography, University of Aberdeen.

Faegri, K. and Iversen, J. (1989). *Textbook of Pollen Analysis*, 4th edn (eds K. Faegri, P.E. Kaland and K. Krzywinski), John Wiley, Chichester.

Fantechi, R. and Ghazi, A. (eds) (1989) *Carbon Dioxide and other Greenhouse Gases: Climatic and Associated Impacts*, Kluwer, Dordrecht.

Fasham, P.J. (1982) The excavation of four ring ditches in central Hampshire. *Proceedings of the Hampshire Field Club and Archaeological Society*, **38**, 19–56.

Firbas, F. (1949) *Spät-und nacheiszeitliche Waldgeschichte Mitteleuropas nördlich der Alpen*, Fischer, Jena.

Fleming, A. (1971) Territorial patterns in Bronze Age Wessex. *Proceedings of the Prehistoric Society*, **37**, 138–66.

Flenley, J.R. (1984) Late Quaternary changes of vegetation and climate in the Malesian mountains. *Erdwissenschaftliche Forschung*, **18**, 261–67.

Flenley, J.R. (1988) Palynological evidence for land use changes in South-East Asia. *Journal of Biogeography*, **15**, 185–97.

Flenley, J.R., King, A.S.M., Teller, J.T., Prentice, M.E., Jackson, J. and Chew, C. (1991) The late Quaternary vegetational and climatic history of Easter Island. *Journal of Quaternary Science*, **6**, 85–115.

Fowler, P.J. (1983) *The Farming of Prehistoric Britain*, Cambridge University Press, Cambridge.

Francis, E. (1987) *The Palynology of the Glencloy Area*, Unpublished Ph.D. Thesis, Queen's University Belfast, Belfast.

Fredskild, B. (1988) Agriculture in a marginal area – south Greenland from the Norse landnam (985 AD) to the present (1985 AD). In *The Cultural Landscape: Past, Present and Future* (eds H.H. Birks, H.J.B. Birks, P.E. Kaland and D. Moe), Cambridge University Press, Cambridge, pp. 381–93.

Frenzel, B. (1966) Climatic change in the Atlantic/Sub-Boreal transition in the Northern Hemisphere: botanical evidence. In *World Climate from 8000 to 0 BC* (ed. J.S. Sawyer), Royal Meteorological Society, London, pp. 99–123.

Fuji, N. (1984) Pollen analysis. In *Lake Biwa* (ed. S. Horie), Junk, Dordrecht, pp. 497–529.

Gaffney, C.F. and Gaffney, C.L. (1988) Some quantitative approaches to site territory and land use from the surface record. In *Conceptual Issues in Environmental Archaeology* (eds D.A. Davidson and E.G. Grant), University Press, Edinburgh, pp. 82–90.

Gamble, C. (1991) The social context for European Palaeolithic art. *Proceedings of the Prehistoric Society* **57**, 3–15.

Gammon, R.H., Sundquist, E.T. and Fraser, P.J. (1985) History of carbon dioxide in the atmosphere. In *Atmospheric Carbon Dioxide and the*

Global Carbon Cycle (ed. J.R. Trabalka), US Department of Energy, Washington DC, pp. 25–62.

Gamper, M. and Oberhaensli, H. (1982) Interpretation von Radiocarbondaten Fossiler Boeden. *Physische Geographie (Zürich)*, **1**, 83–90.

Gandreau, D.C. and Webb, T. III (1985) Late-Quaternary pollen stratigraphy and isochron maps for the northeastern United States. In *Pollen Records of Late-Quaternary North American Sediments* (eds V.M. Bryant and R.G. Holloway), American Association of Stratigraphic Palynologists, Dallas, pp. 245–80.

Garbett, G.G. (1981) The elm decline: the depletion of a resource. *New Phytologist*, **88**, 573–85.

Garton, D. (1987) Buxton. *Current Archaeology*, **103**, 250–3.

Gear, A.J. (1989) *Holocene Vegetation History and the Palaeoecology of* Pinus sylvestris *in Northern Scotland*, Unpublished Ph.D. Thesis, University of Durham.

Gear, A.J. and Huntley, B. (1991) Rapid changes in the range limits of Scots pine 4000 years ago. *Science*, **251** (4993), 544–7.

Gentilcore, R.L. and Donkin, K. (1973) Land surveys of southern Ontario. *Canadian Cartographer*, **10** (Supplement 2).

Gerasimov, I.P. and Chichagova, O.A. (1971) Some problems in the radiocarbon dating of soil. *Soviet Soil Science*, **3**, 519–27.

Geyh, M.A., Roeschmann, G., Wijmstra, T.A. and Middeldorp, A.A. (1983) The unrealiability of ^{14}C dates obtained from buried sandy podzols. *Radiocarbon*, **25**, 409–16.

Geyh, M.A., Röthlisberger, F. and Gellatly, A. (1985) Reliability tests and interpretation of ^{14}C dates from palaeosols in glacier environments. *Zeitschrift für Gletscherkunde und Glacialgeologie*, **21**, 275–81.

Gibbons, M. and Higgins, J. (1988) Archaeology (of Connemara). Derryinver, Rinvyle. Lough Sheeauns area. In *Connemara* (eds M. Warren and W.P. Warren), Field Guide No. 11, Irish Association for Quaternary Studies, Dublin, pp. 8–12 and pp. 55–62.

Gibson, A. (1989) Carneddau, Carno (SN 9989 9978). *Archaeology in Wales*, **29**, 48–9.

Gilet-Blein, N., Marien, G. and Evin J. (1980) Unreliability of ^{14}C dates from organic matter of soils. *Radiocarbon*, **22**, 919–29.

Gimingham, G.H. (1975) *Heathland Ecology*, Oliver & Boyd, Edinburgh.

Girling, M. (1988) The bark beetle *Scolytus scolytus* (Fabricius) and the possible role of elm disease in the early Neolithic. In *Archaeology and the Flora of the British Isles* (ed. M. Jones), Oxford University Committee for Archaeology, Oxford, pp. 34–8.

Goddard, A. (1971) *Studies of the Vegetational Changes Associated with the Initiation of Blanket Peat Accumulation in North-East Ireland*, Unpublished Ph.D. Thesis, Queen's University Belfast, Belfast.

Godwin, H. (1954) Recurrence-surfaces. *Danmarks Geologiska Undersökning*, R II, **80**, 22–30.

Godwin, H. (1960) Radiocarbon dating and Quaternary history in Britain. *Proceedings of the Royal Society* [B], **153**, 287–320.

Godwin, H. (1975) *History of the British Flora: A Factual Basis for Phytogeography*, 2nd edn, Cambridge University Press, Cambridge.

Godwin, H. (1981) *The Archives of the Peat Bogs*, Cambridge University Press, Cambridge.

Godwin, H. and Willis, E.H. (1959) Cambridge University natural radiocarbon measurements I. *American Journal of Science, Radiocarbon Supplement*, **1**, 63–75.

Goh, K.M. and Molloy, B.P.J. (1978) Radiocarbon dating of palaeosols using soil organic matter components. *Journal of Soil Science*, **29**, 567–73.

Good, R. (1931) A theory of plant geography. *New Phytologist*, **30**, 11–171.

Goode, D.A. and Ratcliffe, D.A. (1977) Peatlands. In *A Nature Conservation Review*, Vol. II (ed. D.A. Ratcliffe), Cambridge University Press, Cambridge, pp. 249–87.

Göransson, H. (1983) När börjar neolitikum? *Popular Arkeologi*, **1**, 4–7.

Göransson, H. (1984) Pollen analytical investigations in the Sligo area. In *The Archaeology of Carrowmore. Environmental Archaeology and the Megalithic Tradition at Carrowmore, County Sligo, Ireland* (ed. G. Burenhult), Theses and Papers in North-European Archaeology 14, Stockholm, pp. 154–193.

Göransson, H. (1986) Man and the forests of nemoral broad-leaved trees during the Stone Age. *Striae*, **24**, 143–52.

Göransson, H. (1987a) *Neolithic Man and the Forest Environment around Alvastra Pile Dwelling*, Theses and Papers in North-European Archaeology 20, Stockholm.

Göransson, H. (1987b) Comments on early agriculture in Sweden. On arguing in a circle, on common sense, on the smashing of paradigms, on thistles among flowers, and on other things. *Norwegian Archaeological Review*, **18**, 43–5.

Goransson, H. (1988) Can exchange during Mesolinic time be evidenced by pollen analysis? In *Trade and Exchange in Prehistory* (eds B. Hårdh, L. Larsson, D. Olauson and R. Petre), Acta

Archaeologica Lundensia, Series in 8° No. 16, Lund, pp. 33–40.

Göransson, H. (1991) *Vegetation and Man Around Lake Bjärsjöholmssjon During Prehistoric Time*, LUNDQUA Report 31, Lund.

Gottesfield, A.S. and Gottesfield, L.M.J. (1993) 'Little Ice Age' flood plains in northwestern British Columbia, Canada: extent and preliminary chronology. *The Holocene*, **3** (in press).

Granlund, E. (1932) De svenska Hogmossarnas geologi. *Sveriges Geologiska Undersökning*, **26**, 1–93.

Green, B.H. (1968) Factors influencing the spatial and temporal distribution of *Sphagnum imbricatum* Hornsch ex. Russ. in the British Isles. *Journal of Ecology*, **56**, 47–58.

Green, D.G. and Dolman, G.S. (1988) Fine resolution pollen analysis. *Journal of Biogeography*, **15**, 685–701.

Griffey, N.J. and Matthews, J.A. (1978) Major Neoglacial glacier expansion episodes in southern Norway: evidences from moraine ridge stratigraphy with ^{14}C dates from buried palaeosols and moss layers. *Geografiska Annaler*, **60(A)**, 73–90.

Grime, J.P., Hodgson, J.G. and Hunt, R. (1988) *Comparative Plant Ecology*, Unwin Hyman, London.

Grimm, E.C. (1991) *TILIA and TILIA-GRAPH*, Illinois State Museum, Springfield.

Groenman-van Waateringe, W. (1983) The early agricultural utilization of the Irish landscape: the last word on the elm decline? In *Landscape Archaeology in Ireland* (eds T. Reeves-Smyth and F. Hammond), British Archaeological Reports, British Series 116, Oxford, pp. 217–32.

Grove, J. (1985) The timing of the Little Ice Age in Scandinavia. In *The Climatic Scene* (eds M.J. Tooley and G.M. Sheail), George Allen & Unwin, London, pp. 132–53.

Grove, J. (1988) *The Little Ice Age*, Methuen, London.

Guillet, B. and Robin, A.M. (1972) Interprétation de datations par le ^{14}C d'horizons Bh de deux podzols humo-ferrugineux, l'un formé sous callune, l'autre sous chênaie-hêtraie. *C.R. Hebdomadare Sci. Acad. Sci. Paris*, **274**, 2859–62.

Guthrie, R.D. (1982) Mammals of the mammoth steppe as palaeoenvironmental indicators. In *Paleoecology of Beringia* (eds D.M. Hopkins, J.V. Matthews, Jr., C.E. Schweger and S.B. Young), Academic Press, London, pp. 307–26.

Guyan, W.U. (1955) Das jungsteinzeitliche Moordorf von Thayngen-Weier. *Das Pfahlbauproblem, herausgegeben zum Jubiläum des 100-jährigen Bestehens der Schweizerischen Pfahlbauforschung*, Monographien zur Ur- und Frühgeschichte der Schweiz 11, Basel.

Haberle, S.G., Hope, G.S. and De Fretes, Y. (1991) Environmental change in the Baliem Valley, montane Irian Jaya, Republic of Indonesia. *Journal of Biogeography*, **18**, 25–40.

Hafsten, U. and Solem, T. (1976) Age, origin and palaeoecological evidence of blanket bogs in Nord-Trondelag, Norway. *Boreas*, **5**, 119–41.

Hamilton, A.C. (1982) *Environmental History of East Africa: a Study of the Quaternary*, Academic Press, London.

Hammer, C.U., Clausen, H.B. and Dansgaard, W. (1980) Greenland ice sheet evidence of post-glacial volcanism and its climatic impact. *Nature*, **288**, 230–5.

Hanf, K.I. and Warner, B.G. (1992) European land clearance and *Sphagnum* formation in a kettle-hole mire, southwestern Ontario. In *Peatland Ecosystems and Man: An impact assessment* (eds O.M. Bragg, P.D. Hulme et al.), Department of Biological Sciences, University of Dundee, Dundee, 291–3.

Harrison, S.P. and Digerfeldt, G. (forthcoming) European lakes as palaeohydrological and palaeoclimatic indicators. Submitted to *Quaternary Science Reviews*.

Haslam, C.J. (1987) *Late Holocene Peat Stratigraphy and Climatic Change – A Macrofossil Investigation from the Raised Mires of North Western Europe*, Ph.D. Thesis, University of Southampton.

Hatton, J.M. (1991) *Environmental Change on Northern Dartmoor during the Mesolithic Period*, Unpublished M.Phil. Thesis, University of Exeter.

Hays, J.D., Imbrie J. and Shackleton, N.J. (1976) Variations in the Earth's orbit: Pacemaker of the Ice Age. *Science*, **194**, 1121–32.

Hedges, R.E.M. (1991) AMS Dating: present status and potential applications. In *Radiocarbon Dating: Recent Applications and Future Potential* (ed. J.J.Lowe), Quaternary Proceedings No. 1, Quaternary Research Association, Cambridge, pp. 5–10.

Hedges, R.E.M., Housley, R.A., Law, I.A., Perry, C. and Gowlett, J.A.J. (1987) Radiocarbon dates from the Oxford AMS system: Archaeometry datelist 6. *Archaeometry*, **29**, 289–306.

Hedges, R.E.M., Housley, R.A., Law, I.A., Perry, C. and Hendy, E. (1988) Radiocarbon dates from the Oxford AMS system: Archaeometry datelist 8. *Archaeometry*, **30**, 291–305.

Heidenreich, C.E. (1990) History of the St. Lawrence-Great Lakes area to AD 1650. In *Archaeology of Southern Ontario to AD 1650* (eds C.J. Ellis and N. Ferris), Archaeological Society of Ontario, London, pp. 475–92.

Heim, J. (1962) Recherches sur les relations entre la végétation actuelle et le spectre pollinique recent dans les Ardennes Belges. *Bulletin Societé Royale Botanique Belgique*, **96**, 5–92.

Hemond, H.F. (1980) Biogeochemistry of Thoreau's Bog, Concord, Massachusetts. *Ecological Monographs*, **50**, 507–26.

Hewitt, T.C. (1990) *Land Snails from Danebury*, Unpublished Undergraduate Thesis, University of Wales, Cardiff.

Heybroek, H.M. (1963) Diseases and lopping for fodder as possible causes of a prehistoric decline of Ulmus. *Acta Botanica Neerlandica*, **12**, 1–11.

Hibbert, F.A. and Switsur, V.R. (1976) Radiocarbon dating of Flandrian pollen zones in Wales and northern England. *New Phytologist*, **77**, 793–807.

Hicks, S.P. (1971) Pollen-analytical evidence for the effect of agriculture on the vegetation of north Derbyshire. *New Phytologist*, **70**, 647–67.

Hicks, S.P. (1972) The impact of man on the East Moors of Derbyshire from Mesolithic times. *Archaeological Journal*, **129**, 3–21.

Hicks, S.P. (1985) Problems and possibilities in correlating historical/archaeological and pollen analytical evidence in a northern boreal environment: an example from Kuusamo, Finland. *Fennoscandia Archaeologica*, **11**, 5–84.

Hicks, S.P. (1988) The representation of different farming practices in pollen diagrams from northern Finland. In *The Cultural Landscape: Past, Present and Future* (eds H.H. Birks, H.J.B. Birks, P.E. Kaland and D. Moe), Cambridge University Press, Cambridge, pp. 189–207.

Higham, N. (1986) *The Northern Counties to AD 1000*, Longman, London.

Hill, M.O. (1979) *DECORANA – A FORTRAN program for Detrended Correspondence Analysis and Reciprocal Averaging*, Section of Ecology and Systematics, Cornell University, Ithaca, New York.

Hill, M.O. (1988) *Sphagnum imbricatum* ssp. *austinii* (Sull.(Flatberg)) and ssp. *affine* (Rem. and Card.) in Britain and Ireland. *Journal of Bryology*, **15**, 109–16.

Hill, M.O. and Gauch, H.G. (1980) Detrended correspondence analysis: an improved ordination technique, *Vegetatio*, **42**, 47–58.

Hillman. J. (1976) The dating of Cullyhanna hunting lodge. *Irish Archaeological Research Forum*, **3** (1), 17–20.

Hillman, J., Groves, C.M., Brown, D.M., Baillie, M.G.L., Coles, J.M. and Coles, B.J. (1990) Dendrochronology of the English Neolithic. *Antiquity*, **64**, 210–20.

Hintikka, V. (1963) Über das Grossklima einiger Pflanzenareale in zwei Klimakoordinatensystem Dargestellt. *Annales Botanici Societatis Zoologicae Botanicae Fennicae 'Vanamo'*, **34**(5), 1–64.

Hirons, K.R. and Edwards, K.J. (1986) Events at and around the first and second *Ulmus* declines: palaeoecological investigations in Co. Tyrone, Northern Ireland. *New Phytologist*, **104**, 131–53.

Hodder, A.P.W., De Lange, P.J. and Lowe, D.J. (1991) Dissolution and depletion of ferromagnesian minerals from Holocene tephra layers in an acid bog, New Zealand, and implications for tephra correlation. *Journal of Quaternary Science*, **6**, 195–208.

Holgate, R. (1988) *Neolithic Settlement of the Thames Basin*, British Archaeological Reports, 194, Oxford.

Holland, S. (1975) *Pollen Analytical Studies Concerning Settlements and Early Ecology in Co. Down, N. Ireland*. Unpublished Ph.D. Thesis, Queen's University Belfast, Belfast.

Holloway, R. and Bryant, V.M. (1985) Late-Quaternary pollen record and vegetational history of the Great Lakes region: United States and Canada. In *Pollen Records of Late-Quaternary North American Sediments* (eds V.M. Bryant and R.G. Holloway), American Association of Stratigraphic Palynologists Foundation, Dallas, pp. 205–45.

Hooker, J.P. (1879) The distribution of the North American flora. *American Naturalist*, **8**, 155–70.

Hope, G.S. (1983) The vegetation changes of the last 20,000 years at Telefomin, Papua New Guinea. *Singapore Journal of Tropical Geography*, **4**, 25–33.

Hopkins, D.M., Matthews, J.V. Jr., Schweger, C.E. and Young, S.B. (eds) (1982) *Paleoecology of Beringia*, Academic Press, London.

Housley, R.A., Hedges, R.E.M., Law, I.A. and Bronk, C.R. (1991) Radiocarbon dating by AMS of the destruction of Akrotiri. In *Thera and the Aegean World III*, Vol. 3 (ed. D.A. Hardy), The Thera Foundation, London.

Huntley, B. (1988) Glacial and Holocene vegetation history: Europe, in *Vegetation History* (eds B. Huntley and T. Webb III), Kluwer, Dordrecht, 341–83.

Huntley, B. (1990a) European post-glacial forests: compositional changes in response to climatic change. *Journal of Vegetation Science*, **1**, 507–18.

Huntley, B. (1990b) European vegetation history: palaeovegetation maps from pollen data – 13 000 yr BP to present. *Journal of Quaternary Science*, **5**, 103–22.

Huntley, B. (1990c) Dissimilarity mapping between fossil and contemporary pollen spectra in Europe for the past 13,000 years. *Quaternary Research*, **33**, 360–76.

Huntley, B. (1990d) Studying global change: the contribution of Quaternary palynology. *Global and Planetary Change*, **2**, 53–61.

Huntley, B. (1991) How plants respond to climate change: migration rates, individualism and the consequences for plant communities. *Annals of Botany*, **67** (Supplement 1), 15–22.

Huntley, B. (1992) Rates of change in the European palynological record of the last 13,000 years and their climatic interpretation. *Climate Dynamics*, **6**, 185–91.

Huntley, B. (in press) Pollen–climate response surfaces and the study of climate change. In *Applied Quaternary Research* (ed. J.M. Gray).

Huntley, B. and Birks, H.J.B. (1983) *An Atlas of Past and Present Pollen Maps for Europe 0-13,000 years ago*, Cambridge University Press, Cambridge.

Huntley, B. and Webb, T. III (1989) Migration: species' response to climatic variations caused by changes in the earth's orbit. *Journal of Biogeography*, **16**, 5–19.

Huntley, B., Bartlein, P.J. and Prentice, I.C. (1989) Climatic control of the distribution and abundance of beech (*Fagus* L.) in Europe and North America. *Journal of Biogeography*, **16**, 551–60.

Imbrie, J. and Imbrie, K.P. (1979) *Ice Ages: Solving the Mystery*, Macmillan, London.

Indrelid, S. and Moe, D. (1982) Februk på Hardangervidda i yngre steinalder. *Viking*, **46**, 36–71.

Innes, J.B. (1989) *Fine Resolution Pollen Analysis of Late Flandrian II Peats of North Gill, North York Moors*, Unpublished Ph.D. Thesis, University of Durham.

Innes, J.B. and Simmons, I.G. (1988) Disturbance and diversity: floristic changes associated with pre-elm decline woodland recession in North East Yorkshire. In *Archaeology and the Flora of the British Isles* (ed. M. Jones), University Committee for Archaeology, Oxford, pp. 7–20.

International Study Group (1982) An inter-laboratory comparison of radiocarbon measurements in tree-rings. *Nature*, **298**, 619–23.

IPCC (1990) *Climate Change – The IPCC Scientific Assessment* (eds J.T. Houghton, G.J. Jenkins, and J.J. Ephraums), Cambridge University Press, Cambridge.

Ivanov, K.E. (1981) *Water Movements in Mirelands*, Academic Press, London.

Iversen, J. (1941) Landnam i Denmarks stenalder (Land occupation in Denmark's Stone Age). *Danmarks Geologiske Undersøgelse*, Series II, **66**, 1–68.

Iversen, J. (1944) *Viscum, Hedera* and *Ilex* as climate indicators. *Geologiska Föreningens i Stockholm Förhandlingar*, **66**, 463–83.

Iversen, J. (1949) The influence of prehistoric man on vegetation. *Danmarks Geologiska Undersøgelse*, Series IV, **3**, 1–23.

Iversen, J. (1954) The Late-glacial flora of Denmark and its relation to climate and soil. *Danmarks Geologiske Undersøgelse*, Series II, **80**, 87–119.

Iversen, J. (1956) Forest clearance in the Stone Age. *Scientific American*, **194**, 36–41.

Iversen, J. (1958) The bearing of glacial and interglacial epochs on the formation and extinction of plant taxa. *Uppsala Universiteit Årssk*, **6**, 210–15.

Iversen, J. (1960) Problems of the early Post-glacial forest development in Denmark. *Danmarks Geologiske Undersøgelse*, Series IV, **4**(3), 3–32.

Iversen, J. (1964) Retrogressive vegetational succession in the post-glacial. *Journal of Ecology*, **52**, 59–70.

Iversen, J. (1967) Naturens udvikling siden sidste Isted. In *Danmarks Natur, Vol. 1, Landskabernes Opståen*, Politikens Forlag, Copenhagen, pp. 345–448.

Iversen, J. (1973) The development of Denmark's nature since the last glacial. *Danmarks Geologiske Undersøgelse*, Series V, **7-C**, 1–26.

Jacobi, R.M. (1979) Early Flandrian hunters in the South West. *Proceedings of the Devon Archaeological Society*, **37**, 48–93.

Jacobi, R.M., Tallis, J.H. and Mellars, P.A. (1976) The Southern Pennine Mesolithic and the ecological record. *Journal of Archaeological Science*, **3**, 307–20.

Jacobson, G.L. and Bradshaw, R.H.W. (1981) The selection of sites for palaeovegetational studies. *Quaternary Research*, **16**, 80–96.

Jacobson, G.L., Webb, T. III and Grimm, E.C. (1987) Patterns and rates of vegetation change during the deglaciation of eastern North America. In *The Geology of North America, Volume K-3, North America and Adjacent Oceans During the Last Deglaciation* (eds W.F. Ruddiman and H.E. Wright Jr.), Geological Society of America, Boulder, Colorado, pp. 277–88.

Jäger, J. and Ferguson, H.L. (eds) (1991) *Climate Change: Science, Impacts and Policy. Proceedings of the Second World Climate Conference*, Cambridge University Press, Cambridge.

Jalas, J. and Suominen, J. (eds) (1973) *Atlas Florae Europaeae 2. Gymnospermae (Pinaceae to Ephedraceae)*, Committee for mapping the flora of Europe and Societas Biologica Fennica Vanamo, Helsinki.

Jalas, J. and Suominen, J. (1976) *Atlas Florae Europaeae: Volume 3. Salicaceae to Balanophoraceae*, Societas Biologica Fennica Vanamo, Helsinki.

Janssen, C.R. (1972) The palaeoecology of plant communities in the Dommel Valley, North Brabant, Netherlands. *Journal of Ecology*, **60**, 411–37.

Janssen, C.R. and Ten Hove, H.A. (1971) Some late-Holocene pollen diagrams from the Peel raised bogs (southern Netherlands). *Review of Palaeobotany and Palynology*, **11**, 7–53.

Janusas, S. (1987) *An Analysis of the Historic Vegetation of the Regional Municipality of Waterloo*, Region of Waterloo, Waterloo, Canada.

Jarrige, J. (1981) Chronology of the earlier periods of the Greater Indus as seen from Mehrgarh, Pakistan. In *South Asian Archaeology 1981* (ed. B. Allchin), Cambridge University Press, Cambridge, pp. 21–8.

Jenkinson, D.S. (1975) The turnover of organic matter in agricultural soils. *Welsh Soils Discussion Group Report*, **16**, 91–105.

Jenny, H. (1941) *Factors of Soil Formation, A System of Quantitative Pedology*, McGraw-Hill, New York.

Jeschke, L. and Lange, E. (1987) Zur Landschafts – und Vegetationsgeschichte im Gebiet der Sternberger Seen in Nordwesten der DDR. *Flora*, **179**, 317–334.

Jessen, K. (1949) Studies in late Quaternary deposits and flora-history of Ireland. *Proceedings of the Royal Irish Academy [B]*, **52**, 85–290.

Johnson, E.A., Fryer, G.I. and Heathcott, M.J. (1990) The influence of man and climate on frequency of fire in the interior wet belt forest, British Columbia. *Journal of Ecology*, **78**, 403–12.

Johnson, S. (1989) *English Heritage Book of Hadrian's Wall*, Batsford, London.

Johnson, W.C. and Webb, T. III (1989) The role of blue jays (*Cyanocitta cristata* L.) in the postglacial dispersal of fagaceous trees in eastern North America. *Journal of Biogeography*, **16**, 561–71.

Jones, R., Benson-Evans, K. and Chambers, F.M. (1985) Human influence upon sedimentation in Llangorse Lake, Wales. *Earth Surface Processes and Landforms*, **10**, 227–35.

Jones, R.L. (1976) The activities of Mesolithic man: further palaeobotanical evidence from north-east Yorkshire. In *Geoarchaeology. Earth Science and the Past* (eds D.A. Davidson and M.L. Shackley), Duckworth, London, pp. 355–67.

Jowsey, P.C. (1966) An improved peat sampler. *New Phytologist*, **65**, 245–8.

Kaland, P.E. (1986) The origin and management of Norwegian coastal heaths as reflected by pollen analysis. In *Anthropogenic Indicators in Pollen Diagrams* (ed. K.-E. Behre), A.A. Balkema, Rotterdam, pp. 19–36.

Kapp, R.O. (1978) Presettlement forests of the Pine river Watershed (central Michigan) based on original land survey records. *Michigan Botanist*, **17**, 3–15.

Keble Martin, W. and Fraser, G.T. (eds) (1939) *Flora of Devon*, T. Buncle & Co. Ltd, Arbroath.

Keppie, L. (1986) *Scotland's Roman Remains*, Donald, Edinburgh.

Kerney, M.P., Preece, R.C. and Turner, C. (1980) Molluscan and plant biostratigraphy of some Late Devensian and Flandrian deposits in Kent. *Philosophical Transactions of the Royal Society [B]*, **291**, 1–43.

Kershaw, A.P. (1986) Climatic change and Aboriginal burning in north-east Australia during the last two glacial/interglacial cycles. *Nature*, **322**, 47–9.

Kershaw, A.P., D'Costa, D.M., McEwen Mason, J.R.C. and Wagstaff, B.E. (1991) Palynological evidence for Quaternary vegetation and environments of mainland southeastern Australia. *Quaternary Science Reviews*, **10**, 391–404.

Kononova, M.M. (1961) *Soil Organic Matter, Its Nature, Its Role in Soil Formation and in Soil Fertility*, Pergamon Press, Oxford.

Kononova, M.M. (1975) Humus of virgin and cultivated soils. In *Soil Components Volume I: Organic Components* (ed. J.E. Gieseking), Springer Verlag, Berlin.

Kratz, T.K. (1988) A new method for estimating horizontal growth of the peat mat in basin-filling peatlands. *Canadian Journal of Botany*, **66**, 826–8.

Kratz, T.K. and De Witt, C.B. (1986) Internal factors controlling peatland–lake ecosystem development. *Ecology*, **67**, 100–7.

Kuhn, T.S. (1970) *The Structure of Scientific Revolutions*, University of Chicago Press, Chicago.

Kullman, L. (1992) Orbital forcing and tree-limit history: hypothesis and preliminary interpretation of evidence from Swedish Lappland. *The Holocene*, **2**(1), 131–37.

Küster, H. (1988) The history of the landscape around Auerberg, southern Bavaria – a pollen analytical study. In *The Cultural Landscape: Past, Present and Future* (eds H.H. Birks, H.J.B. Birks, P.E. Kaland and D. Moe), Cambridge University Press, Cambridge, pp. 301–10.

Küster, H. (1989) Pollen analytical evidence for the beginning of agriculture in south central Europe. In *The Beginnings of Agriculture* (eds A. Milles, D. Williams and N. Gardner), British

Archaeological Reports, International Series S496, pp. 137–47.

Kutzbach, J.E. and Gallimore, R.G. (1988) Sensitivity of a coupled atmosphere/mixed-layer ocean model to changes in orbital forcing at 9000 years BP *Journal of Atmospheric Science*, **93**, 803–21.

Kutzbach, J.E. and Guetter, P.J. (1986) The influence of changing orbital parameters and surface boundary conditions on climate simulations for the past 18000 years. *Journal of Atmospheric Science*, **43**, 1726–59.

Kutzbach, J.E. and Street-Perrott, F.A. (1985) Milankovitch forcing of fluctuations in the level of tropical lakes from 18 to 0 kyr BP *Nature*, **317**, 130–4.

LaMarche, V.C. Jr. and Hirschboeck, K.K. (1984) Frost rings in trees as records of major volcanic eruptions. *Nature*, **307**, 121–6.

Lamb, H.F., Eicher, U. and Switsur, V.R. (1989) An 18,000 year record of vegetation, lake level and climatic change from Tigalmamine, Middle Atlas, Morocco. *Journal of Biogeography*, **16**, 65–74.

Lamb, H.H. (1977a) *Climate: Past, Present and Future*, Vol. II, Methuen, London.

Lamb, H.H. (1977b) The late Quaternary history of the climate of the British Isles. In *British Quaternary Studies: Recent Advances* (ed. F.W. Shotton), Clarendon Press, Oxford, 283–98.

Lamb, H.H. (1979) Climatic variation and changes in the wind and ocean circulation: the Little Ice Age in the northeast Atlantic. *Quaternary Research*, **11**, 1–20.

Leggett, J. (ed.) (1990) *Global Warming – the Greenpeace Report*, Oxford University Press, Oxford.

Leuschner, von H.H. (1991) Subfossil trees. *Proceedings of the Lund Conference, September 1990*, (forthcoming).

Leuschner, von H.H. and Delorme, A. (1984) Verlangerung der Gottingen Eichenjahrringchronologien fur Nord-und Suddeutschland bis zum Jahr 4008 v.Chr. *Forstarchiv*, **55**, 1–4.

Leuschner, von H.H. and Delorme, A. (1988) Tree-ring work in Gottingen – absolute oak chronologies back to 6255 BC. *Pact*, II.5 Wood and Archaeology, 123–32.

Lewis, F.J. (1905) The plant remains in Scottish Peat Mosses. Part I. The Scottish Southern Uplands. *Transactions of the Royal Society of Edinburgh*, **41**, 699–723.

Lewis, F.J. (1906) The plant remains in Scottish Peat Mosses. Part II. The Scottish Highlands. *Transactions of the Royal Society of Edinburgh*, **45**, 335–60.

Lewis, F.J. (1907) The plant remains in Scottish Peat Mosses. Part III. The Scottish Highlands and the Shetland Islands. *Transactions of the Royal Society of Edinburgh*, **46**, 33–70.

Lewis, F.J. (1911) The plant remains in Scottish Peat Mosses. Part IV. The Scottish Highlands and Shetland, with an appendix on the Icelandic peat deposits. *Transactions of the Royal Society of Edinburgh*, **47**, 793–833.

Li, X. and Liu, J. (1988) Holocene vegetational and environmental changes at Mt. Luoji, Sichuan. *Acta Geographica Sinica*, **43**, 44–51.

Limbrey, S. (1975) *Soil Science and Archaeology*, Academic Press, London.

Lindsay, R.A., Charman, D.J. Everingham, F., O'Really, R.M., Palmer, M.A., Rowell, T.A. and Stroud, D.A. (1988) *The Flow Country: the Peatlands of Caithness and Sutherland*, NCC, Peterborough.

Livingstone, D.A. (1967) Postglacial vegetation of the Ruwenzori mountains in equatorial Africa. *Ecological Monographs*, **37**, 25–52.

Livingstone, D.A. (1971) A 22,000 year pollen record from the plateau of Zambia. *Limnology and Oceanography*, **16**, 349–56.

Lorius, C., Barkov, N.I., Jouzel, J., Korotkevich, Y.S., Kotlyakov, V.M. and Raynaud, D. (1988) Antarctic ice core: CO_2 and climatic change over the last climatic cycle. *Eos*, **69**, 681–4.

Lotter, A.F. (1991) Absolute dating of the Late-Glacial period in Switzerland using annually laminated sediments. *Quaternary Research*, **35**, 321–30.

Lowe, J.J. (1977) *Pollen Analysis and Radiocarbon Dating of Late Glacial and Early Flandrian Deposits in Southern Perthshire*. Unpublished Ph.D. Thesis, University of Edinburgh.

Lowe, J.J. (1978) Radiocarbon-dated Late glacial and early Flandrian pollen profiles from the Teith Valley, Perthshire, Scotland. *Pollen et Spores*, **20**, 367–97.

Lowe, J.J. (1982) Three Flandrian pollen profiles from the Teith Valley, Perthshire, Scotland. I. vegetational history. *New Phytologist*, **90**, 355–70.

Lowe, J.J. (1991) Stratigraphic resolution and radiocarbon dating of Devensian Late glacial sediments. In *Radiocarbon Dating: Recent Applications and Future Potential* (ed. J.J. Lowe), Quaternary Proceedings No. 1, Quaternary Research Association, Cambridge, pp. 19–26.

Lowe, J.J. and Walker, M.J.C. (1984) *Reconstructing Quaternary Environments*, Longman, London.

Lowe, J.J. and Walker, M.J.C. (1991) Vegetational history of the Isle of Skye. II. The Flandrian. In *The Quaternary of the Isle of Skye: Field Guide* (eds

C.K. Ballantyne, D.I. Benn, J.J. Lowe and M.J.C. Walker), Quaternary Research Association, Cambridge, pp. 119–42.

Lundqvist, B. (1962) Geological radiocarbon datings from the Stockholm station. *Sveriges Geologiska Undersokning*, **C 589**, 3–23.

Lynch, A. (1981) *Man and Environment in Southwest Ireland, 4000 BC–AD 800: a Study of Man's Impact on the Development of Soil and Vegetation*, British Archaeological Reports, British Series, 85, Oxford.

Lynn, C.J. (1987) Deer Park Farms, Glenarm, Co. Antrim. *Archaeology in Ireland*, **1** (1), 11–15.

MacDonald, G.M., Larsen, C.P.S., Szeicz. J.M. and Moser, K.A. (1991) The reconstruction of boreal forest fire history from lake sediments: a comparison of charcoal, pollen, sedimentological and geochemical indices. *Quaternary Science Reviews*, **10**, 53–71.

MacMillan, C. (1893) On the occurrence of *Sphagnum* atolls in central Minnesota. *Minnesota Botanical Studies Bulletin*, **9**, 2–13.

Maguire, D.J. and Caseldine, C.J. (1985) The former distribution of forest and moorland on northern Dartmoor. *Area*, **17**, 193–203.

Mallik, A.U., Gimingham, C.H. and Rahman, A.A. (1984) Ecological effects of heather burning. I. Water infiltration, moisture retention and porosity of surface soil. *Journal of Ecology*, **72**, 767–76.

Mallory, J.P. and Warner, R.B. (1988) The date of Haughey's Fort. *Emania*, **5**, 36–40.

Malmros, C. (1986) A Neolithic road built of wood at Tibirke, Zealand, Denmark. Contribution to the history of coppice management in the Sub-Boreal period. *Striae*, **24**, 153–56.

Maloney, B.K. (1985) Man's impact on the rainforests of West Malesia: the palynological record. *Journal of Biogeography*, **12**, 537–58.

Maloney, B.K. (1990) Grass pollen and the origins of rice agriculture in North Sumatra. *Modern Quaternary Research in Southeast Asia*, **11**, 135–61.

Maloney, B.K. and McAlister, J.J. (1990) Khok Phanom Di, central Thailand: chemical analysis of pollen core KL2 and AMS radiocarbon dates from cores KL2 and BMR2. *Geoarchaeology*, **5**, 375–82.

Mamakowa, K. (1968) Lille Bukken and Lerøy – two pollen diagrams from Western Norway. *Årbok Universitet i Bergen Naturvitenskapelig*, **4**, 5–42.

Manby, T.G. (1976) The excavation of the Kilham long barrow, East Riding of Yorkshire. *Proceedings of the Prehistoric Society*, **42**, 111–59.

Manning, S. (1990) The Thera eruption: The third congress and the problem of the date. *Archaeometry*, **32** (1), 91–100.

Mannion, A.M. (1991) *Global Environmental Change*, Longman, London.

Martin, P.S. (1984) Prehistoric overkill: the global model. In *Quaternary Extinctions, a Prehistoric Revolution* (eds P.S. Martin and R.G. Klein), University of Arizona Press, Tucson, pp. 354–403.

Mason, Sir J. (1991) Models of global climatic change. *Proceedings of the EURISY Symposium on the Earth's Environment, Venice, Italy 10–11 April 1991*, **ESA SP-337**, 47–56.

Matthews, J.A. (1976) 'Little Ice Age' palaeotemperatures from high-altitude tree growth in S. Norway. *Nature*, **264**, 243–5.

Matthews, J.A. (1977) Glacier and climatic fluctuations inferred from tree-growth variations over the last 250 years, central southern Norway. *Boreas*, **6**, 1–24.

Matthews, J.A. (1980) Some problems and implications of ^{14}C dates from a podzol buried beneath an end moraine at Haugabreen, southern Norway. *Geografiska Annaler*, **62**(A), 185–208.

Matthews, J.A. (1981) Natural ^{14}C age/depth gradient in a buried soil. *Naturwissenschaften*, **68**, 472–4.

Matthews, J.A. (1982) Soil dating and glacier variations: a reply to Wibjörn Karlen. *Geografiska Annaler*, **64**(A), 15–20.

Matthews, J.A. (1984) Limitations of ^{14}C dates from buried soils in reconstructing glacier variations and Holocene climate. In *Climatic Changes on a Yearly to Millennial Basis: Geological. Historical and Instrumental Records* (eds N.A. Mörner and W. Karlén), Reidel, Dordrecht, pp. 281–90.

Matthews, J.A. (1985) Radiocarbon dating of surface and buried soils: principles, problems and prospects. In *Geomorphology and Soils* (eds K.S. Richards., S. Ellis and R.R. Arnett), George Allen & Unwin, London, pp. 269–88.

Matthews, J.A. (1987) Some aspects of the ^{14}C dating of buried soil horizons. In *Soils and the Time Factor* (ed. P. Stevens), Welsh Soils Discussion Group Report No. 24, Bangor, pp. 13–30.

Matthews, J.A. (1991a) The late Neoglacial ('Little Ice Age') glacier maximum in southern Norway: new ^{14}C-dating evidence and climatic implications. *The Holocene*, **1**(3), 219–33.

Matthews, J.A. (1991b) Deposits indicative of Holocene climatic fluctuations in the timberline areas of Northern Europe: some physical proxy data sources and research approaches. *Paläoklimaforschung* (in press).

Matthews, J.A. and Caseldine, C.J. (1987) Arctic-alpine Brown Soils as a source of palaeoenvironmental information: further ^{14}C dating and palynological evidence from Vestre Memurubreen, Jotunheimen, Norway. *Journal of Quaternary Science*, **2**, 59–71.

Matthews, J.A. and Dresser, P.Q. (1983) Intensive ^{14}C dating of a buried palaeosol horizon. *Geologiska Föreningens i Stockholm Förhandlingar*, **105**, 59–63.

Matthews, J.A., Innes, J.L. and Caseldine, C.J. (1986) ^{14}C dating and palaeoenvironment of the historic 'Little Ice Age' glacier advance of Nigardsbreen, southwest Norway. *Earth Surface Processes and Landforms*, **11**, 369–75.

Mattson, S. and Lönnemark, H. (1939) The pedography of hydrologic podzol series I. *Annals of the Agricultural College of Sweden*, **7**, 185–227.

Mayer, L.J. (1977) Palynological studies in the western arm of Lake Superior. *Quaternary Research*, **7**, 14–44.

McAndrews, J.H. (1988) Human disturbance of North American forests and grasslands: the fossil pollen record. In *Vegetation History* (eds B. Huntley and T. Webb III), Kluwer, Dordrecht, pp. 673–97.

McAndrews, J.H. and Boyko-Diakonow (1989) Pollen analysis of varved sediments at Crawford Lake, Ontario: evidence of Indian and European farming. In *Quaternary of Geology of Canada and Greenland* (ed. R.J. Fulton), ch. 7. (*Geology of Canada*, **1**, Geological Survey of Canada).

McCormack, R.F. (1991) *Further Palaeoecological Investigations Towards the Reconstruction of Vegetation and Land Use History at Derryinver, Rinvyle, N.W. Connemara*, Unpublished B.Sc. Thesis, University College Galway, Galway.

McGlone, M.S. (1983) Polynesian deforestation of New Zealand: a preliminary synthesis. *Archaeology in Oceania*, **18**, 11–25.

McVean, D.N. and Ratcliffe, D.A. (1962) *Plant Communities of the Scottish Highlands*, Monographs of the Nature Conservancy, HMSO, Edinburgh.

Meadows, M.E. (1988) Late Quaternary peat accumulation in southern Africa. *Catena*, **15**, 459–72.

Meier, M.F. (1965) Glaciers and climate. In *The Quaternary of the United States* (eds H.E. Wright Jr. and D.G. Frey), Princeton University Press, Princeton, pp. 795–805.

Mellars, P.A. (1976) Fire ecology, animal populations and man: a study of some ecological relationships in prehistory. *Proceedings of the Prehistoric Society*, **42**, 15–45.

Mellor, A. (1984) *An Investigation of Pedogenesis on Selected Neoglacial Moraine Ridge Sequences, Jostedalsbreen and Jotunheimen, Southern Norway*, Ph.D. Thesis, University of Hull.

Mercer, R.J. (1981) *Grimes Graves, Norfolk: Excavations 1971–72*, Vol. 1, HMSO, London.

Merryfield, D.L. and Moore, P.D. (1974) Prehistoric human activity and blanket peat initiation on Exmoor. *Nature*, **250**, 439–41.

Metcalfe, S.E., Street-Perrott, F.A., Brown, R.B., Hales, P.E., Perrott, R.A. and Steininger, F.M. (1989) Late Holocene human impact on lake basins in central Mexico. *Geoarchaeology*, **4**, 119–41.

Middeldorp, A.A. (1982) Pollen concentration as a basis for indirect dating and quantifying net organic and fungal production in a peat bog ecosystem. *Review of Palaeobotany and Palynology*, **37**, 225–82.

Mighall, T. and Chambers, F.M. (1989) The environmental impact of iron-working at Bryn y Castell hillfort, Merioneth. *Archaeology in Wales*, **29**, 17–21.

Miller, N.G. and Futyma, R.P. (1987) Palaeohydrological implications of Holocene peatland development in northern Michigan. *Quaternary Research*, **27**, 297–311.

Milles, A., Williams, D. and Gardner, N. (eds) (1989) *The Beginnings of Agriculture*, British Archaeological Reports, International Series S496, Oxford.

Mills, C.Wright. (1959) *The Sociological Imagination*, Oxford University Press, New York.

Mitchell, G.F. (1953) Vegetational environmental studies. *Advancement of Science*, **36**, 430–2.

Mitchell, G.F. (1956) Post-Boreal pollen diagrams from Irish raised bogs. *Proceedings of the Royal Irish Academy*, [B], **57**, 225–82.

Mitchell, G.F. (1965) Little Bog, Tipperary: an Irish agricultural record. *Journal of the Royal Society of Antiquaries of Ireland*, **95**, 121–32.

Mitchell, G.F. (1976) *The Irish Landscape*, Collins, London.

Mitchell, G.F. (1986) *The Shell Guide to Reading the Irish Landscape*, Country House, Dublin.

Moar, N.T. (1969) Late Weichselian and Flandrian pollen diagrams from south-west Scotland. *New Phytologist*, **68**, 433–67.

Molloy, K. (1989) *Palaeoecological Investigations Towards the Reconstruction of Prehistoric Human Impact in N.W. Connemara, Western Ireland*, Unpublished Ph.D. Thesis, University College Galway, Galway.

Molloy, K. and O'Connell, M. (1991) Palaeoecological investigations towards the reconstruction of woodland and land-use history at Lough Sheeauns, Connemara, western Ireland. *Review of Palaeobotany and Palynology*, **67**, 75–113.

Moore, P.D. (1968) Human influence upon vegetational history in north Cardiganshire. *Nature*, **217**, 1006–7.

Moore, P.D. (1973) The influence of prehistoric cultures upon the initiation and spread of blanket bog in upland Wales. *Nature*, **241**, 350–3.

Moore, P.D. (1975) Origin of blanket mires. *Nature*, **256**, 267–9.

Moore, P.D. (1980) Resolution limits of pollen analysis as applied to archaeology. *MASCA Journal*, **1**, 118–20.

Moore, P.D. (1986a) Hydrological changes in mires. In *Handbook of Holocene Palaeoecology and Palaeohydrology* (ed. B.E. Berglund), John Wiley, Chichester, pp. 91–107.

Moore, P.D. (1986b) Unravelling human effects. *Nature* **321**, 204.

Moore, P.D. (1987) Man and mire, a long and wet relationship. *Transactions of the Botanical Society of Edinburgh*, **45**, 77–95.

Moore, P.D. (1988) The development of moorlands and upland mires. In *Archaeology and the Flora of the British Isles* (ed. M. Jones), University Committee for Archaeology, Oxford, pp. 116–22.

Moore, P.D. and Bellamy, D.J. (1973) *Peatlands*, Elek Science, London.

Moore, P.D. and Chater, E.H. (1969) The changing vegetation of west-central Wales in the light of human history. *Journal of Ecology*, **57**, 361–79.

Moore, P.D. and Webb, J.A. (1978) *An Illustrated Guide to Pollen Analysis*, Hodder and Stoughton, London.

Moore, P.D., Merryfield, D.L. and Price, M.D.R. (1984) The vegetation and development of blanket mires. In *European Mires* (ed. P.D. Moore), Academic Press, London, pp. 203–35.

Moore, P.D., Evans, A.R. and Chater, M. (1986) Palynological and stratigraphic evidence for hydrological changes in mires associated with human activity. In *Anthropogenic Indicators in Pollen Diagrams* (ed. K.-E. Behre), A.A. Balkema, Rotterdam, pp. 209–20.

Moore, P.D., Webb, J.A. and Collinson, M.E. (1991) *An Illustrated Guide to Pollen Analysis*, 2nd edn, Blackwell Scientific, Oxford.

Morley, R.J. (1982) A palaeoecological interpretation of a 10,000 year pollen record from Danau Padang, Central Sumatra, Indonesia. *Journal of Biogeography*, **9**, 151–90.

Morrison, M.E.S. (1968) Vegetation and climate in the uplands of south-western Uganda during the later Pleistocene period. I. Muchoya Swamp, Kigezi District. *Journal of Ecology*, **56**, 363–84.

Morrison, M.E.S. and Hamilton, A.C. 1974. Vegetation and climate in the uplands of south-western Uganda during the later Pleistocene period. II. Forest clearance and other vegetational changes in the Rukiga Highlands during the past 8,000 years. *Journal of Ecology*, **62**, 1–31.

Nesje, A. and Johannessen, T. (1992) What were the primary forcing mechanisms of high-frequency Holocene climate and glacier variations? *The Holocene* **2**(1), 79–84.

Nesje, A., Kvamme, M., Rye, N. and Løvlie, R. (1991) Holocene glacial and climatic history of the Jostedalsbreen region, western Norway: evidence from lake sediments and terrestrial deposits. *Quaternary Science Reviews*, **10**, 87–114.

Newell, P.J. (1988) A buried wall in peatland by Sheshader, Isle of Lewis. *Proceedings of the Society of Antiquaries of Scotland*, **118**, 79–93.

Newey, W.W. (1968) Pollen analyses from southeast Scotland. *Transactions of the Botanical Society of Edinburgh*, **40**, 424–34.

Newnham, R.M. and Lowe, D.J. (1991) Holocene vegetation and volcanic activity, Auckland isthmus, New Zealand. *Journal of Quaternary Science*, **6**, 177–93.

Nilssen, E.J. (1988) Development of the cultural landscape in the Lofoten area, north Norway. In *Cultural Landscape: Past, Present and Future* (eds H.H. Birks, H.J.B. Birks, P.E. Kaland, D. Moe), Cambridge University Press, Cambridge, pp. 369–80.

Nilsson, T. (1935) Die pollenanalytische zonen gliederung der spat-und post-glazialen Bildungen Schonens. *Geologiska Föreningens i Stockholm Förhandlinger*, **57**, 385–62.

Nilsson, T. (1961) Ein neues Standardpollendiagram aus Bjärsjöholmssjön in Schonen. *Lunds Universitets Årsskrift*, N.F. 2, **56**, 1–34.

Nordhagen, R. (1954) Om barkebrød og treslaget alm i kulturhistoriskbelysning. Ethnobotanical studies on barkbread and the employment of wych-elm under natural husbandry. *Danmarks Geologiske Undersøgelse*, II, **80**, 262–308.

O'Connell, M. (1986) Reconstruction of local landscape development in the Post-Atlantic based on palaeoecological investigations at Carrownaglogh prehistoric field system, County Mayo, Ireland. *Review of Palaeobotany and Palynology*, **49**, 117–76.

O'Connell, M. (1987) Early cereal-type pollen records from Connemara, western Ireland and their possible significance. *Pollen et Spores*, **29**, 207–24.

O'Connell, M. (1990a) Origins of Irish lowland and blanket bog. In *Ecology and Conservation of*

Irish Peatlands (ed. G.J. Doyle), Royal Irish Academy, Dublin, pp. 49–71.

O'Connell, M. (1990b) Early land use in North-East County Mayo – the palaeoecological evidence. *Proceedings of the Royal Irish Academy [C]*, **90**, No. 9, 259–79.

O'Connell, M., Molloy, K. and Bowler, M. (1988) Post-glacial landscape evolution in Connemara, western Ireland with particular reference to woodland history. In *The Cultural Landscape: Past, Present and Future* (eds H.H. Birks, H.J.B. Birks, P.E. Kaland and D. Moe), Cambridge University Press, Cambridge, pp. 267–87.

Oerlemans, J. (ed.) (1989) *Glacier Fluctuations and Climatic Change*, Kluwer, Dordrecht.

Oldfield, F. (1960a) Pollen analysis and man's role in the ecological history of the southeast Lake District. *Geografiska Annaler*, **45**, 23–40.

Oldfield, F. (1960b) The coastal mud-bed at Mouligna, Bidart, and the age of the Asturian industry in the Pays Basque. *Pollen et Spores*, **2**, 57–70.

Oldfield, F. (1960c) Studies in the post-glacial history of British vegetation: Lowland Lonsdale. *New Phytologist*, **59**, 192–217.

Oldfield, F. (1963) Pollen analysis and man's role in the ecological history of the southeast Lake District. *Geografiska Annaler*, **45**, 23–40.

Oldfield, F., Thompson, R. and Barber, K.E. (1978) Changing fallout of magnetic particles recorded in recent ombrotrophic peat sections. *Science*, **199**, 679–80.

Oliver, C.D. (1981) Forest development in North America following major disturbances. *Forest Ecology and Management*, **3**, 153–68.

Ó Nualláin, S. (1984) A survey of the stone circles in Cork and Kerry. *Proceedings of the Royal Irish Academy [C]* **84**, 1–77.

Ó Nualláin, S. (1988) Stone rows in the south of Ireland. *Proceedings of the Royal Irish Academy [C]*, **88**, 179–256.

O'Sullivan, A. (1991) *Historical and Contemporary Effects of Fire on the Native Woodland Vegetation of Killarney, S.W. Ireland*, Unpublished Ph.D. Thesis, Trinity College, Dublin.

O'Sullivan, P.E. (1974) Two Flandrian pollen diagrams from the east-central Highlands of Scotland. *Pollen et Spores*, **16**, 33–57.

O'Sullivan, P.E. (1975) Early and middle-Flandrian pollen zonation in the eastern Highlands of Scotland. *Boreas*, **4**, 197–207.

Overbeck, F. (1975) *Botanisch-geologische moorkunde*, K. Wachholtz Verlag, Neumünster.

Overbeck, F., Munnich, K.O., Aletsee, L. and Averdieck, F.R. (1957) Das Alter des 'Grenzhorizonts' norddeutcher Hochmoore nach Radiocarbon-Datierungen. *Flora*, **145**, 337–71.

Owen, R.B., Barthelme, J.W., Renaut, R.W. and Vincens, A. (1982) Palaeolimnology and archaeology of Holocene deposits north-east of Lake Turkana, Kenya. *Nature*, **298**, 523–9.

Parry, M. (1990) *Climatic Change and World Agriculture*, Earthscan, London.

Passmore, J. (1974) *Man's responsibility for Nature*, Duckworth, London.

Pears, N.V. (1968) Post-glacial tree lines of the Cairngorm Mountains, Scotland. *Transactions of the Royal Society of Edinburgh*, **40**, 361–94.

Pears, N.V. (1970) Post-glacial tree-lines of the Cairngorm Mountains: some modifications based on radiocarbon dating. *Transactions of the Botanical Society Edinburgh*, **40**, 536–44.

Pears, N.V. (1972) Interpretation problems in the study of tree-line fluctuations. In *Research Papers in Forest Meteorology* (ed. J.A. Taylor), Cambrian News, Aberystwyth, pp. 31–45.

Pearson, G.W. (1979) Precise ^{14}C measurement by liquid scintillation counting. *Radiocarbon*, **21**, 1–21.

Pearson, G.W. (1986) Precise calendrical dating of known growth period samples using a 'curve fitting' technique. *Radiocarbon*, **28**, 292–9.

Pearson, G.W. and Stuiver, M. (1986) High-precision calibration of the Radiocarbon Time Scale, 500–2500 BC. *Radiocarbon*, **28**, 839–62.

Pearson, G.W., Pilcher, J.R., Baillie, M.G.L. and Hillman, J. (1977) Absolute radiocarbon dating using a low altitude European tree-ring calibration. *Nature*, **270**, 25–28.

Pearson, G.W., Pilcher, J.R., Baillie, M.G.L., Corbett, D.M. amd Qua, F. (1986) High-precision ^{14}C measurement of Irish oaks to show the natural ^{14}C variations from AD 1840–5210 BC. *Radiocarbon*, **28**, 911–34.

Peglar, S.M. (1979) A radiocarbon-dated pollen diagram from Loch of Winless, Caithness, north-east Scotland. *New Phytologist*, **82**, 245–63.

Peglar, S.M., Fritz, S.C. and Birks, H.J.B. (1988) Vegetation and land-use history at Diss, Norfolk, England. *Journal of Ecology*, **77**, 203–22.

Pennington, W. (1947) Lake sediments: pollen diagrams from the bottom deposits of the north basin of Windermere. *Philosophical Transactions of the Royal Society of London [B]*, **233**, 137–75,

Pennington, W. (1965) The interpretation of some Post-glacial vegetation diversities at different Lake District sites. *Proceedings of the Royal Society of London*, [B], **161**, 310–23.

Pennington, W. (1970) Vegetation history of northwest England: a regional synthesis. In *Studies in the Vegetational History of the British Isles* (eds D. Walker and R.G. West), Cambridge University Press, Cambridge, pp. 41–79.

Pennington, W. (1975) The effect of Neolithic man on the environment in North West England: the use of absolute pollen diagrams. In *The Effect of Man on the Landscape: the Highland Zone* (eds J.G. Evans, S. Limbrey and H. Cleere), CBA Research Report 11, Council for British Archaeology, London, pp. 74–86.

Pennington, W. (Mrs T.G. Tutin) (1986) Lags in adjustment of vegetation to climate caused by the pace of soil development. *Vegetatio*, **67**, 105–18.

Perry, I. and Moore, P.D. (1987) Dutch elm disease as an analogue of neolithic elm decline. *Nature*, **326**, 72–3.

Persson, C. (1971) Tephrochronological investigation of peat deposits in Scandinavia and on the Faroe Islands. *Sveriges Geologiska Undersokning*, C **565**, 1–33.

Phillips, F.M., Zreda, M.G., Smith, S.S., Elmore, D., Kubik, P.W. and Sharma, P. (1990) Cosmogenic chlorine-36 chronology for glacial deposits at Bloody Canyon, Eastern Sierra Nevada. *Science*, **248**, 1529–32.

Piggott, S. (1954) *The Neolithic Cultures of the British Isles*, Cambridge University Press, London.

Pigott, C.D. (1958) Biological flora of the British Isles (*Polemonium caeruleum* L.). *Journal of Ecology*, **46**, 507.

Pilcher, J.R. (1969) Archaeology, palaeoecology and ^{14}C dating of the Beaghmore stone circle site. *Ulster Journal of Archaeology*, **32**, 73–91.

Pilcher, J.R. (1973) Pollen analysis and radiocarbon dating of a peat on Slieve Gallion, Co. Tyrone, N. Ireland. *New Phytologist*, **72**, 681–9.

Pilcher, J.R. (1990) Radiocarbon calibration and radiocarbon accuracy: why they matter to Quaternary scientists. Conference on Radiocarbon dating – recent applications and future potential. *Quaternary Newsletter*, **61**, 41 (abstract).

Pilcher, J.R. (1991) Radiocarbon dating – a users' guide. In *Geochronological Dating, a Technical Manual of the Quaternary Research Association* (ed. P. Smart), QRA, Cambridge, pp. 1–26.

Pilcher, J.R. and Baillie, M.G.L. (1978) Implications of a European radiocarbon calibration. *Antiquity*, **52**, 217–22.

Pilcher, J.R. and Larmour, R. (1982) Late-glacial and post-glacial vegetational history of the Meenadoan nature reserve, County Tyrone. *Proceedings of the Royal Irish Academy* [B], **82**, 277–95.

Pilcher, J.R. and Smith, A.G. (1979) Palaeoecological investigations at Ballynagilly, a Neolithic and Bronze Age settlement in Co. Tyrone, N. Ireland. *Philosophical Transactions of the Royal Society of London* [B], **286**, 345–69.

Pilcher, J.R., Smith, A.G., Pearson, G.W. and Crowder, A. (1971) Land clearance in the Irish Neolithic: new evidence and interpretation. *Science*, **172**, 560–2.

Pilcher, J.R., Hillman, J., Baillie, M.G.L., and Pearson, G.W. (1977) A long sub-fossil oak tree-ring chronology from the North of England. *New Phytologist*, **79**, 713–29.

Pilcher, J.R., Baillie, M.G.L., Schmidt, B. and Becker, B. (1984) A 7272-year tree-ring chronology for western Europe. *Nature*, **312**, 150–2.

Piperino, D.R. (1990) Aboriginal agriculture and land usage in the Amazon Basin, Ecuador. *Journal of Archaeological Science*, **17**, 665–77.

Ponel, Ph., de Bealieu, J.-L., and Tobolski, K. (1992) Holocene palaeoenvironment at the timberline in the Taillefer Massif, French Alps: a study of pollen, and plant and insect macrofossils. *The Holocene*, **2**, 117–30.

Popper, K.R. (1963) *Conjectures and Refutations: the Growth of Scientific Knowledge*, Routledge & Kegan Paul, London.

Porter, S.C. (1981) Glaciological evidence of Holocene climatic change. In *Climate and History: Studies in Past Climates and Their Impact on Man* (eds T.M.L. Wigley, M.J. Ingram and G. Farmer), Cambridge University Press, Cambridge, pp. 82–110.

Pott, R. (1989) The effects of wood pasture on vegetation. *Plants Today*, **2**, 170–5.

Potzger, J.E. (1956) Pollen profiles as indicators in the history of lake filling and bog formation. *Ecology*, **37**, 476–83.

Preece, R.C. (1980) The biostratigraphy and dating of the tufa deposit at the Mesolithic site at Blashenwell, Dorset, England. *Journal of Archaeological Science*, **7**, 345–62.

Preece, R.C. and Robinson, J.E. (1984) Late Devensian and Flandrian environmental history of the Ancholme Valley, Lincolnshire: molluscan and ostracod evidence. *Journal of Biogeography*, **11**, 319–52.

Preece, R.C., Coxon, P. and Robinson, J.E. (1986) New biostratigraphic evidence of the Postglacial colonization of Ireland and for Mesolithic forest disturbance. *Journal of Biogeography*, **13**, 487–509.

Prentice, I.C. (in press) Climate change and long-term vegetation dynamics. In *Vegetation Dynamics Theory* (eds D.C. Glenn-Lewin, R.K. Peet and T.T. Veblen), Chapman & Hall, London.

Prentice, I.C. and Leemans, R. (1990) Pattern and process and the dynamics of forest structure: a simulation approach. *Journal of Ecology*, **78**, 340–55.

Prentice, I.C., Bartlein, P.J. and Webb, T. III (1991) Vegetation change in eastern North America since the last glacial maximum: a response to continuous climatic forcing. *Ecology*, **72**, 2038–56.

Price, M.D.R. and Moore, P.D. (1984) Pollen dispersion in the hills of Wales: a pollen shed hypothesis. *Pollen et Spores*, **26**, 127–36.

Proctor, M.C.F. (1989) Notes on mire vegetation on Dartmoor. *Report and Transactions of the Devonshire Association*, **121**, 129–51.

Pyle, D.M. (1989) Ice core acidity peaks, retarded tree growth and putative eruptions. *Archaeometry*, **31**, 88–91.

Rackham, O. (1980) *Ancient Woodland: its History, Vegetation and Uses in England*, Edward Arnold, London.

Rackham, O. (1986) *The History of the Countryside*, Dent, London.

Rackham, O. (1988a) Wildwood. In *Archaeology and the Flora of the British Isles* (ed. M. Jones), University Committee for Archaeology, Oxford, pp. 3–6.

Rackham, O. (1988b) Trees and woodland in a crowded landscape – the cultural landscape of the British Isles. In *The Cultural Landscape: Past, Present and Future* (eds H.H. Birks, H.J.B. Birks, P.E. Kaland and D. Moe), Cambridge University Press, Cambridge, pp. 53–77.

Ratcliffe, D.A. and Walker, D. (1958) The Silver Flowe, Galloway, Scotland. *Journal of Ecology*, **46**, 407–45.

Raynaud, D., Chapellaz, J., Barnola, J.M., Korotkevich, Y.S. and Lorius, C. (1988) Climatic and CH_4 cycle implications of glacial–interglacial change in the Vostok ice core. *Nature*, **333**, 655–7.

Regnéll, J. (1989) *Vegetation and Land Use During 6000 Years. Palaeoecology of the Cultural Landscape at Two Lake Sites in Southern Skane, Sweden*, LUNDQUA thesis 27, Lund.

Reille, M. and de Beaulieu, J.L. (1988) History of the Wurm and Holocene vegetation in western Velay (Massif Central, France): a comparison of pollen analysis from three corings at Lac du Bouchet. *Review of Palaeobotany and Palynology*, **54**, 233–48.

Rendell, H. Worsley, P., Green, F. and Parks, D. (1991) Thermoluminescence dating of the Chelford Interstadial. *Earth and Planetary Science Letters*, **103**, 182–189.

Richards, J. (1990) *The Stonehenge Environs Project*, English Heritage, London.

Ridley, H.N. (1930) *The Dispersal of Plants Throughout the World*, L. Reeve, Ashford, Kent.

Rigg, G.B. (1916) A summary of bog theories. *Plant World*, **19**, 310–25.

Roberts, N. (1989) *The Holocene – an Environmental History*, Blackwell, Oxford.

Robinson, D. (1983) Possible Mesolithic activity in the west of Arran: evidence from peat deposits. *Glasgow Archaeological Journal*, **10**, 1–6.

Robinson, D. (1987) Investigations into the Aukhorn peat mounds, Keiss, Caithness: pollen, plant macrofossil and charcoal analyses. *New Phytologist*, **106**, 195–200.

Robinson, D. and Dickson, J.H. (1988) Vegetational history and land use: a radiocarbon-dated pollen diagram from Machrie Moor, Arran, Scotland. *New Phytologist*, **109**, 223–251.

Robinson, D. and Rasmussen, P. (1989) Botanical investigations at the neolithic lake village at Weier, north east Switzerland: leaf hay and cereals as animal fodder. In *The Beginning of Agriculture* (eds A. Milles, D. Williams and N. Gardner), British Archaeological Reports, International Series S496, Oxford, pp. 149–63.

Robinson, M. (1988) Molluscan evidence for pasture and meadowland on the flood plain of the upper Thames basin. In *The Exploitation of Wetlands* (eds P. Murphy and C. French), British Archaeological Reports, Oxford, pp. 101–12.

Robinson, T. (1990) *Connemara. Part 1: Introduction and Gazetteer. Part 2: a One-inch Map*, Folding Landscape, Roundstone.

Rodwell, J.S. (ed.) (1991) *British Plant Communities: Volume 1 – Woodlands and Scrub*, Cambridge University Press, Cambridge.

Rosenqvist, I.T. (1977) *Acid Soil – Acid Water*, Ingeniørforlaget, Oslo.

Rossignol, M. and Pastouret, L. (1971) Analyse pollinique de niveaux sapropéliques postglaciaires dans une carotte en Méditerranée orientale. *Review of Palaeobotany and Palynology*, **11**, 227–38.

Rowley-Conwy, P. (1981) Slash and burn in the temperate European Neolithic. In *Farming Practice in British Prehistory* (ed. R. Mercer), Edinburgh University Press, Edinburgh, pp. 85–96.

Rowley-Conwy, P. (1982) Forest grazing and clearance in temperate Europe with special reference to Denmark: an archaeological view In *Archaeological Aspects of Woodland Ecology* (eds M. Bell and S. Limbrey), British Archaeological Reports, International Series S146, Oxford, pp. 199–215.

Ruddiman, W.F. and McIntyre, A. (1976) Northeast Atlantic palaeoclimatic changes over the past 6000 years. In *Investigation of Late Quaternary Palaeoceanography and Palaeoclimatology* (eds R.M. Cline and J.D. Hays), pp. 111–146. (*Geological Society of American Memoirs*, 145).

Rymer, L. (1977) A Late-glacial and early Postglacial pollen diagram from Drimnagall, north Knapdale, Argyllshire. *New Phytologist*, **79**, 211–21.

Saldarriaga, J.G. and West, D.C. (1986) Holocene fires in the Northern Amazon Basin. *Quaternary Research*, **26**, 358–66.

Samuelsson, G. (1910) Scottish peat mosses: a contribution to the knowledge of the Late-Quaternary vegetation and climate of northwestern Europe. *Bulletin of the Geological Institute, University of Uppsala*, **10**, 197–260.

Sauer, J.D. (1988) *Plant Migration: the Dynamics of Geographic Pattering in Seed Plant Species*, University of California Press, Berkeley.

Scharpenseel, H.W. (1971a) Radiocarbon dating of soils: problems, troubles, hopes. In *Palaeopedology: Origin, Nature and Dating of Paleosols* (ed. D.H. Yaalon), International Society of Soil Scientists and Israel Universities Press, Jerusalem, pp. 77–88.

Scharpenseel, H.W. (1971b) Radiocarbon dating of soils. *Soviet Soil Science*, **3**, 76–83.

Scharpenseel, H.W. (1972) Natural radiocarbon measurment of soil and organic matter fractions and on soil profiles of different pedogenesis. *Proceedings 8th International Radiocarbon Dating Conference, Lower Hutt, New Zealand*, **2**, 382–93.

Scharpenseel, H.W. (1975) Natural radiocarbon measurements on humic substances in the light of carbon cycle estimates. In *Humic Substances: Their Structure and Function in the Biosphere* (eds D. Povoledo and H.L. Goltermann), Centre for Agricultural Publishing and Documentation, Wageningen.

Scharpenseel, H.W. (1977) The search for biologically inert and lithogenic carbon in recent soil organic matter. In *Soil Organic Matter Studies*, Vol. 2, (ed. International Atomic Energy Agency), IAEA, Vienna, pp. 193–201.

Scharpenseel, H.W. (1979) Soil fraction dating. In *Proceedings 9th International Radiocarbon Conference Los, Angeles* (eds R. Berger and H.E. Suess), pp. 277–83.

Scharpenseel, H.W. and Schiffmann, H. (1977a) Radiocarbon dating of soils: a review. *Zeitschrift Pflanzenernährung Dündung Bodenkunde*, **140**, 159–74.

Scharpenseel, H.W. and Schiffmann, H. (1977b) Soil radiocarbon analysis and soil dating. *Geophysical Surveys*, **3**, 143–56.

Schneekloth, H. (1965) Die Rekurrenzflache im Grosse Moor bei Gifhorn, eine Zeitgleiche Bilding? *Geologisches Jahrbuch*, **83**, 477–96.

Schnitzer, M., Lowe, L.E., Dormaar, J.F. and Martel, Y. (1981) A procedure for the characterization of soil organic matter. *Canadian Journal of Soil Science*, **61**, 517–19.

Schofield, A.J. (1987) The role of palaeoecology in understanding variations in regional survey data. *Circaea*, **5**, 33–42.

Scott, E.M., Aitchison, T.C., Harkness, D.D. and Baxter, M.S. (1990) An overview of all three stages of the international radiocarbon comparison. *Radiocarbon*, **32** (3), 309–19.

Scott, L. (1987) Late Quaternary forest history in Venda, southern Africa. *Review of Palaeobotany and Palynology*, **53**, 1–10.

Sears, P.B. (1935) Glacial and postglacial vegetation. *Botanical Gazette*, **1**, 37–51.

Sernander, R. (1908) On the evidences of Postglacial changes of climate furnished by the peat-mosses of Northern Europe. *Geologiska Föreningens i Stockholm Förhandlingar*, **30**, 467–78.

Seymour, W.P. (1985) *The Environmental History of the Preseli Region of South-West Wales over the Past 12,000 Years*, Unpublished Ph.D. Thesis, University of Wales.

Shackleton, N.J. (1987) Oxygen isotopes, ice volume and sea level. *Quaternary Science Reviews*, **6**, 183–90.

Shackleton, N.J. and Opdyke, N.D. (1973a) Quaternary isotope and palaeomagnetic stratigraphy of equatorial Pacific core V28–238: Oxygen isotope temperatures and ice volume on a 10^5 year and 10^6 year scale. *Quaternary Research*, **3**, 39–55.

Shackleton, N.J. and Opdyke, N.D. (1973b) Oxygen-isotope and paleomagnetic stratigraphy of Pacific core V28–239, Late Pliocene to latest Pleistocene. In *Investigation of Late Quaternary Paleogeography and Paleoecology* (eds R.M. Cline and J.D. Hays), Geological Society of America Memoir 145, Boulder, pp. 449–64.

Sharma, C. (1985) On the late Quaternary vegetational history in Himachal Pradesh-3. Parasram Tal. *Geophytology*, **15**, 206–18.

Sharma, C. and Chauhan, M.S. (1988) Studies in the late Quaternary vegetational history in Himachal Pradesh-4. Rewalsar Lake II. *Pollen et Spores*, **30**, 395–408.

Sharma, C. and Singh, G. (1972a) Studies in the late Quaternary vegetational history in Himachal Pradesh-1. Khajiar Lake. *The Palaeobotanist*, **21**, 144–62.

Sharma, C. and Singh, G. (1972b) Studies in the late Quaternary vegetational history in Himachal Pradesh-2. Rewalsar Lake. *The Palaeobotanist*, **21**, 321–38.

Silvester, B. (1989) Carneddau, Carno (SN 99 99). *Archaeology in Wales*, **29**, 67.

Simmons, I.G. (1964) Pollen diagrams from Dartmoor. *New Phytologist*, **63**, 165–80.

Simmons, I.G. (1969a) Evidence for vegetation changes associated with Mesolithic man in Britain. In *The Domestication and Exploitation of Plants and Animals* (eds P.J. Ucko and G.W. Dimbleby), Duckworth, London, pp. 111–19.

Simmons, I.G. (1969b) Pollen diagrams from the North York Moors. *New Phytologist*, **68**, 807–27.

Simmons, I.G. (1975) The ecological setting of Mesolithic man in the Highland Zone. In *The Effect of Man on the Landscape: the Highland Zone*, (eds J.G. Evans, S. Limbrey and H. Cleere), CBA Research Report 11, Council for British Archaeology, London, pp. 57–63.

Simmons, I.G. and Innes, J.B. (1981) Tree remains in a North York Moors peat profile. *Nature*, **294**, 76–8.

Simmons, I.G. and Innes, J.B. (1985) Late Mesolithic land use and its impact in the British uplands. In *The Biogeographical Impact of Land Use Change: Collected Essays* (ed. R.T. Smith), Biogeographical Monographs 2, Biogeography Study Group, Leeds, pp. 7–17.

Simmons, I.G. and Innes, J.B. (1987) Mid-Holocene adaptations and later Mesolithic forest disturbance in Northern England. *Journal of Archaeological Science*, **14**, 385–403.

Simmons, I.G. and Innes, J.B. (1988a) The later Mesolithic period (6000–5000 bp) on Glaisdale Moor, North Yorkshire. *Archaeological Journal*, **145**, 1–12.

Simmons, I.G. and Innes, J.B. (1988b) Late Quaternary vegetational history of the North York Moors. VIII Correlation of Flandrian II litho- and pollen-stratigraphy at North Gill, Glaisdale Moor. *Journal of Biogeography*, **15**, 249–72.

Simmons, I.G. and Innes J.B. (1988c) Late Quaternary vegetational history of the North York Moors. IX. Numerical analysis and pollen concentration analysis of Flandrian II peat profiles from North Gill, Glaisdale Moor. *Journal of Biogeography*, **15**, 273–97.

Simmons, I.G. and Innes, J.B. (1988d) Late Quaternary vegetational history of the North York Moors. X. Investigations on East Bilsdale Moor. *Journal of Biogeography*, **15**, 299–324.

Simmons, I.G. and Tooley, M.J. (eds) (1981) *The Environment in British Prehistory*, Duckworth, London.

Simmons, I.G., Dimbleby, G.W. and Grigson, C. (1981) The Mesolithic. In *The Environment in British Prehistory* (eds I.G. Simmons and M.J. Tooley), Duckworth, London, pp. 82–124.

Simmons, I.G., Rand, J.I. and Crabtree, K. (1983) A further pollen-analytical study of the Blacklane peat section on Dartmoor, England. *New Phytologist*, **94**, 655–67.

Simmons, I.G., Turner, J. and Innes, J.B. (1989) An application of fine-resolution pollen analysis to later Mesolithic peats of an English upland. In *The Mesolithic in Europe* (ed. C. Bonsall), John Donald, Edinburgh, pp. 206–17.

Singh, G. (1971) The Indus Valley culture seen in the context of post-glacial climatic and ecological studies in northwest India. *Archaeology and Physical Anthropology in Oceania*, **6**, 177–89.

Singh, G. and Geissler, E.A. (1985) Late Cainozoic history of vegetation, fire, lake levels and climate, at Lake George, New South Wales, Australia. *Philosophical Transactions of the Royal Society of London* [B], **311**, 379–447.

Singh, G. and Smith, A.G. (1966) The post-glacial marine transgression in N. Ireland – conclusions from estuarine and 'raised beach' deposits: a contrast. *The Palaeobotanist*, **15**, 230–4.

Singh, G. and Smith, A.G. (1973) Postglacial vegetational history and relative land-and sea-level changes in Lecale, Co. Down. *Proceedings of the Royal Irish Academy*, **73**, (B 1), 1–51.

Singh, G., Joshi, R.D., Chopra, S.K. and Singh, A.B. (1974) Late Quaternary history of vegetation and climate of the Rajasthan Desert, India. *Philosophical Transactions of the Royal Society of London* [B], **267**, 467–501.

Smart, P.L. and Frances, P.D. (eds) (1991) *Quaternary Dating Methods – a User's Guide*, Quaternary Research Association Technical Guide No. 4, QRA, Cambridge.

Smart, T.L. and Hoffman, E.S. (1988) Environmental interpretation of archaeological charcoal. In *Current Palaeoethnobotany* (eds H. Hastorf and V. Popper), University of Chicago Press, Chicago and London, pp. 167–205.

Smith, A.G. (1985a) Pollen analytical investigations of the mire at Fallahogy TD, Co. Derry. *Proceedings of the Royal Irish Academy*, **59** (B 16), 329–43.

Smith, A.G. (1958b) Two lacustrine deposits in the south of the English Lake District. *New Phytologist*, **57**, 363–86.

Smith, A.G. (1958c) Post-glacial deposits in South Yorkshire and North Lincolnshire. *New Phytologist*, **57**, 19–49.

Smith, A.G. (1959) The mires of south-western Westmorland: stratigraphy and pollen analysis. *New Phytologist*, **58**, 105–27.

Smith, A.G. (1961a) The Atlantic–Sub-boreal transition. *Proceedings of the Linnean Society of London*, **172**, 38–49.

Smith, A.G. (1961b) Cannons Lough, Kilrea, Co. Derry: stratigraphy and pollen analysis. *Proceedings of the Royal Irish Academy*, **61** (B 20), 369–83.

Smith, A.G. (1964) Problems in the study of the earliest agriculture in Northern Ireland. *Report of the VI International Congress on the Quaternary, INQUA* (Warsaw 1961), **II**, 461–471.

Smith, A.G. (1965) Problems of inertia and threshold related to post-glacial habitat changes. *Proceedings of the Royal Society [B]*, **161**, 331–42.

Smith, A.G. (1970a) Late-and post-Glacial vegetational and climatic history of Ireland: a review. In *Irish Geographical Studies in Honour of E. Estyn Evans* (eds N. Stephens and R.E. Glasscock), Department of Geography, The Queen's University of Belfast, Belfast, pp. 65–88.

Smith, A.G. (1970b) The influence of Mesolithic and Neolithic man on British vegetation. In *Studies in the Vegetational History of the British Isles* (eds D. Walker and R.G. West), Cambridge University Press, Cambridge, pp. 81–96.

Smith, A.G. (1975) Neolithic and Bronze Age landscape changes in Northern Ireland. In *The Effect of Man on the Landscape: The Highland Zone* (eds J.G. Evans, S. Limbrey and H. Cleere), Research Report No. 11, Council for British Archaeology, London, pp. 64–74.

Smith, A.G. (1981a) Palynology of a Mesolithic–Neolithic site in County Antrim, Northern Ireland. *Proceedings of the IV Palynology Conference Lucknow*, **3**, 248–57.

Smith, A.G. (1981b) The Neolithic. In *The Environment in British Prehistory* (eds I.G. Simmons and M.J. Tooley), Duckworth, London, pp. 125–209.

Smith, A.G. (1984) Newferry and the Boreal Atlantic Transition. *New Phytologist*, **98**, 35–55.

Smith, A.G. and Cloutman, E.W. (1988) Reconstruction of Holocene vegetation history in three dimensions at Waun Fignen Felen, an upland site in South Wales. *Philosophical Transactions of the Royal Society [B]*, **322**, 159–219.

Smith, A.G. and Collins, A.E.P. (1971) The stratigraphy, palynology and archaeology of diatomite deposits at Newferry, Co. Antrim, Northern Ireland. *Ulster Journal of Archaeology*, **34**, 3–26.

Smith, A.G. and Goddard, I.C. (1991) A 12 500 year record of vegetational history at Sluggan Bog, Co. Antrim, N. Ireland (incorporating a pollen zone scheme for the non-specialist). *New Phytologist*, **118**, 167–87.

Smith, A.G. and Green, C.A. (in press) A three-dimensional approach to the study of topogenous peat development and Late-Flandrian vegetation history: Cefn Glas, South Wales, UK. *The Holocene*.

Smith, A.G. and Morgan, L.A. (1989) A succession to ombrotrophic bog in the Gwent Levels, and its demise: a Welsh parallel to the peats of the Somerset Levels. *New Phytologist*, **112**, 145–67.

Smith, A.G. and Pilcher, J.R. (1973) Radiocarbon dates and vegetational history of the British Isles. *New Phytologist*, **72**, 903–14.

Smith, A.G. and Willis, E.H. (1961–62) Radiocarbon dating of the Fallahogy landnam phase. *Ulster Journal of Archaeology*, **24–25**, 16–24.

Smith, A.G., Baillie, M.G.L., Hillam, J., Pilcher, J.R. and Pearson, G.W. (1972) Dendrochronological work in progress in Belfast: the prospects for an Irish post-glacial tree-ring sequence. *Proceedings of the 8th International Conference on Radiocarbon Dating*, A 92–6.

Smith, A.G., Gaskell Brown, C., Goddard, I.C., Goddard, A., Pearson, G.W. and Dresser, P.Q. (1981) Archaeology and environmental history of a barrow at Pubble, Loughermore Townland, County Londonderry. *Proceedings of the Royal Irish Academy*, **81** (C 2), 29–66.

Smith, A.G., Pilcher, J.R. and Pearson, G.W. (1971) Belfast radiocarbon dates IV. *Radiocarbon*, **13**, 450–67.

Smith, A.G., Whittle, A., Cloutman, E.W. and Morgan, L.A. (1989) Mesolithic and Neolithic activity and environmental impact on the south-east Fen-edge in Cambridgeshire. *Proceedings of the Prehistoric Society*, **55**, 207–49.

Smith, B.M. (1985) *A Palaeoecological Study of Raised Mires in the Humberhead Levels*, Unpublished Ph.D. Thesis, University of Wales.

Smith, C.J. (1980) *Ecology of the English Chalk*, Academic Press, London.

Smith, R.T. and Taylor, J.A. (1989) Biopedological processes in the inception of peat formation. *International Peat Journal*, **3**, 1–24.

Snell, E.A. (1989) Recent wetland loss in southern Ontario. In *Wetlands: Inertia or Momentum* (eds M.J. Bardecki and N. Patterson), Federation of Ontario Naturalists, Don Mills, pp. 183–97.

Solem, T. (1986) Age, origin and development of blanket mires in Sor-Trondelag, Central Norway. *Boreas*, **15**, 101–15.

Solem, T. (1989) Blanket mire formation at Haramsoy, More og Romsdal, Western Norway. *Boreas*, **18**, 221–35.

Sowunmi, M.A. (1981) Late Quaternary environmental changes in Nigeria. *Pollen et Spores*, **23**, 125–48.

Spray, M. (1981) Holly as fodder in England. *Agricultural History Review*, **29**, 97–110.

Stevenson, A.C. (1985) Studies in the vegetational history of S.W. Spain, II. Palynological investigations at Laguna de las Madres, S.W. Spain. *Journal of Biogeography*, **12**, 293–314.

Stevenson, A.C. and Moore, P.D. (1988) Studies in the vegetational history of S.W. Spain, IV. Palynological investigations of a valley mire at El Acebron, Huelva. *Journal of Biogeography*, **15**, 339–61.

Stockmarr, J. (1975) Retrogressive forest development as reflected in a mor pollen diagram from Manteringbos, Drenthe, the Netherlands. *Palaeohistoria*, **XVII**, 37–51.

Stout, J.D., Goh, K.M. and Rafter, T.A. (1981) Chemistry and turnover of naturally occurring resistant organic compounds in soil. In *Soil Biochemistry*, Vol. 5 (eds E.A. Paul and J.N. Ladd), Marcel Dekker, New York, pp. 1–73.

Street-Perrott, F.A. (1991) General circulation (GCM) modelling of palaeoclimates: a critique. *The Holocene*, **1**, 74–80.

Stuijts, I., Newsome, J.C. and Flenley, J.R. (1988) Evidence for Late Quaternary vegetational change in the Sumatran and Javan highlands. *Review of Palaeobotany and Palynology*, **55**, 207–16.

Stuiver, M. and Pearson, G.W. (1986) High-precision calibration of the radiocarbon timescale, AD 1950–500 BC. *Radiocarbon*, **28**, 805–38.

Stuiver, M. and Reimer, P.J. (1986) A computer programme for radiocarbon age calibration. *Radiocarbon*, **28**, 1022–30.

Sturlodottir, S.A. and Turner, J. (1985) The elm decline at Pawlaw mire: an anthropogenic interpretation. *New Phytologist*, **99**, 323–9.

Succow, M. (1988) *Landschaftsökologische Moorkunde*, Gebrüder Borntraeger, Berlin.

Suess, H. (1970) Bristlecone pine calibration of the radiocarbon timescale from 5200 BC to the present. In *Radiocarbon Variation and Absolute Chronology* (ed I.U. Olsson), Almquist and Wiksell, Stockholm, pp. 303–12.

Sun, X., Wu. Y., Qiao, Y. and Walker, D. (1986) Late Pleistocene and Holocene vegetation history at Kunming, Yunnan Province, southwest China. *Journal of Biogeography*, **123**, 441–76.

Sutherland, D.G. (1984) Modern glacier characteristics as a basis for inferring former climates with particular reference to the Loch Lomond Stadial. *Quaternary Science Reviews*, **3**, 291–310.

Sutton, D.G. (1987) A paradigmatic shift in Polynesian prehistory: implications for New Zealand. *New Zealand Journal of Archaeology*, **9**, 135–55.

Svensson, G. (1988) Bog development and environmental conditions as shown by the stratigraphy of Storre Mosse mire in S. Sweden. *Boreas*, **17**, 89–111.

Swadling, P. and Hope, G.S. (1991) Environmental change in New Guinea since human settlement. In *The Naive Lands – Prehistory and Environmental Change in the Southwest Pacific* (ed. J.R. Dodson), Longmans-Cheshire, Melbourne, pp. 13–42.

Swain, A.M. (1973) A history of fire and vegetation in northeastern Minnesota as recorded in lake sediments, *Quaternary Research*, **3**, 383–96.

Swain, A.M. (1978) Environmental changes during the last 200 years in north-central Wisconsin: analysis of pollen, charcoal and seeds from varved lake sediments. *Quaternary Research*, **10**, 55–68.

Swan, J.M.A. and Gill, A.M. (1970) The origins, spread and consolidation of a floating bog in Harvard Pond, Petersham, Massachusetts. *Ecology*, **51**, 829–40.

Talbot, M.R., Livingstone, D.A., Palmer, P.G., Maley, J., Melack, J.M., Delibrias, G. and Gulliksen, S. (1984) Preliminary results from sediment cores from Lake Bosumtwi, Ghana. *Palaeoecology of Africa and the Surrounding Islands*, **16**, 173–92.

Tallis, J.H. (1964a) Studies on southern Pennine peats. I. The general pollen record. *Journal of Ecology*, **52**, 324–53.

Tallis, J.H. (1964b) The pre-peat vegetation of the southern Pennines. *New Phytologist*, **63**, 363–73.

Tallis, J.H. (1975) Tree remains in southern Pennine peats. *Nature*, **256**, 482–4.

Tallis, J.H. (1991) *Plant Community History*, Chapman & Hall, London.

Tallis, J.H. and Switsur, V.R. (1973) Studies on southern Pennine peats. VI. A radiocarbon-dated pollen diagram from Featherbed Moss, Derbyshire. *Journal of Ecology*, **61**, 743–51.

Tallis, J.H. and Switsur, V.R. (1990) Forest and moorland in the South Pennine Uplands in the mid-Flandrian period. II. The Hillslope Forests. *Journal of Ecology*, **78**, 857–83.

Tamm, C.O. and Holmen, H. (1967) Some remarks on soil organic matter turnover in Swedish podzol profiles. *Meddelelser Norske Skogsforsøksvesen*, **85**, 67–88.

Tansley, A.G. (1939) *The British Islands and Their Vegetation*, Cambridge University Press, London.

Tarnocai, C., Adams, G.D., Glooschenko, et al. (1988) The Canadian wetland classification system. In *Wetlands of Canada* (coordinated by C.D.A. Rubec), Polyscience Publications Inc., Montreal, P.Q., pp. 415–27.

Tauber, H. (1965) Differential pollen dispersal and the interpretation of pollen diagrams. *Danmarks Geologiske Undersøgelse II*, **89**, 1–69.

Taylor, C.C. (1984) *Village and Farmstead: a History of Rural Settlement in England*, Philip, London.

Taylor, J.A. and Smith, R.T. (1972) Climatic peat – a misnomer? *Proceedings of the 4th International Peat Congress, Helsinki*, Vol. 1, pp. 471–84.

Taylor, J.A. and Smith, R.T. (1980) The role of pedogenic factors in the initiation of peat formation and in the classification of mires. *Proceedings of the 6th International Peat Congress, Duluth*, pp. 109–18.

Thirgood, J.V. (1981) *Man and the Mediterranean Forest – A History of Resource Depletion*, Academic Press, London.

Thomas, K.D. (1982) Neolithic enclosures and woodland habitats on the South Downs in Sussex, England. In *Archaeological Aspects of Woodland Ecology* (eds M. Bell and S. Limbrey), British Archaeological Reports, Oxford, pp. 147–70.

Thomas, K.D. (1985) Land snail analysis in theory and practice, in *Palaeobiological Investigations: Research Design, Methods and Data Analysis* (eds D.D Gilbertson and N.G.A. Ralph), British Archaeological Reports, Oxford, pp. 131–56.

Thomas, K.D. (1989) Vegetation of the British chalklands in the Flandrian period: a response to Bush. *Journal of Archaeological Science*, **16**, 549–53.

Thompson, R. and Oldfield, F. (1986) *Environmental Magnetism*, Allen & Unwin, London.

Thoreau, H.D. (1837–1848) *Journal*, Vols 1 and 2, Dover Publications, New York.

Thorley, A. (1981) Pollen analytical evidence relating to the vegetation history of the Chalk. *Journal of Biogeography*, **8**, 93–106.

Tinsley, H.M. (1981) The Bronze Age. In *The Environment in British Prehistory* (eds I.G. Simmons and M.J. Tooley), Duckworth, London, pp. 210–49.

Tolonen, M. (1978) Palaeoecology of annually laminated sediments at Lake Ahvenainen, S. Finland. I. Pollen and charcoal analyses and their relation to human impact. *Annales Botanicae Fennici*, **15**, 177–208.

Transeau, E.N. (1903) On the geographic distribution and ecological relations of the bog plant societies of northern North America. *Botanical Gazette*, **36**, 401–20.

Troels-Smith, J. (1954) Ertebøllekultur–Bondekultur. Resultater af de sidste 10 aars undersøogelser i Aamosen, Vestsjaelland (Ertebølle Culture – Farmer Culture. Results of the past ten years' excavations in Aamosen Bog, West Zealand. *Aarbøger for Nordisk Oldkyndighed og Historie 1953*, pp. 5–62.

Troels-Smith, J. (1960) Ivy, mistletoe and elm. Climate indicators/fodder plants. *Danmarks Geologiske Undersøgelse*, IV, **4**, 1–32.

Troels-Smith, J. (1984) Stall-feeding and field-manuring in Switzerland about 6000 years ago. *Tools and Tillage*, **5**, 13–25.

Trueman, A.E. (1971) *Geology and Scenery in England and Wales* (revised by J.B. Whittow and J.R. Hardy), Penguin, London.

Tsukada, M. (1966) Late postglacial absolute pollen diagram in Lake Nojiri. *The Botanical Magazine (Tokyo)*, **79**, 179–84.

Tsukada, M. (1967a) Pollen succession, absolute pollen frequency and recurrence surfaces in central Japan. *American Journal of Botany*, **54**, 821–31.

Tsukada, M. (1967b) Vegetation in subtropical Formosa during the Pleistocene glaciations and the Holocene. *Palaeogeography, Palaeoclimatology, Palaeoecology*, **3**, 49–64.

Tsukada, M. (1981) The last 12,000 years – the vegetation history of Japan II. New pollen zones. *Japanese Journal of Ecology*, **31**, 201–15.

Tsukada, M. (1983) Man and vegetation in prehistoric Japan. *Kodansha Encyclopedia of Japan*, **5**, 92–96. (Kodansha Ltd, Tokyo).

Tsukada, M. and Deevey, E.S. (1967) Pollen analyses from four lakes in the southern Maya area of Guatemala and El Salvador. In *Quaternary Ecology* (eds E.J. Cushing and H.E. Wright), Yale University Press, New Haven, pp. 303–31.

Tsukada, M. and Stuiver, M. (1966) Man's influence on vegetation in central Japan. *Pollen et Spores*, **8**, 309–13.

Turner, J. (1962) The *Tilia* decline: an anthropogenic interpretation. *New Phytologist*, **61**, 328–41.

Turner, J. (1964) The anthropogenic factor in vegetational history I: Tregaron and Whixall Mosses. *New Phytologist*, **63**, 73–89.

Turner, J. (1965) A contribution to the history of forest clearance. *Proceedings of the Royal Society of London [B]*, **161**, 343–54.

Turner, J. (1970) Post-Neolithic disturbance of British vegetation. In *Studies in the Vegetational History of the British Isles* (eds D. Walker and R.G. West), Cambridge University Press, Cambridge, pp. 97–116.

Turner, J. (1979) The environment of northeast England during Roman times as shown by pollen analysis. *Journal of Archaeological Science*, **6**, 285–290.

Turner, J. (1981) The Iron Age. In *The Environment in British Prehistory* (eds I.G. Simmons and M.J. Tooley), Duckworth, London, pp. 250–81.

Turner, J. (1986) Principal components analyses of pollen data with special reference to anthropogenic indicators. In *Anthropogenic Indicators in Pollen Diagrams* (ed. K.-E. Behre), A.A. Balkema, Rotterdam, pp. 221–32.

Turner, J. and Greig, J. (1975) Some Holocene pollen diagrams from Greece. *Review of Palaeobotany and Palynology*, **20**, 171–204.

Turner, J. and Peglar, S.M. (1988) Temporally-precise studies of vegetation history. In *Vegetation History* (eds B. Huntley and T. Webb III), Kluwer, Dordrecht, pp. 753–77.

Tzavaras, J. (1991) *Molluscan Analysis of Ditch Deposits at Millbarrow, a Neolithic Long Barrow at Winterbourne Monkton, North Wiltshire*. Unpublished Undergraduate Thesis, University of Wales Cardiff, Cardiff.

van Den Brink, L.M. and Janssen, C.R. (1985) The effect of human activities during cultural phases on the development of montane vegetation in the Serra da Estrela, Portugal. *Review of Palaeobotany and Palynology*, **44**, 193–215.

van Der Hammen, T. (1962) Palinologia de la region de 'Laguna de Los Bobos'. *Revista Academia Colombiona de Ciencias Exactas, Fisicas y Naturales*, **11**, 359–61.

van Der Hammen, T. and Gonzalez, E. (1965) A pollen diagram from 'Laguna de la Herrera' (Sabana de Bogota). *Leidse Geologische Mededelingen*, **32**, 183–91.

van Geel, B. (1972) Palynology of a section from the raised bog Wietmarscher Moor, with special reference to fungal remains. *Acta Botanica Neerlandica*, **21**, 261–84.

van Geel, B. (1978) A palaeoecological study of Holocene peat bog sections in Germany and the Netherlands. *Review of Palaeobotany and Palynology*, **25**, 1–120.

van Geel, B. and Middeldorp, A.A. (1988) Vegetational history of Carbury Bog (Co. Kildare, Ireland) during the last 850 years and a test of the temperature indicator value of $^2H/^1H$ measurements of peat samples in relation to historical sources and meteorological data. *New Phytologist*, **109**, 377–92.

van Geel, B. and Mook, W.G. (1989) High-resolution ^{14}C dating of organic deposits using natural atmospheric ^{14}C variations. *Radiocarbon*, **31**, 151–5.

van Zeist, W. (1954) A contribution to the problem of the so-called Grenzhorizont. *Palaeohistoria*, **3**, 220–4.

van Zeist, W. (1959) Studies on the post Boreal vegetational history of south-eastern Drenthe (Netherlands). *Acta Botanica Neerlandica*, **8**, 156–85.

van Zeist, W. (1964) A palaeobotanical study of some bogs in western Brittany, (Finistère), France. *Palaeohistoria*, **10**, 157–80.

van Zeist, W. and Bottema, S. (1977) Palynological investigations in Western Iran. *Palaeohistoria*, **19**, 19–85.

van Zeist, W. and Bottema, S. (1982) Vegetational history of the eastern Mediterranean and the Near East during the last 20,000 years. In *Palaeoclimates, Palaeoenvironments and Human Communities in the Eastern Mediterranean Region in Later Prehistory* (eds J.L. Bintliff and W. van Zeist), British Archaeological Reports, International Series S133, Oxford, pp. 277–321.

van Zeist, W. and van Der Spoel-Walvius, M.R. (1980) A palynological study of the Late-glacial and the Postglacial in the Paris Basin. *Palaeohistoria*, **22**, 67–109.

van Zeist, W. and Woldring, H. (1980) Holocene vegetation and climate of northwestern Syria. *Palaeohistoria*, **22**, 111–25.

van Zeist, W., Woldring, H. and Stapert, D. (1975) Late Quaternary vegetation and climate of southwestern Turkey. *Palaeohistoria*, **17**, 53–143.

van Zinderen Bakker E.M. (1989) Middle Stone age palaeoenvironment at Florisbad (South Africa). *Palaeoecology of Africa and the Surrounding Islands*, **20**, 133–54.

Vasari, Y. (1977) Radiocarbon dating of the Late glacial and Early Flandrian vegetational succession in the Scottish Highlands and Isle of Skye. In *Studies in the Scottish Late Glacial Environment* (eds J.M. Gray and J.J. Lowe), Pergamon, Oxford, pp. 143–62.

Vasari, Y. and Vasari, A. (1968) Late-and Post-glacial macrophytic vegetation in the lochs of northern Scotland. *Acta Botanica Fennica*, **80**, 1–120.

Viereck, L.A. (1966) Plant succession and soil development on gravel outwash of the Muldrow Glacier, Alaska. *Ecological Monographs*, **36**, 181–99.

Vlaming, de V. and Proctor, V.W. (1968) Dispersal of aquatic organisms: viability of seeds recovered from the droppings of captive killdeer and mallard ducks. *American Journal of Botany*, **55**, 20–6.

Vorren, K.-D. (1986) The impact of early agriculture on the vegetation of Northern Norway – a discussion of anthropogenic indicators in biostratigraphical data. In *Anthropogenic Indicators in Pollen Diagrams* (ed. K.-E. Behre), A.A. Balkema, Rotterdam, pp. 1–18.

Vuorela, I. (1985) Comments on early agriculture in Scandinavia. Source areas of pollen spectra in Southern Finland. *Norwegian Archaeological Review*, **18**, 97–9.

Wainwright, G.J. (1979) *Mount Pleasant, Dorset: Excavations 1970–1971*, Society of Antiquaries, London.

Walker, D. (1955) Studies in the post-glacial history of British vegetation XIV. Skelsmergh Tarn and Kentmere, Westmorland. *New Phytologist*, **54**, 222–54.

Walker, D. (1956) A site at Stump Cross, near Grassington, Yorkshire, and the age of the Pennine microlith industry. *Proceedings of the Prehistoric Society*, **22**, 23–8.

Walker, D. (1966) The late Quaternary history of the Cumberland lowland. *Philosophical Transactions of the Royal Society of London [B]*, **251**, 1–120.

Walker, D. and Hope, G.S. (1982) Late Quaternary vegetation history. In *Biogeography and Ecology of New Guinea* (ed. J.L. Gressitt), Junk, The Hague, pp. 263–85.

Walker, D. and Walker, P.M. (1961) Stratigraphic evidence of regeneration in some Irish bogs. *Journal of Ecology*, **49**, 169–85.

Walker, M.F. and Taylor, J.A. (1976) Post-neolithic vegetation changes in the western Rhinogau, Gwynedd, North Wales. *Transactions of the Institute of British Geographers*, **1**, 323–45.

Walker, M.J.C. (1982) Early- and mid-Flandrian environmental history of the Brecon Beacons, South Wales. *New Phytologist*, **91**, 147–65.

Walker, M.J.C. (1984) Pollen analysis and Quaternary research in Scotland. *Quaternary Science Reviews*, **3**, 369–404.

Walker, M.J.C. and Harkness, D.D. (1990) Radiocarbon dating the Devensian Lateglacial in Britain: new evidence from Llanilid, South Wales. *Journal of Quaternary Science*, **5**, 135–44.

Walker, M.J.C. and Lowe, J.J. (1979) Postglacial environmental history of Rannoch Moor. II. Pollen diagrams and radiocarbon dates from the Rannoch Station and Corrour areas. *Journal of Biogeography*, **6**, 349–62.

Walker, M.J.C. and Lowe, J.J. (1981) Postglacial environmental history of Rannoch Moor. III. Early and mid-Flandrian pollen-stratigraphic data from sites on western Rannoch Moor and near Fort William. *Journal of Biogeography*, **8**, 475–91.

Walker, M.J.C. and Lowe, J.J. (1982) Lateglacial and early Flandrian chronology of the Isle of Mull, Scotland. *Nature*, **296**, 558–61.

Walker, M.J.C. and Lowe, J.J. (1987) Flandrian environmental history of the Isle of Mull, Scotland. III. A high-resolution pollen profile from Gribun, western Mull. *New Phytologist*, **106**, 333–47.

Walker, M.J.C. and Lowe, J.J. (1990) Reconstructing the environmental history of the Last Glacial–Interglacial transition: evidence from the Isle of Skye, Inner Hebrides, Scotland. *Quaternary Science Reviews*, **9**, 15–49.

Walker, M.J.C., Ballantyne, C.K., Lowe, J.J. and Sutherland, D.G. (1988) A reinterpretation of the Lateglacial environmental history of the Isle of Skye, Inner Hebrides, Scotland. *Journal of Quaternary Science*, **3**, 135–46.

Ward, R.G.W., Haggart, B.A. and Bridge, M.C. (1987) Dendrochronological studies of bog pine from the Rannoch Moor area, western Scotland. In *Applications of Tree-Ring Studies* (ed. R.G.W. Ward), British Archaeological Reports, International Series S333, Oxford, pp. 215–25.

Ward, S.D., Jones, A.D. and Manton, M. (1972) The vegetation of Dartmoor. *Field Studies*, **3**, 505–33.

Warner, B.G. (1988) Geological and palaeoecological aspects of *Sphagnum* bogs in Ontario. In *Wetlands: Inertia or Momentum* (eds M.J. Bardecki and N. Patterson), Federation of Ontario Naturalists, Don Mills, pp. 329–38.

Warner, B.G., Kubiw, H.J. and Hanf, K.I. (1989) An anthropogenic cause of quaking mire formation in southwestern Ontario. *Nature*, **340**, 380–84.

Wasylikowa, K. (1986) Plant macrofossils preserved in prehistoric settlements compared with anthropogenic indicators in pollen diagrams. In *Anthropogenic Indicators in Pollen Diagrams* (ed. K.-E. Behre), A.A. Balkema, Rotterdam, pp. 173–85.

Waterman, W.G. (1926) Ecological problems from the *Sphagnum* bogs of Illinois. *Ecology*, **7**, 255–72.

Watkins, N.D. (1971) Polarity events and the problem of 'the reinforcement syndrome'. *Comments on Earth Sciences: Geophysics*, **2**, 36–42.

Waton, P.V. (1982) Man's impact on the chalklands: some new pollen evidence. In *Archaeological Aspects of Woodland Ecology* (eds M. Bell and S. Limbrey), British Archaeological Reports, Oxford, pp. 75–91.

Waton, P.V. (1986) Palynological evidence for early and permanent woodland on the chalk of central Hampshire. In *The Scientific Study of Flint and Chert* (eds G. de G. Sieveking and M.B. Hart), Cambridge University Press, Cambridge, pp. 169–74.

Watts, W.A. (1973) Rates of change and stability in vegetation in the perspective of long periods of time. In *Quaternary Plant Ecology* (eds H.J.B. Birks and R.G. West), Blackwell, Oxford, pp. 195–206.

Watts, W.A. and Bradbury, J.P. (1982) Palaeoecological studies at Lake Patzcuaro on the west-central Plateau and at Chalco in the Basin of Mexico. *Quaternary Research*, **17**, 56–70.

Webb, S.L. (1986) Potential role of passenger pigeons and other vertebrates in the rapid Holocene migrations of nut trees. *Quaternary Research*, **26**, 367–75.

Webb, T. III (1986) Is vegetation in equilibrium with climate? How to interpret late-Quaternary pollen data. *Vegetatio*, **67**, 75–91.

Webb, T. III (1988) Eastern North America. In *Vegetation History* (eds B. Huntley and T. Webb III), Kluwer, Dordrecht, pp. 385–414.

Webb, T. III., Bartlein, P.J. and Kutzbach, J.E. (1987) Climatic change in eastern North America during the past 18000 years; comparisons of pollen data with model results. In *North America and Adjacent Oceans During the Last Deglaciation, Decade of North American Geology*, Geological Society of America, Boulder, v. K-3, pp. 447–62.

Weber, C.A. (1900) Uber die Moore, mit besonderer Berucksichtigung der zwischen Underweser und Unterelbe liegenden. *Jahresbericht der Manner von Morgenstern*, **3**, 3–23.

Welinder, S. (1985) Comments on early agriculture in Scandinavia. *Norwegian Archaeological Review*, **18**, 94–6.

Welinder. S. (1989) Mesolithic forest clearance in Scandinavia. In *The Mesolithic in Europe* (ed. C. Bonsall), John Donald, Edinburgh, pp. 362–6.

Welten, M. (1958) Pollenanalytische Untersuchung alpiner Bodenprofile: historische Entwicklung des Bodens und säkulare Sukzession der örtlichen Pflanzengesellschaften. *Veröffentlichungen Geobotanischen Instituts, Eidgenössische Technische Hochschule, Rübel in Zürich*, **37**, 330–45.

West, D.C., Shugart, H.H. and Botkin, D.B. (eds) (1981) *Forest Succession. Concepts and Application*, Springer-Verlag, New York.

West, R.G. (1956) The Quaternary deposits at Hoxne, Suffolk. *Philosophical Transactions of the Royal Society of London [B]*, **239**, 265–356.

West, R.G. (1968) *Pleistocene Geology and Biology – with Especial Reference to the British Isles*, Longmans, London.

West, R.G. (1980) Pleistocene forest history of East Anglia. *New Phytologist*, **85**, 571–622.

West, R.G. (1991) *Pleistocene Palaeoecology of Central Norfolk*. Cambridge University Press, Cambridge.

West, R.G. and McBurney, C.M.B. (1954) The Quaternary deposits at Hoxne, Suffolk, and their archaeology. *Proceedings of the Prehistoric Society*, **20**, 131–54.

Whittington, G., Edwards, K.J. and Cundill, P.R. (1990) *Palaeoenvironmental investigations at Black Loch in the Ochil Hills of Fife, Scotland. (O'Dell Memorial Monograph,* 22), University of Aberdeen, 64pp.

Whittington, G., Edwards, K.J. and Cundill, P.R. (1991a) Palaeoecological investigations of multiple elm declines at a site in north Fife, Scotland. *Journal of Biogeography*, **18**, 71–87.

Whittington, G., Edwards, K.J. and Caseldine, C.J. (1991b) Late- and post-glacial pollen-analytical and environmental data from a near-coastal site in north-east Fife, Scotland. *Review of Palaeobotany and Palynology*, **68**, 65–85.

Whittle, A. (1978) Resources and population in the British Neolithic. *Antiquity*, **52**, 34–42.

Whittle, A. (1990a) A model for the Mesolithic – Neolithic transition in the upper Kennet valley, north Wiltshire. *Proceedings of the Prehistoric Society*, **56**, 101–10.

Whittle, A. (1990b) Prolegomena to the study of the Mesolithic–Neolithic transition in Britain and Ireland. In *Rubané et Cardial* (eds D. Cahen and M. Otte), Etudes et Recherches Archéologiques de l'Université de Liège 39, Liège, pp. 209–27.

Wigley, T.M.L (1988) The climate of the past 10 000 years and the role of the sun. In *Secular Solar and Geomagnetic Variations in the Last 10 000 Years* (eds F.R. Stephenson and A.W. Wolfendale), Kluwer, Dordrecht, pp. 209–24.

Wigley, T.M.L. and Kelly, P.M. (1990) Holocene climatic change, ^{14}C wiggles and variations in solar irradiance. *Philosophical Transactions of the Royal Society of London [A]*, **330**, 547–60.

Wijmstra, T.A. and van Der Hammen, T. (1966) Palynological data on the history of tropical savannas in northern South America. *Leidse Geologische Mededlingen*, **38**, 71–90.

Wijmstra, T.A., Hoekstra, S., De Vries, B.J. and van der Hammen, T. (1984) A preliminary study of periodicities in percentage curves dated by pollen density. *Acta Botanica Neerlandica*, **33**, 547–55.

Wilcox, D.A. and Simonin, H.A. (1988) The stratigraphy and development of a floating peatland, Pinhook Bog, Indiana. *Wetlands*, **8**, 75–91.

Williams, C.T. (1985) *Mesolithic exploitation patterns in the central Pennines – a palynological study of Soyland Moor*. British Archaeological Reports, 139, Oxford.

Williams, W. (1977) *The Flandrian Vegetational History of the Isle of Skye and Morar Peninsula*. Unpublished Ph.D. Thesis, University of Cambridge.

Wiltshire, P.E.J. and Moore, P.D. (1983) Palaeovegetation and palaeohydrology in upland Britain. In *Background to Palaeohydrology* (ed. K.J. Gregory), John Wiley, Chichester, pp. 433–51.

Wimble, G.A. (1986) *The Palaeoecology of Lowland Coastal Raised Mires of South Cumbria*, Unpublished Ph.D. Thesis, University of Wales.

Winkler, M.G. (1988) Effect of climate on development of two *Sphagnum* bogs in south-central Wisconsin. *Ecology*, **69**, 1032–43.

Woodman, P.C. (1985) Prehistoric settlement and environment. In *The Quaternary History of Ireland* (eds K.J. Edwards and W.P. Warren), Academic Press, London, pp. 251–78.

Woods, K.D. and Davis, M.B. (1989) Palaeoecology of range limits: beech in the Upper Peninsula of Michigan. *Ecology*, **70**, 681–96.

Wright, H.E. and Barnosky, C.W. (1984) Introduction to the English edition. In *Late Quaternary Environments of the Soviet Union* (eds A.A. Velichko, H.E. Wright and C.W. Barnosky), University of Minnesota Press, Minneapolis, Minnesota, pp. xiii-xxii.

Yasuda, Y. (1978) Prehistoric environment in Japan. Palynological approach. *Science Reports of the Tohoku University, 7th Series (Geography)*, **28**, 115–281.

Yasuda, Y. (1981-2) Influence of prehistoric and historic man on Japanese vegetation. *Researches related to the UNESCO's Man and the Biosphere Programme in Japan*, pp. 35–47.

Zackrisson, O. (1977) Influence of forest fires on the north Swedish boreal forest. *Oikos*, **29**, 22–32.

Zahn, R. (1992) Deep ocean circulation puzzle. *Nature*, **356**, 744–6.

Zolitschka, B. (1988) Spätquartäre Sedimentationsgeschichte des Meerfelder Maares(Westeifel)–Mikrostratigraphie jahreszeitlich geschichteter Seesediment. *Eiszeitalter und Gegenwart*, **38**, 87–93.

Zolitschka, B. (1989) Jahreszeitlich geschichtete Seesedimente aus dem Holzmaar und dem Meerfelder Maar. *Zeitschrift deutsche geologische Gesellschaft*, **140**, 25–33.

Zvelebil, M. (ed.) (1986) *Hunters in Transition*, Cambridge University Press, Cambridge.

Index

All references in **bold** type represent figures and those in *italics* are tables.

Aaby, B. 52, 53, 147, 235
  and Tauber, H. 51, 235
Aaltonen, V.T. 93–4
Aborigines 109–10
Accelerator mass spectrometer (AMS) 27, 29, 165, 250
Acid rain 16–19, 259
Acorns 210–11
Afforestation 171
Africa *105*, 109
Agriculture 106, 107, 142, 145, 183
  arable 183, 228
  Carneddau 173
  China 108
  Connemara National Park 200, 201
  Cumbria 228, 230–1
  Derryinver Hill 199–200
  early 120, 168
  and forest destruction 108–10
  medieval 184, 185
  Mesolithic-Neolithic 113–14, 141
  mixed 139
  Roman 183
  shifting 139, 140
  valley 108
  see also Cereal cultivation; Forest farming; Grazing
Alder (*Alnus*) 9–10, 23–4, 52, 70, 113, 257
  Bonfield and North Gills 114–15
  Carneddau region 181–2
  Flandrian expansion 181
  at Lismore Fields 161, 167
  in peat bogs 48–9
  spread patterns 253
  as wetness indicator 74

Alder carr 167, 168, 169
Aletsee, L. *et al.* 50
Alignments, stone **189** 188, 200
  soil profiles 192–3, 194–5
Alluviation 155, 156
Alnus, *see* Alder (*Alnus*)
Animals 120, 141, 142, 252, 257
Antonine wall 231–2, 235
Apparent mean residence time (AMRT) 84–6
Aquatic plants 209–10
Archaeological surveys 151–2, 153, 171–2
  and molluscan analysis **157**, *157*
Archaeology, environmental 261
Arran, Isle of (Scotland) 74, 170
Artefacts 122–3, 152, 154, 220
Ascott-under-Wychwood (Oxfordshire) **150**, 152
Asia, south-east *105*
Assemblages 151, 152
  fossil 161
  macrofossil 235
  molluscan 154, 155, 157
  open-country 153, 155
  pit 152, 155–6
Astronomical forcing theory (Milankovitch) 255, 259
Atlantic Ocean influence 214
Atlantic time
  Boreal transition 23–4
  sub-Boreal transition 8
Atmospheric composition 213, **259**
Australia *105*, 107, 109
Avebury (Wiltshire) **150**, 152, 154, **155**, 156

Averdieck, F.R., *et al.* 50

Baillie, M.G.L. 39
  and Brown, D.M. 36–8
  et al. 36
Bank barrows 154
Barber, K.E. 50, 51, 55
  on stratigraphy and climate 227, 228, 230–1, 235, 236
Barrow ditches 155
Barthlein, P.J., *et al.* 210
Bartholin, T.S., *et al.* 36
Baur, M. 151
Bedrock basins 244
Beetles 209–10, 257–8
Belfast
  master chronology (BLC7000) 38, 41
  Palaeoecology Laboratory 27
  Radiocarbon Laboratory 24
Bellamy, D.J., and Moore, P.D. 224
Bell, M. 156
Bennett, K.D., *et al.* 112
Berglund, B. 142–3
Beringia 251
Birch (*Betula*) 68–70, 112, 138–9, 209
  Corylus sequence 140
  tree limit 84, 95
Birks, H.H. 48
Birks, H.J.B. 73, 257
  and Huntley, B. 211
Blacklane Brook (Dartmoor) 122
Black Mountains (S. Wales) 7, 224
Black Ridge Brook (Dartmoor) 122, 124–6, **125**, 130
  Pinswell comparison 132–3

Blashenwell (Dorset) **150**, 152, 153
Bloak Moss (Ayrshire) 233
Blytt, A. 48
Blytt–Sernander scheme 47–9, 52, 73, 74, 256
Bogs and mires 7, 24, **25**, 50, 113, 145
  blanket 54–5, 219–26, *256*
  Hadrianic-Antonine zone **232**, 233
  Mälaran Valley (Sweden) 63–5
  mire extension model 224
  peat 35
    proxy climatic data source 47–56
  quaking mats 239–47
  raised 55, 228, 235–7
  stratigraphy 80
  surface wetness 236, **237**
  tephra layers 255
  see also Floating bogs
Bolton Fell Moss (Cumbria) 228–30, **229**, 231–2, 233, 235
  surface wetness **237**, 237
Border raids 230, 231
Bradshaw, R.H.W., and Jacobson, G.L 233
Brenninkmeijer, C.A.M., *et al.* 51
Bridge, M.C.
  et al. (1987) 77
  et al. (1990) 74, 77, 80
Briffa, K.R. *et al.* 36
British Isles
  human impact publications 105
  suggested population graph **254**
Bronze Age 8, 36–7, 155–6, 173, 182–3, 228
Brown Earths (soils) 88, **89**, 91–2, 92, 94–7, 150, 209
  arctic-alpine 84
Burgess, C. 36
Burning, *see* Fire
Bush, M.B. 112, 153
  and Flenley, J.R. 153
Buxton (Derbyshire), riverine site 159–70
Bysjön, Lake (Sweden) 146

Cairngorm Mountains (Scotland) 77, 80
Cairns 173, 182, 187

Calibration 29, 53, 90
Calluna 130
Cambridge Botany School 6
Canada 108, 239–47
Canadian Shield 244
Carbon
  'bomb' 90, 92
Cardiff Radiocarbon Dating Laboratory 83–4, 86–8, **87**
Carneddau Hengwm (N. Wales) 223
Carneddau (mid-Wales) 171–85, 223, **172**
Carrs 153, 167–9
Carsegowan Moss 233, 235, **237**
Carter, B.A. 151
Carya 210
Case, H.J., *et al.* 220, 222
Caseldine, C.J. 95
  et al., (1986) 90
  and Maguire, D.J. 124, 126
  and Matthews, J.A. 91, 95
Cattle 140, 142
Causewayed camps 154
Central America *105–6*
Cereal cultivation 106
  and forest farming 139, 140, 141, 143–6
  Lismore Fields 168, 169–70
Cerealia-type pollen 140, 141, 144–6, 166, 168
Chalklands, molluscan evidence study 149–58, **150**
Chandler, C., *et al.* 120
Chapman, S.B. 220, 224
Charcoal **63**, 107–10, 166
  and alder pollen 115
  analysis 80, 126–9, **127–9**
  blanket peat basal layer 224
  Bonfield Gill Head 114, 115
  drainage effects 222
  and fire importance 119
  in lake sediments 112, 216
  Lismore Fields **163**, 163, 167, 169
  Lochstrathy diagram **79**
  Melampyrum association 130
  Pinswell 126–9, **127–9**, 133
  stratigraphies 9–10
Cheney, P., *et al.* 120
Cherhill (Wiltshire) **150**, 152, 155
China 103, 108
Chronology 7, 24, 103–6, **104**, 107

Cist burial **189**, 190
Clark, R.L. 126
Clashgour (Rannoch Moor, Scotland) 77, **78**, 80, **81**
Clearance horizons 151
Climate
  change
    human impact concern 258–9
    and vegetational response 251–7
Climatic tension 58
Cloutman, E.W., and Smith, A.G. 146, 224, 225
Clwyd-Powys Archaeological Trust 171
COHMAP team 73
Colluviation 156
Coneybury (Wiltshire) **150**, 154
Connemara (Ireland) 187
  National Park 200, 201
Contamination
  'bomb' carbon 90, 92
  root 126–9
Conway, V.M. 220, 225
Coppicing 142–3, 144–5, 222–3
Corban Lough (Co. Fermanagh) 34–5, 40
Corylus, see Hazel (*Corylus*)
Crannog 34–5
Crowder, A. *et al.* 7
Cumbria 51, 228–31, 235
Cundill, P.R. *et al.* 114
Curves
  charcoal 107
  hydroclimatic **237**, 237

Dachnowski, A.P. 245
Dags Mosse (Sweden) **137**
Damman, A.W.H., and French, T.W. 242
Danebury (Hampshire) **150**, 155–6
Dartmoor 112, 119, 121–33, 220
Data
  paleobotanical 74–5
  proxy climatic 36, 47–56, 52–5, 75, 83–4
Databases 19, 21
Dating 29, 250–1
  techniques 52–3
  see also Radiocarbon dating

Index

Davis, M.B. and Woods, K.D. 58–60
Decomposition, aerobic 54
DECORANA, plots at Lismore Fields 165, **166**, 166, **167**
Deevey, E.S. 17, 244
Defoliation 38–41
Deforestation 72, 228, 239, 240, 246–7
see also Forest and woodland
Deglaciation 214, 252–3
Dendrochronology 27, 33–41, 250–1
Denmark 51, 60, 130, 138, 142, 145
leaf-foddering 140–2
Derbyshire 159–70
Derryinver Hill (Co. Galway) 187–201, **189**
Devon 122–3
De Vries, B.J. *et al.* 52
Dickinson, W. 50
Digerfeldt, G. 146
Dimbleby, G.W. 95–6, 220, 221
et al. 220, 222
Dispersal 253–4
Ditch assemblages 152
Dorset 151–2, 154–5
Draved Forest experiment 146
Dresser, P.Q. 25–7
Drumlin 188
Dubois, A.D. and Ferguson, D.K. 80
Dumfriesshire 235
Dupont index 256
Dupont, L.M. 51, 52, 256
Durham, County 228
Durrington Walls (Wiltshire) **150**, 154
Dust veils, volcanic 35
Dutch elm disease 139, 257

East Anglia 112, 119
Easton Down (Hampshire) **150**, 155
Eckstein, D. *et al.* 36
Edwards, K.J. 112, 113–14, 115
et al. 114
Elm (*Ulmus*) 48–9, 70, 209
coppiced 145
decline 7, 24, **26**, 72, 114, 139, 257
and blanket mire correlation 224

dates 28
initial 141–2
length of 146
mid-Holocene 8, 135
pollarding effect 140
northern limits 212
pollen-climate response surface 214
Embanked enclosure **189**, 189, **190**, 200
soil profile 196–8
Empetrum 70, **76**
End moraines 83–4, **85**, 88–97
Environmental management 260–1
Europe **105**, 210, 211
north-west 257
forest farming 135–47
glaciated regions 209
western 57, 60–3, **61**
Expansion-regression model 142–3, 146

Fagaceae 211, 215
Fagus grandifolia 57, 58–60
Fallahogy (Co. Down) 6, 7
Falsification, deductive 17–18, 21
Far East *105*
Farming, *see* Agriculture
Fauna, *see* Animals
Fellend Moss (Cumbria) 228, 234
Ferguson, D.K., and Dubois, A.D. 80
Fern Lake (Ohio) **242–3**
Field sampling 161
Field systems 155, 173, 184
Fieldwalking 156
Filtration effect 144
Fire 9, 65, 112–3, 146–7
deliberate use 106, 109, 140, 207, 222, 223
Black Ridge Brook 125
early Holocene 213
Lismore Fields 169
Mesolithic 121–2, 224
North Gill 118–19
Pinswell 130
natural 119–20, 213, 216
Flandrian period 72, 182–4
at Carneddau 178–82
chronozones 112–13
Flenley, J.R. and Bush, M.B. 153

Flint
artefacts 122–3, 152, 154
mines 154
Floating bogs 243, 244
in Great Lakes basin (N. America) 239–47
growth conditions 241–2
Floating shrub mat **242–3**
Flooding 240–1
Fodder 140–2, 144–5
Forest farming 135–47, 155, 170
Forest and woodland 135, 252, 257
on base-rich soils 208–9
blanket mire replacement 225–6
boreal-nemoral 63–5
Carneddau 178–84
clearance 9, 131, 153, 155, 156, 228
and agriculture 108–10
Bronze Age 182–3, 228
Dartmoor 124–5
Derryinver Hill 201
methods 222–3
military use 233
pollen analysis 110
pre-peat 221
Roman 233
Romano-British 183
see also Landnam model (Iversen)
coppiced 142–3, 144–5, 222–3
deciduous 63, 68–72, 74, 112–13, 120
mixed 119, 138–9, 150
Flandrian 72, 112–13, 182–4
Holocene 9, 253–5
mixed 119, 138–9, 181
openings 115–18, 119, 120
Pinswell 129–32
Prästholmen 63
prehistoric modification 223–5
regeneration
Bonfield and North Gills 115
Derryinver Hill 201
Scotland **71**, 74
simulation models 58
temperate 57, 253–5
utilization model **143**, 143–6
see also Fire; Trees
Fowler, P.J. 228
Fozy Moss 233, 234

Fraximus 209
French, T.W. and Damman, A.W.H. 242
Frenzel, B. 50
Frontier zone, Hadrianic-Antonine 232–5
FRPA (fine-resolution pollen analysis) 114, 115–18
Fruits, dispersal distance 211
Fuel, cutting wood for 222, 223
Fulvic acids 88

Galloway (Scotland) 235, 236
Gandreau, D.C., and Webb, T. III 28
Gear, A.J., and Huntley, B. 77–80
General circulation models (GCMs) 213, 215
Germany 34, 35, 50, 168
Giants' Hills 2 long barrow (Skendleby) **150**, 153–4
Gimingham, C.H., *et al.* 222
Girdling, *see* Ring-barking
Glacial maximum, last 251–3
'Glacial refugia' hypothesis 210, 211, 215
Glaciers 213
  Norway 83–4, 88–97
Glasson Moss 233, 235
Global change
  projected temperature rises 260
  research agenda 19, 21
  warming 259
Global circulation models (GCM) 259
Glyceria-hollow (Denmark) 138, 141–2
Godwin, H. 8–9, 24, 52, 211, 219–20
  and Willis, E.H. 24
Göransson, H. 143–6
Gramineae (grasses) 161–3, 167–8, 169
  pollen curves 230, 233, 234–5, **235**
Grampian Highlands (Scotland) 68, 70
Granlund, E. 50
Grasses, *see* Gramineae (Grasses)
Grass Lake (Ontario) 244
Grassland 154, 155, 169
  wet 130–2

Grazing 107, 130, 131–2, 142–3, 222–3
  pollen indicators 140
  pressures 135
Great Lakes drainage basin (North America) 239–47, **240**
Great Wold Valley (Yorkshire) **150**, 153
Greenhouse effect 19, 213, 259
Greenland ice cores 35, 255
Gribun (Isle of Mull) 75

Hadrian's Wall 231–2, 234
Haggart, B.A.
  et al. (1987) 77
  et al. (1990) 74, 77, 80
Hampshire 155, 156
Haramsoy (Norway) 225
Haslam, C.J. 50, 235, 237, 256
Hatton, J.M. 124, 126
Haugabreen glacier (Norway) 84, 91
  palaeopodsol 88, **89**, 91, **93**, 93–4, 95, **96**
Haughey's Fort (Co. Armagh) 36–8
Hazel (*Corylus*) 9, 10, 52, 70, 106, 207
  abundance 214, 216–17, 257
  hypotheses for 208–16
  coppiced 145
  fire-tolerance 112, 213, 214
  groves 139
  hurdle track (Walton Heath) 145
  nuts 211, 212, 216
  Pinswell woodland 130–2
  scrub 198
  in Wales 179–80
  in western Britain 180–1
Hebrides (Scotland)
  Inner 70, 77
  Outer 225
Heim, J. 165
Hekla-3 eruption 27, 31, 37
Hemlock decline 28
Hemond, H.F. 245, 246
Henges **150**, 154, 156
Herbs 130, 133
Herding, *see* Grazing
Highlands (Scotland) 70, 74, 77, **78**, 80
Hill-forts 36–7, 155
Hillsborough Fort (Co. Down) 39

Hintikka, V. 214
Hoekstra, S. *et al.* 52
Holly (*Ilex aquifolium*) 142
Holocene 9–10
  climate 67–8, 213
    and palaeobotanical records 75–82
  environmental change 227–38
  Scottish vegetational history 68–72
Hoxnian Interglacial **104**, 110
Hughes, M.K. *et al.* 36
Human impact
  evidence 107, 257–8
  Carneddau 171–85
  Chalkland molluscs 149–58
  in floating bogs 245–7
Humic acids 88
Hunter-gatherers 113, 170, 212–13
Huntley, B. 119
  and Birks, H.J.B. 211
  and Gear, A.J. 77–80
Hypotheses
  Corylus abundance 208–16
  testing 10, 17–18, 255–6

Ice
  caps 90, 213–14
  sheets 214
Iceland 255
Illuviation concept, ascending 93–4, **94**
Immigration
  human 212
  tree taxa 74, 119
  see also Migration
India 105
Indicators
  burnt habitat 130
  proxy climatic 74, 75
  Roman forest clearance 233
  vegetation change 107
Inertia 8–9
  climatic 57
  vegetational 57–8
Innes, J.B. 115
  et al. 115, 153
  and Simmons, I.G. 120
Innes, J.L., *et al.* 90
'Interglacial cycle' concept 208–9
Interstadials, and oceanic record correlation 252

Investigation
  scientific
    deductive 13–14, **14**, 16–19, 21
    inductive 14–16, 18, 20
Ireland 6, **25**, 63–5, 107, **150**, 153
  Co. Galway 187–201
  dendrochronology 33–41
  Neolithic and Bronze Age 8
  Northern (Ulster) 38–41, *40*, **41**, 135, **136**, 138, 224
    Co. Armagh 35, 36–8
    Co. Down 39
    Co. Fermanagh 34–5
    oak chronology in 33–4
    see also Belfast
  pollen diagrams 24
  southern 170
  vegetational history 8
  western 209, 220, 220–1
    blanket mires 219–20
Iron Age 155–6, 183
Iron pan 222
Isopoll mapping 253
Isotopic analysis 51
Italy 226
Iversen, J. 138–9, 141–2, 144–7, 221, 222, 255
Ivy 142

Jacobi, R.M. 122
  et al. 225
Jacobson, G.L. and Bradshaw, R.H.W. 233
Java 109
Jays 211
Jessen, K. 23–4
Johnson, W.C., and Webb, T. III 211
Jones, P.D. *et al.* 36
Jope, M. 24, 33–4
Jostedalsbreen ice cap 85, *90*, 90
Jotunheimen/Jostedalsbreen (Norway) region 84, **85**, 88
Juncus 197–8
Juniper (*Juniperus*) 68–70, 75, **76**

Kaland, P.E. 225
Karlen, W. *et al.* 36
Kennet, River (Wiltshire) 153, 155, 156
Kent 151, 152
Keppie, L. 231

Kerney, M.P., *et al.* 151
Kesselmoor, *see* Floating bogs
Kettle-hole basins 239–47
Kilham long barrow **150**, 153
Killarney National Park (Ireland) **63**
Lake District 122, 220
Lakes 24, 225
  acidification 16–17, 259
  sediments 112, 119, 139, 215, 216
  water levels 77, 145, 214
Lamb, H.H. 220, 231
Land abandonment (Avebury) 154, **155**
Landnam model (Iversen) 107, **138**, 138–40, 141, 142, 144, 146
  at Fallahogy 6, 7
  early Neolithic 200
Landscape, management 260–1
Land use 220
  Carneddau 183–4
  Derryinver Hill 187–201
  Mesolithic 224, 225
Laurentide ice sheet 214
Leaf-foddering model 140–2, 144–5
Leaf hay 141
Letham Moss 233, 235
Lewis, Isle of (Scotland) 225–6
Lewis, F.J. 48
Lightning strikes 119–20, 213, 216
Lincolnshire 151, 153–4
Lismore Fields (Derbyshire) 160–70
'Little Ice Age' 88, 90, 92–3, 184, 231, 256–7
Little Turnbull Lake (Ontario) 244
Llangorse Lake (Wales) 183
Local pollen assemblage zones (LPAZ) 112, 163–5
  Carneddau 174–8, *175*
  Derryinver Hill 191
Lochstrathy (Scotland) **79**
Loess 209
Lomond, Loch (Scotland) 77
Long barrows 152–3, 154
  ditches 155
Longhouses 173
Lotter, A.F. 215

Lotus uliginosus 130
Lowe, J.J. *et al.* 74, 77, 80
Lundqvist, B. 50
Lycopodium **76**
Lynch, A. 188

Macrofossils
  analysis 51, **237**, *237*, 256
  Derryinver Hill 191–8
  pine **81**, 74, 77, 80
  plant 72, 211
Maguire, D.J., and Caseldine, C.J. 124, 126
Maguire family (of Fermanagh) 34–5
Maiden Castle (Dorset) 155, 157
  bank barrow **150**, 154
  causewayed camp **150**, 154
Mälaran Valley (Sweden) 58, 63–5, **64**
Mallik, A.U. *et al.* 222
Mallory, J.P. and Warner, R.B. 36–7
Mammals, *see* Animals
Mantingerboos (Denmark) 130
Maps 8, 180
Massachusetts (USA) 245
Matthews, J.A. 92–3
  and Caseldine, C.J. 91, 95
  et al. 90
Megaliths 187, **189**
Mellars, P.A. *et al.* 225
Merryfield, D.L.
  et al. 224
  and Moore, P.D. 224
Mesolithic period 111–2, 112–13, 122, 212
  chalkland vegetation 152–3
  Dartmoor communities 122–3
  human influence 72, 150–1, 152–3, 258
    Neolithic transition 113–14, 153
  people 8, 9, 10, 122
Michigan, Lake (USA) 58–60
Microfossils 51
Middeldorp, A.A. 52
  and van Geel, B. 236
Migrational lag 57, 58, 60, 209–11, 254
Migration (spread) xix, 8, 27, 215, 253–4
Milankovitch astronomical forcing theory 255, 259

Millbarrow long barrow (Wiltshire) **150**, 154, 155
Mills, C. Wright 15
Mires, *see* Bogs and mires
Mistletoe 142
Mitchell, G.F. 50, 220
  et al. 220, 222
Moel y Gerddi (Wales) 138, 181
Mollands (Scotland) **69**
Molloy, K., and O'Connell, M. 191
Molluscs 149–58, **157**
Monuments 154, 173, 182, 187–90, **188**
Mook, W.G. *et al.* 51
Moore, P.D. 122, 220, 221, 223, 224–5
  and Bellamy, D.J. 224
  et al. 224
Moorland 121–33
Mor 129, 130
Moraines 83–4, **85**, 88–97
Morrison, M.E.S. *et al.* 220, 222
Moss 88, 90
  see also Sphagnum
Mount Pleasant (Dorset) **150**, 154, 155
Muds 25, 29
Mull, Isle of (Scotland) 68, 75, 77
Munnich, K.O. *et al.* 50
Myriophyllum 75–7
Myths 15–16, 18

Nature conservation 260–1
Navan Fort (Co. Armagh) 35
Near East 108–9
Neolithic period 8, 150–1, **152**, 153–5
Netherlands 51, 135, 145
New Guinea 105, 109
Newlands Cross (Co. Dublin) **150**, 153
New Zealand 105, 109, *255*
Nigardsbreen glacier **85**, *90*
Nilsson, T. 50
Non-mire pollen sum (NMP) 228–30
North America 57, 58–60, 105, 108–9
  eastern 120, 215, 216–17, 253
  forest fires 119, 120
  Great Lakes drainage basin 239–41

Northumberland 224, 233
North Yorkshire Moors 122, 142, 153
  Bonfield Gill 114–15
  North Gill **114**, 114–19
Norway **85**, 138, 225
  glaciers 83–4, **85**, 88–97

Oak (*Quercus*) 35, 70, 112, 209, 212, 214
  at Derryinver Hill 200
  dramatic growth reduction (DGR) **39**, 39–41, *40*, **41**
  felling dates 37–8
  northern limits 212
  tree chronology 3–4
Ocean-core records 251, 252
O'Connell, M. 53, 219
  and Molloy, K. 191
Oldfield, F. 227
Oliver's Bog (Ontario) 244, 246
Ontario (Canada) 226, 239, 244, 246
Ó Nualláin, S. 188
Open habitat 115–18, 155, 167–8
Open pools 231
Overbeck, F. *et al.* 50
Oxfordshire **150**, 152
Oxford University 165
Oxygen-isotope stages, ocean-core **251**, 251, 252

Palaeoecology **49**
  hypothesis testing 257–8
  quaternary period 13–21
Palaeoenvironments, global 214
Palaeoexperiments 18–19
Palaeopodsols, *see* Podsols
Palaeosols, *see* Soils
Paludification 118
Passmore, J. 16
Pastoralism 108, 141, 183, 185, 198, 200, 222
  Cumbria 230
Pears, N.V. 77
Pearson, G.W.
  et al. 7
  and Stuiver, M. 90
Peat *30*, 160, 224, 256
  blanket 25, 113, 119, 219–26
  origins 121–2
  Pinswell 126, 129–32, 132
  Scotland 74, 77, **79**, 80

bogs, *see* Bogs and mires
Bonfield Gill 114–15
Carneddau 174, 184
components 25–7, 28
cutting 173, 184
inception and fire 122–3
initiation 119, 129–32, 198–9
North Gill 114–15, 118–19
profiles 53, 174, 224
stratigraphy 48, 227, 256
Peatlands, threat to 228
Pedogenesis 8, 209
Peglar, S.M., *et al.* 112
Pennines 122, 132, 220, 225
Pennington, W. 209, 220, 227
Permafrost 212
Philosophy 24
pH variations 17
Phytogeography 8, 24
Picea abies 58, 60, 252
Pickering, Vale of (Yorkshire) 7
Pilcher, J.R.
  et al. 7, 36
  and Smith, A.G. 7–8, 9, 25, 28, 48–9
Pine (*Pinus*) **63, 79**, 181, 200, 216
  decline 28, 29–32
  Holocene range limits 77–80
  in Scotland 48, 70, 72
  stumps 48, 74
Pinswell (Dartmoor) 123, 126–9
Pinus strobus (white pine) 240
Pinus sylvestris 57, 60–3, **61**
Pit assemblages 152, 155–6
Plantago lanceolata 183
Plant migrations 27
Plateaux 153, 156
Pleistocene period 257
Ploughing 220–1, 222
Podsolization 220–2
Podsols 84, 88, **89**, 91
  Haugabreen 88, *89*, 91, **93**, 93–4, 95, **96**
Polemonium caeruleum 168
Pollarding 40, 140, 141, 222–3
Pollard, J. 155
Pollen 6, 10, 119, 211
  analysis 6–7, 110, 223, 257
  Carneddau 173, 174
  Carneddau Hengwm 223
  Derryinver Hill 191–8
  Hadrianic–Antonine zone 233

Index

Lismore Fields 161–5, **162**, *163*, 165
  soil 95, 97
cereal-type 140, 141, 144–6, 166, 168
climate response surfaces 210
  curves 142–3, 146, 230, 233
  data 6, 68, 119, 173
  diagrams 76, 78, 108, **138**, **229**
    Carneddau **176**, **177**, **178**, **179**, **180**
    Derryinver Hill **192**, **194**, **196**, **199**
    Hadrianic–Antonine zone 233–4
    human impact chronology 103, 106
    Lismore Fields **162**, *163*, 163
    Lochstrathy **79**
    North Gill profile **116**, **117**, **118**
    Walton Moss **234**
  indicators 140, 143, 144
  Pinswell *126*, 126–9
  profiles 28, *163*, 163, **194**
  stratigraphy 69, 72–5, 95
  zones 7–8, 8–9, 48–9, 72–3
    boundaries 7, 24, **26**
Pollen assemblage zones (PAZ) 201
Pomatias elegans 151
Popper, K.R. 10, 13, **14**, 15–16, 17
Population decline 145
Postbridge (Dartmoor) 123, 133
Post-glacial, divisions *48*, 48
Post track (Somerset Levels) 35
Precipitation 77, 80, 235–6
  /evaporation ratio 219, 226
Precision 6, 7, 53, 250
Preece, R.C. 151, 153
  et al. 151
  and Robinson, J.E. 151
Prehistoric communities 220
Prentice, I.C. 60, 65
  et al. 210
Price, M.D.R., *et al.* 224
Proudfoot, V.B., *et al.* 220, 222
Proxy climate records 36, 256
Publications, human impact chronology *105*, 106

Q-7971 sample 37–8
Quaking mires (Canada) 226
Quaternary period 24
  palaeoecology 13–21
Quercus, see Oak (*Quercus*)

Rackham, O. 112, 119, 141, 213
Radiocarbon dating 7, 103, 216, 234, 250
  calibration **30**, 31–2, *53*, 53, 263–4
  Carneddau *174*
  conventions xix–xx
  Corylus migration rate 215
  in dendrochronology 27–9
  Derryinver Hill *198*, 198–200
  errors 28–9, *30*
  from peat 52–3
  Lismore Fields **165**, 165
  and palynology 23–32
  soil 91, 93, 95, 97
Rahman, A.A., *et al.* 222
Ralston, I. and Edwards, K.J. 112
Rasmussen, P. and Robinson, D. 141
Reconstruction 52, 151
  Carneddau environment 184–5
  human activity 187
  palaeobotanical 82, 97
  palaeoclimatic 36, 60, 74, 210
  palaeovegetational 7, 166–9, 187, 252
Records
  palaeobotanical 75–82
  proxy climatic 36, 256
Regional pollen assemblage zones (RPAZ) 174–8, *175*
Renvyle Peninsula (Co. Galway, Ireland) 187–201, **188**
Research 19, 20, 250
  acid rain 16–19
  chalkland techniques 150, **157**, 157–8
  climate change 20–1
  deductive mode 13–14, **14**, 16–19
  inductive mode 14–16
  molluscan analysis 156–8
  strategy, Carneddau 173–7
Revertance, climatic 75–7, **76**

Rice 108
Ridges **85**
Ring-barking (girdling) 139, 141, 144, 145, 222
River gravels 35
River valleys 153, 155, 156, 159–70
Robinson, D., and Rasmussen, P. 141
Robinson, J.E., and Preece, R.C. 151
Roman invasion 228, 230, 231–2
Roots, plant 53, 92, 126–9
Round barrows 155
Rowley-Conwy, P. 140, 141
Rumex acetosella 130, 131

Sagabreen glacier (Norway) **85**, 88
Sample fractionation 250
Samuelsson, G. 48
Sandy soils 228
Santorini eruption 35
Scandinavia 214, 230, 257
Scharpenseel, H.W. 91
Schneekloth, H. 50
Schweingruber, F.H. *et al.* 36
Schwingmoor, *see* Floating bogs
Scolytus scolytus 257–8
Scotland 48, 112, 115, 209, 219, 225
  isolating climatic factors 67–82
  Roman invasion 231
  vegetational history 68–72
Sea, global temperatures 213
Seasons, climatic 214
Sediments
  aeolian 95
  bog *255*
  lake 112, 119, 139, 215, 216
Sernander, R. 48
  see also Blytt–Sernander scheme
Sheeauns, Lough (Connemara, Ireland) 200–1
Sheep 142
Shifting cultivation 139, 140
Shropshire 181, 183, 238
Shrubs
  Lismore Fields **163**
  migration **68**, 68–70, 73–4
Signatures, clearance 233
Simmons, I.G. 220
  et al. 116, 153

Simmons, I.G. *contd*
 and Innes, J.B. 120
Simonson, W.D., *et al.* 112
Skye, Isle of (Scotland) 68, 75, **76**, 77
Slash and burn 140
Slopes 153, 154, 156
Smith, A.G. 5–11, 24, 33–4, 110, 170
 and Cloutman, E.W. 7, 146, 224, 225
 and Pilcher, J.R. 7–8, 9, 25, 28, 48–9
 and Willis, E.H. 7
 et al. 7
 on blanket peat initiation 122, 226, 227
 on Blytt-Sernander scheme 52
 on *Corylus* abundance 207, 213, 215, 217
 on dendrochronology 27
 on fire incidence 121–2
 Mesolithic period thesis 111
 on species immigration 112
 on threshold and inertia 8–9, 55, 57–8, 65, 74, 75, 255
 on vegetation research 150–1
Smith, B.M. 235
Smith, R.T. and Taylor, J.A. 118, 221–2
Snails 154
Snowdonia (Wales) 181
Soils 119, 154, 155, 209, 221–2, 228
 dating problems 84–8, *86*
 Derryinver Hill 191
 development and succession hypothesis 208–9
 erosion 109, 183
 Haugabreen and Vestre Memurubreen 93–5
 Lismore Fields 170
 sample pretreatment 84–8, **87**
 time since formation 88–90, *90*, 90–2
 see also Brown Earths; Podsols
Solar radiation 213, 214
Somerset Levels 34, 35, 50, 145
South America *105–6*
South Downs (Sussex) 156

South Street (Wiltshire) **150**, 152, 154
Sphagnum 51, 55, 236
 mats 239–47
Spores, reference slides 6
Spread (migration) xix, 8, 27, 215, 253–4
Steng Moss (Northumbria) 233
Stimulus concept 15, 16
Stone circles 173, 188
Stonehenge (Wiltshire) **150**, 154, 156
Stoneman, R. 256
Stone walls, pre-bog 188, **189**, 189, 193–4
Storbreen glacier (Norway) **85**, 88
Stuiver, M., and Pearson, G.W. 90
Successions 68, 75
 bog plant 241
 conventional 117
 and soil development hypothesis 208–9
'Suck-in' effect 36
Sudden Bog (Ontario) **244**
Suess, H. 27
Sumatra 107, 109
Surface water 16, 240–1
Svensson, G. 50
Sweden 50, 57–8, **137**, 142, 145
 boreal-nemoral ecotone 63–5
Sweet track (Somerset Levels) 34, 35, 145
Switzerland 141, 145

Tallis, J.H. 220
 et al. 225
 and Switsur, V.R. 132
Tauber, H., and Aaby, B. 51, 235
Taxa, thermophilous 209, 210, 211, 213
Taylor, J.A., and Smith, R.T. 118, 221–2
Tephra 27, 32, 53, *254*, *255*
Thickthorn Down (Dorset) **150**, 155
Thomas, P., *et al.* 120
Threshold theory (Smith) 9, 55, 57–8, 65, 74, 75
Tilia 65, 141–2, 182–3
Time-lagged responses 73–4

Timescales 258–9
Time-series analysis 52
Tombs, megalithic 187, 188, **189**
Traband, L., *et al.* 120
Tree line 77, 95, 138
Tree rings 27–9, 35, 80, 215, 257
Trees 8–9, *141*, 147, 208, 216
 alder 23–4
 Bonfield Gill Head stumps 114
 felling and burning 109, 138, 222
 immigration 58–60, **68**, 68–70, 73–4, 112, 119
 Lismore Fields **163**, *164*
 see also individual species
Treethrow pits 152
Troels-Smith, J. 140, 141, 146
Tsuga (hemlock) decline 257
Tufa **150**, 152, 153
Tulach, The (Connemara, Ireland) 187–8
Tully Mountain (Connemara, Ireland) 187
Turner, C., *et al.* 151
Turner, J. 227, 228, 230
 and Davies, C. 228
 et al. 116, 153

Ulmus, *see* Elm (*Ulmus*)
Ulster, *see* Ireland, Northern
Ulster crannog case study 34–5
Unidentifiable organic matter (UOM) 236
Upland sites 108, 159
USA 57, 58–60, 108

Vallonia costata 151, 152
van der Hammen, T. *et al.* 52
van Geel, B. 51
 et al. 51
 and Middeldorp A.A. 236
van Zeist, W. 50
Vegetation 142, 151, 222
 chalkland 149–50
 change 24, 121–2, 124, **125**
  Carneddau region 171–85
  Dartmoor 124–6, **125**
  Pinswell 129–32, **131**
 climatic forcing 254–5
 history 7–8, 95, **96**
  Derryinver Hill 187–201
  Lismore Fields 165–9

Vestre Memurubreen 95–7
  human impact 58, 103–6, **104**, 112, 220, 225
  Mesolithic and Neolithic 150–1
  succession **68**, 75
  time-lagged responses 73–4
Vestre Memurubreen glacier (Norway) 84
  buried brown soil 88, **89**, 91–2, 92, 94–7
Volcanic activity 35, **254**, *254*, 255
Volcanic ash (tephra) 27, 32, 53, **254**, 255

Wales 83–4, 86–8, 138, 146, 223
  Carneddau region 171–85
  north 221, 223
  south 112, 122
  uplands 183, 225
Walker, D. 227, 228
Wall construction 198, 201
Walton Moss (Cumberland) 233, **234**, *234*, 235
  macrofossil results **236**, 236
Ward, R.G.W. *et al.* 77
Warner, R. 34
Water budget 245
Waterlogging **221**, 221–2, **222**, 224, 226
  hydrological model 224
Waterman, W.G. 241–2, 246
Waun-Fignen-Felen (Wales) 7, 112, 122, 146, 224
Webb, T. III
  et al. 210
  and Johnson, W.C. 211
Weber, C.A. 50
Weeds, anemophilus 106
Weir's Lough (N. Ireland) **136**
Welten, M. 95–7
Wester Ross (Scotland) 77
Wetland forms 242–3
Whittington, G. *et al.* 114
Whixall Moss (Shropshire) 183
Wiggle matching 27, **30**, 30–1, 53, 250
Wigley, T.M.L. *et al.* 36
Wijmstra, T.A. *et al.* 52
Williams, D., *et al.* 120

Willis, E.H., and Smith, A.G. 7
Willow Garth (Yorkshire) **150**, 153
Wilsford Shaft (Wiltshire) **150**, 150
Wiltshire **150**, 152, 154, 155, 156
Wimble, G.A. 235
Windmill Hill (Wiltshire) 154
Wisconsinian stage 251–3
Woodhenge (Wiltshire) **150**, 154
Woodland, *see* Forest and woodland
Woods, K.D. and Davis, M.B. 58–60
Wye, River (Derbyshire) 159–60

Yorkshire 7, 112, 114–19, 122, 142, 153
Yorkshire Wolds 112
Ytterholmen (Sweden) 63, **64**

Zetterberg, P. *et al.* 36
Zolitschka, B. 215